Inventive Geniuses Who Changed the World

John Bailey

Inventive Geniuses Who Changed the World

Fifty-Three Great British Scientists and Engineers and Five Centuries of Innovation

 Springer

John Bailey
Evenley, UK

ISBN 978-3-030-81383-3   ISBN 978-3-030-81381-9   (eBook)
https://doi.org/10.1007/978-3-030-81381-9

This Springer imprint is published by the registered company Springer Nature Switzerland AG
The registered company address is: Gewerbestrasse 11, 6330 Cham, Switzerland

Timeline for the key achievements of outstanding British scientists and engineers

Seventeenth century	1628		Circulation of blood identified (William Harvey)	
		Corpuscular theory of matter (Robert Boyle)	1661	
	1665		Micrographia/study of the miniature world and cells (Robert Hooke)	
		Universal gravitation theory (Isaac Newton)	1667	
	1668		Scientific explanation of colour (Isaac Newton)	
		First-generation steam engine (Thomas Savery)	1698	
Eighteenth century	1701		Seed drill (Jethro Tull)	
		Second-generation steam engine (Thomas Newcomen)	1712	
	1733		Flying shuttle loom for weaving wider fabrics (John Kay)	
		Latent heat required to induce a change in state (Joseph Black)	1762	
	1765		Spinning "jenny" raised productivity of cotton manufacture (James Hargreaves)	
		Man-made canal (James Brindley and Duke of Bridgewater)	1765	
	1769		Third-generation steam engine, reducing coal need by 60% (James Watt)	
		Flush toilet with S-bend	1775	
	1780		"Mouth broom"/toothbrush (William Addis)	
		Air is not a unitary element (Henry Cavendish)	1780s	
	1786		Corn threshing machine (Andrew Meikle)	
		Different kinds of air (Joseph Priestley)	1786	
	1798		Smallpox vaccination (Edward Jenner)	

N.B. Not all discoveries/inventions/theories can be accurately attributed to any particular person, nation or date

(continued)

Timeline (continued)

Nineteenth century	Fourth generation of steam engine that was mobile (Richard Trevithick) 1808	1802
	Use of electrolysis to isolate new metals (Humphrey Davy) 1812	Atoms exist. First table of atomic weights (John Dalton) 1808
	Miners' safety lamp (Humphrey Davy) 1810–1820	Parts of an ichthyosaur skeleton found (Mary Anning) 1815
	Principle of electric motor (Michael Faraday) 1823	Macadamised turnpike roads (John McAdam) 1821
	Portland cement (Joseph Aspdin) 1829	Liquefaction of gases/principle of refrigeration (Michael Faraday) 1824
	First public intercity railway line (George Stephenson) 1831	Rocket steam locomotive (Robert Stephenson) 1830
	5-year expedition on HMS Beagle begins (Charles Darwin) 1834	Principles of AC generator/transformer (Michael Faraday) 1831
	Faraday protective cage (Michael Faraday) 1837	Concept of a general-purpose, programmable computer (Charles Babbage) 1836
	Makings of a computer programmer (Ada Lovelace) 1843	Telegraphic signalling (William Cooke and Charles Wheatstone) 1843
		Iron hull, steam propulsion and screw propeller (Isambard Brunel)

(continued)

Timeline (continued)

UK patent for vulcanisation of rubber (Thomas Hancock) 1845	1843
	Pneumatic tyre for horse-drawn carriages (Robert Thomson) 1849–1854
First observational epidemiological study showed cholera is a water-borne disease (John Snow)	
One-piece water closet (Joseph Jennings) 1855	1852
	The Lady with the Lamp (Florence Nightingale) 1856
The making of steel without fuel (Henry Bessemer) 1858	
	Evolution and natural selection (Charles Darwin and Alfred Wallace) 1860
Applying probability to gas behaviour (James Clerk Maxwell) 1860s	
	Cistern with floating ballcock (Thomas Crapper) 1866
Trans-Atlantic electric telegraph cable (William Thomson) 1867	
	Antiseptic surgical procedures (Joseph Lister) 1873
Unifying theory of electricity, magnetism and light (James Clerk Maxwell) 1875	
	London's integrated sewage system (Joseph Bazalgette) 1876
Telephone for transmitting vocal sounds (Alexander Bell) 1880	
	Greenwich Mean Time for train time tabling 1888
Dunlop's tyre patent rescinded 1894	
	First public demonstration of a radio (Oliver Lodge) 1897
Properties of cathode rays/electrons (Joseph Thomson)	

(continued)

Twentieth century	1900	Transmutation of atoms (Ernest Rutherford)
	Nuclear model of the atom (Ernest Rutherford)	1909
	1912	Prototype of modern X-ray diffractometers (William Henry Bragg)
	Structural analysis of crystals (William Lawrence Bragg)	1913
	1914	Atomic number/periodic law (Henry Moseley)
	Identified 212 of 254 naturally occurring, stable isotopes (Francis Aston)	1919
	1919	Artificial nuclear transformation (Ernest Rutherford)
	Proton as a nuclear particle (Ernest Rutherford)	1920
	1928	Anti-microbial penicillin (Alexander Fleming)
	Proof of the existence of neutrons (James Chadwick)	1932
	1932	First particle accelerator to split the atom (John Cockcroft and Ernest Walton)
	Molecular structure of nucleosides and nucleotides via chemical synthesis and degradation (Alexander Todd)	1930s and 1940s
	1937	J. J. Thomson's son jointly awarded Nobel Prize for diffraction of electrons by crystals (George Thomson)
	Partition chromatography (Archer Martin and Richard Synge)	1941
	1941	Maiden flight of propeller-less plane with jet engine (Frank Whittle)

(continued)

	Breaking the Enigma code (Alan Turing)	1941
	1942	Decryption of Lorenz ciphers (William Tutte)
	World's second electronic and digital computer with programmability—Colossus (Tommy Flowers)	1944
	1950s	Structure of the globular protein, myoglobin (John Kendrew)
	DNA's complex double-helical structure (Francis Crick and James Watson)	1953
	1955	Reagent to facilitate amino acid sequencing of complex proteins (Frederick Sanger)
	First nuclear reactor for domestic use (Calder Hall 1)	1956
	1977	Sequencing technique for nucleotides of DNA (Frederick Sanger)
	In vitro fertilisation (Patrick Steptoe and Robert Edwards)	1977
	1984	DNA genetic fingerprinting (Alec Jeffreys)
	DNA profiling (Alec Jeffreys)	1985
	1990	Enablers for the World Wide Web (Timothy Berners-Lee)
Twenty-first century	Methodologies for forensic investigations, including soil science (Lorna Dawson)	2011

Talent hits a target no one else can hit; Genius hits a target no one else can see.

Arthur Schopenhauer, 1818

Preface

Scope

From the cradles of civilisation to the twenty-first century, great scientific and engineering minds have changed the way we understand ourselves and the universe we inhabit. Certain, elite individuals have made monumental contributions to specific scientific and engineering fields, changing our world forever and propelling humanity forward.

This publication explores the scientific and inventive endeavours of some of our most noteworthy British scientists and engineers. Fifty of them are deceased; three are still alive and continue to make weighty contributions to the scientific world.

They all have one thing in common, namely commanding abilities which they applied to make a profound influence on mankind, thereby attaining a foremost place in the history of science and its practical application.

The manuscript looks at their childhood, education, notable achievements and decorations. It endeavours to make some of their complex theories more comprehensible and accessible. It aspires to set these things in the sociopolitical context of their place in history, sometimes wandering from their own activities to those of others, to round off their story.

In the introductory chapter, a sociopolitical context is set and the national and personal attributes of, and characteristics for, scientific and engineering greatness are defined. The second chapter explores scientific, agricultural and industrial "revolutions", trying to pinpoint the reasons why Great Britain was such fertile ground for the world's First Industrial Revolution. It asks why did it occur; why did it occur when it did; and why did it occur in Great Britain before other nations?

An account of each scientist or engineer is assigned to one of 14 chapters. In numerical order, these chapters deal with: (3) The Steam Age; (4) New Forms of Transport and its Infrastructure; (5) Holistic Approach to Public Health via Sanitation Technology; (6) Polymaths; (7) Natural Science; (8) History of the Atom; (9) Life Sciences; (10) Electricity, Magnetism and Light; (11) Palaeontology and Evolution; (12) X-ray Crystallography and Biomolecules; (13) Nucleosides, Nucleotides and DNA; (14) Science of Key Building Materials; (15) Communication, Telephone, Computers, WWW, the Digital Age; and (16) Solving Crime via Forensic Science. Chapter 17 guides the reader forward to consider current scientific and technological challenges and potential opportunities that lie ahead. How satisfying it would be if some of my readers were to rise to the challenge and become the great scientists or engineers of the future.

Examples of our scientific and engineering prowess are given in the 14 chapters. Each chapter begins with a précis. The author hopes that these prefatory tasters will whet the readers' appetite for a more detailed account of the luminaries appearing under the appropriate chapter heading. To dispel the adage that "a little learning is a dangerous thing", the author encourages the reader to devour more.

How Were These Fifty-Three Eminent British Scientists and Engineers Chosen?

Many of us have an unquenchable thirst for knowledge. The UK has been blessed with a growing number of relentlessly curious people having an insatiable desire to better understand the universe in which we live, hoping to uncover nature's secrets and mysteries. In the last 450 years, few other countries can claim such prowess and achievement in science and engineering.

I have singled out 53 exceptional British scientists and engineers whose search for fundamental truths has powered scientific and engineering endeavour. Fifty-three is, indeed, an odd number to choose. The target was fifty, but another three were so exceptional that they could not be omitted. There are many more deserving of inclusion. Some have been mentioned in the text because they laid the theoretical or practical groundwork for those selected.

The choice was based not only on their ingenuity and brilliance in their chosen field, but also on the extent to which their contribution had profound socio-economic consequences, improving the lives of mankind. I have chosen neither pure theorists (e.g. S. W. Hawking) nor solely pragmatists. Instead, I have concentrated on those who, by scientific and/or engineering wizardry, satisfied a fundamental human need, bringing about an extraordinary change to the way ordinary people conduct their lives.

Those chosen had time to explore and develop their exceptional thoughts devoid of a prescribed pathway. In the main, they did not have a route bull-dozed by their parents. It is noteworthy that, before they reached adolescence, ten of them either lost one parent or both or they were abandoned by both parents. Only four were from privileged backgrounds, and only three attended famous public schools. Three of them had no formal primary or secondary education. Twenty-five of them were not university graduates.

Living in a pluralist society and stable democracy, they had the freedom to imagine, exchange ideas, experiment, hypothesise, invest and fail, before success brought them to public attention. In previous centuries, it was the private sector, and not the state, that played the major role in key advances such as steam engines, canals, vaccinations, railways, cementitious materials, steel, electrical mechanisation, textiles, the telegraph, telephone and other forms of mass communication. Likewise, in the twentieth century, the computer, antibiotics, IVF and forensic science, all had their roots in academia or the private sector.

Sources of Information

In the main, I have used sources freely available on the Internet, supported by TV and radio documentaries, articles in newspapers and periodicals, together with my own knowledge and reference books.

Chapter lengths vary depending on the availability of biographical information.

In seeking validation of my synopses, I have contacted five of the contemporary scientists and engineers or their offspring, inviting comment on my draft scripts.

Inevitably, however, there will be errors of fact, for which I apologise. Should a second edition be forthcoming, I will endeavour to correct any errors of fact or misinformation about which I am notified.

No book can ever be finished. The moment one turns away from it, one becomes aware of omissions and shortcomings.

Acknowledgements

The project started as a casual outlet for a retiree during the Brexit debate. It became all-consuming during the COVID-19 quarantine period which made it easy for me to lock myself in my office. It did mean that I neglected my wife and family and I thank them for their forbearance.

I am extremely grateful to Prof. Alec Jeffreys and Prof. Lorna Dawson for sighting my draft document on their achievements and making valuable comments on my synopses.

I also appreciate helpful clarifications from Prof. Andrew Steptoe (son of Patrick Steptoe); Jenny Joy and, sister, Caroline Roberts (daughters of Robert Edwards); and Ian Whittle (son of Frank Whittle) on their respective fathers' story.

Every effort has been made by the author to ensure the accuracy of the information contained in this book. The opinions expressed herein are those of the author and not the publishing company.

Evenley, UK   John Bailey, BSc PhD (Cantab)
May 2021

Contents

1

Introduction

Many forms of Government have been tried and will be tried in this world of sin and woe. No one pretends that democracy is perfect or all-wise. Indeed, it has been said that democracy is the worst form of Government except for all those other forms that have been tried from time to time… (Winston S Churchill, 11 November 1947)

1.1 Background to the Manuscript

Listening to the endless debate about Brexit prompted me to think about what it is that makes us British, and ask, what have our pre-eminent scientists, engineers and inventors contributed to the world, as we know it?

I believe that, overwhelming, our success as a nation results from our freedom of expression, being essentially independent citizens, living in a pluralist society, and stable democracy. After over 750 years of service, the British parliament is one of the oldest representative assemblies in the world. Having lived under democratic governance for so long, we assume that personal freedom is an indispensable part of everyday life. In this regard, we have often led the way in establishing religious and racial freedom, and the protection of vulnerable groups.

Our laws and customs are derived mainly from our Christian heritage. Our individual civil rights and liberties, in an old, established country, have been won steadily and incrementally since Magna Carta, in 1215. Alongside rules and laws established by the legislature is common law, made by judges.

© The Author(s), under exclusive license to Springer Nature
Switzerland AG 2022
J. Bailey, *Inventive Geniuses Who Changed the World*,
https://doi.org/10.1007/978-3-030-81381-9_1

We do not have a codified constitution but an evolving array of acts of parliament, court judgements and conventions. In the main, changes to our democracy have been brought about, not by revolution, but by reform in an orderly, relatively stable, and rational way. Free citizens fight with words and not with swords. An ability to speak freely sits alongside an ability to think freely.

We have enjoyed freedom of speech and universal suffrage for longer than most other countries. Since the United Kingdom does not have a written constitution, the principle of parliamentary sovereignty exists which means that Parliament can create or end any law. Judicial independence means that, in the main, the judiciary is kept away from other branches of government, private or partisan interests.

The free exchange of ideas and opinions is the lifeblood of a liberal society. Our right to argue, challenge and, potentially, change minds, is key to a tolerant society. Our political system facilitates independent and radical thoughts, some straying outside the 'box', to question all, and to be creative. Our ability to have visionary dreams, particularly during the Victorian era, propelled us to a futuristic world.

1.2   National and Personal Attributes of, and Characteristics for, Scientific and Engineering Greatness

Knowledge can only be finite, while ignorance must necessarily be infinite. Science is not a body of theological truths, rather it is a means of inquiry to which disputation and debate are key. Since the data available at any given moment are not an unambiguous set of immutable facts, the understanding of evidence must always be provisional. Thus, the science of today provides a tentative explanation of the things around us and offers the best knowledge upon which we can currently rely. Its universal laws and hypotheses attempt to describe the structural properties of reality but are open, of course, to refutation and, therefore, adaptation. Science is a human activity in which prevailing theories are systematically challenged and advanced.

Super-intelligence alone is no guarantee of monumental achievement in any walk of life. Intelligence remains a complex human trait, being a dynamic mix of 'nature' and 'nurture'. Success in the scientific and engineering worlds requires creativity which is an amalgam of curiosity, open mindedness, imagination, and inventiveness.

Our cultural heritage continues to inspire us, having reshaped, not just the UK, but other nations as well. No country on the planet has given as much to the world, in terms of language, writers, the arts, science and technology, sport, political and judicial structures, as has the UK. We are a global trading power with the sixth largest economy in the world (based on gross domestic product in 2019). For centuries, our wealth has come from our ambition, innovative and entrepreneurial skills, and global reach.

The most dangerous people in any society can be its rulers. In the last century, many European countries fell under the spell of extreme political ideologies. We, however, are intolerant of undemocratic forces, such as Nazism, Fascism, Communism, and dictatorships—going to war to uphold liberty and free expression. On several occasions, we have rescued Europe from a tyrannical power, not because we were initially attacked, but because the sovereignty of a friendly nation had been violated. Our union of four nations, therefore, has been shaped by what we have done, and suffered, together. We chose political freedom because it offers a dignified form of human co-existence.

In the past half-millennium, we have had a noble tradition of welcoming genuine victims of persecution such as French Huguenots, European Jews, Ugandan Asians and perhaps next, Hong Kong Chinese. We should celebrate the values and systems that bind us, including tolerance, the rule of law, and democratic assent.

We are an amalgam of peoples and countries with diverse ethnicities and cultural backgrounds which have enriched us all. We have a long—often uneasy—history whose shared experience, at times of grave danger and ultimate success has come to represent one of the most powerful and enduring symbols of union, freedom, liberty, and democracy anywhere in the world. Stoicism and patriotic mutuality are our major strengths. We have an innate doggedness with quiet, yet invincible courage. We are valiant, steadfast and self-effacing.

We used to be accepting of individualism. We generated peace-time explorers (e.g., Cook, Rhodes and Livingstone) as well as great war-time admirals and vice-admirals (e.g., Nelson and Drake) field marshals (Wellington) and leaders (Celtic warrior, Boudica, Queen Elizabeth I and Churchill), the latter a Nobel Literature Laureate who 'mobilised the English language and sent it into battle'. Our explorers delivered overseas colonies, starting 500 years ago with Newfoundland, in 1497, and ending with the hand-over of Hong Kong, in 1997.

1.3   Dissemination of Information and the Importance of Education

In 1479, William Caxton brought printing to Britain, giving us a new process for disseminating information and circulating literature. The flexible, English language has one of the richest vocabularies of any tongue. It is ready to absorb words from other languages, adding to its versatility. Some of the world's greatest writers (e.g., Chaucer, Shakespeare, Austen, Scott, Dickens, Bronte sisters, Stevenson, Kipling, Rowling etc.) have recorded in English prose and poetry how we behave as individuals and as part of a wider society. Former hostage, Terry Waite, said, "Good language, like good music, has the capacity to breathe harmony into the soul".

Because English has become the world's favourite *lingua franca*, such literary giants are read internationally. Printing equipment facilitated this reporting process for posterity, whilst the recent digital age has accelerated the transmission of news and information to the masses. English has become the international business and social language.

William Forster's Elementary Education Act of 1870 was the first of several of acts of parliament, passed between 1870 and 1893, to create compulsory, publicly funded education for children. The 1880 Act made school attendance mandatory for all children aged between five and ten years of age, with the leaving age progressively raised since then. Literacy and numeracy improved, so, as well as leading the world industrially, Britain started to lead the globe in scientific thought. For today's generation, we have developed good primary and secondary educational facilities, not only private, but state schools as well.

It is noteworthy that at least three of the scientists chosen by the author had no formal education. For instance, George Stephenson, pioneer of steam locomotives and the first inter-city railway, was born to illiterate parents and learnt his engineering skills on-the-job. He, together with self-taught palaeontologist Mary Anning and self-educated natural philosopher (scientist), Michael Faraday, should be inspirations to girls and boys of humble origins who wish to become scientists or engineers.

After the death of the last Tudor monarch—Elizabeth, Queen of England—James VI of Scotland was proclaimed King of England, uniting the two crowns. When he addressed the English parliament in 1603, he asked, "*Hath He not made us all in One Island, compassed with One Sea and of itself by Nature so indivisible?*". James was an ardent unionist and, in 1604, he assumed the name and style of King of Great Britain. He was the first monarch to advocate for British political union, but this did not happen for over a century—during the reign of Queen Anne.

Whilst not necessarily a treaty of affection to everyone, the Act of Union in 1707 unified the English and Scottish parliaments, resulting in the creation of the United Kingdom of Great Britain. It is a relationship bound by covenant, not contract, and has proved to be one of the most successful and enduring political, social and economic unions in the World. Scientists and inventors from all four nations have been responsible for some major global advances. Ten of the world-famous Scottish innovators are reported herein. In the manuscript, their order of appearance is: Watt, Macadam, Dunlop, Cummings, Lord Kelvin (Irish-Scottish), Fleming, Maxwell, Lord Todd, Bell, and Dawson.

In 1721, Robert Walpole was made the de facto first Prime Minister of Great Britain. Since then, there have been 76 prime ministerships held by 56 different prime ministers (to 15 April 2020). Forty-six of those were educated at fee-paying schools or home educated. Thirty-three were educated at one of three prestigious public schools, namely Eton College (19), Harrow (8) and Westminster School (6).

Only ten PMs were educated at non-fee-paying schools, and seven of these held office between 1964 and 2019. Indeed, between 1964 and 1997, State grammar schools provided the UK with five successive prime ministers, before a public-school alumnus was restored, ironically by a Labour prime minister—Scottish-born, Tony Blair. Twenty-seven of our fifty-six different prime ministers were educated at Oxford University, fourteen at the University of Cambridge and four at Scottish universities.

This publication, however, demonstrates that a private education is not a passport to high achievement in the worlds of science and engineering. Of the fifty-three great scientists and engineers chosen by this author, only three (viz John Harrington, Robert Hooke, and Robert Boyle) attended one of the three elite educational establishments mentioned above. Like all talented people, the others had the right blend of intellectual equipment and work ethic.

We have some of the world-leading universities and these are not necessarily London-centric. the Times Higher Education (THE) World University Ranking's Table for the life sciences, in 2020, registers four UK universities in the top-25. For specialization in the life sciences, in 2020, there was only one university (Wageningen) in the top-25 that is located on the continent of Europe. In the physical sciences, there are three UK and four continental universities in the top-25.

Universities are one of our national assets. The UK is the second most popular university destination in the world, resulting in 'soft power' benefits.

In 2015, according to the Higher Education Policy Institute (HEPI), fifty-five world leaders (presidents, prime ministers, or monarchs) from fifty-one counties had attended higher-level educational institutes in the UK.

The University of Oxford is the oldest university in the English-speaking world. A history of teaching here started in about 1096 and the university was founded in the twelfth century. Since 1901, when the first Nobel Prize was awarded, fifty-one of its university community have become Nobel laureates. Oxford graduates are prominent in politics, the legal profession and serve as opinion formers in the media.

The Times Higher Education World University Rankings 2021 rated more than 1,500 universities, from 93 countries and regions, based on five core missions, namely, teaching, research, citations, international outlook and income from industry. For five consecutive years, Oxford University has achieved the top ranking.

The University of Cambridge is the second oldest English-speaking university, founded in 1209. It is especially famous for its contribution to mathematics and science, due largely to the long-term achievements of some of its elite graduates. These include Sir Isaac Newton, James Clerk Maxwell, Lord Kelvin, Sir Francis Bacon, as well as Nobel Prize winner, Lord Rayleigh.

The University of Cambridge is one of the finest universities in the world, boasting a total of 110 Nobel laureates—from numerous nations—among its affiliates (viz alumni, academics who carried out research at the University, visiting fellowships, lectureships etc.). This is equivalent to 11% of all Nobel Laureates awarded between 1901 and 2020. Its Cavendish Laboratory has been described as the greatest research centre for physics in the world. Twenty-nine scientists, from various countries who, at some time in their career, conducted research at the Cavendish, have been awarded the illustrious Nobel Prize.

Frequent reference is made to the Nobel Prize because it is the most distinguished and coveted global honour—the *ne plus ultra* for scientists. It can be awarded to a single individual or shared between two or a maximum of three laureates. Numerous scientists from Oxbridge and other British universities and research institutes have been the recipients of Nobel and Economic Science Prizes. Between 1901 and 2020, 603 prizes were shared amongst 962 Nobel laureates.

In the three Nobel categories of a scientific nature (namely, chemistry, physics and physiology or medicine) there were 624 Laureates who achieved the epitome of scientific glory between 1901 and 2020. Thirty two percent of these were American-born, 14% German/Prussian, 13% were born in United Kingdom, dropping to the 4th highest—5%, of French lineage. When the other three Nobel categories (viz Literature, Economics and Peace) are included, there are 962 Laureates and UK overtakes Germany in percentage terms.

Twelve of the scientists featured in this manuscript are Nobel laureates. One of them—Frederick Sanger—is the only person to have received two Nobel Prizes in Chemistry. William L Bragg, at 25, is the second youngest Nobel Laureate, the youngest being Malala Yousafzai who received the Peace prize at the age of seventeen.

Throughout history, women have been denied formal education and deterred from advancing professionally. To date, in 2020, a Nobel prize has been awarded 58 times to women. There are, however, only 57 different female laureates since Polish-born Marie Curie was honoured in both the physics and chemistry categories. Four of the female Laureates are British, one of them being Dorothy Hodgkin, a scientist, who received the Nobel Prize in Chemistry, in 1964.

The Royal Society of London for Improving Natural Knowledge, founded in 1663, is the oldest, independent, scientific academy in existence. Its members, together with those of the Royal Institution, have designed experiments to entice nature to reveal its secrets and mysteries. The Philosophical Transaction of the Royal Society—the oldest scientific journal in continuous publication in the world—reported on some of their discoveries.

Our universities and learned societies have been bastions of free speech, where received wisdom has been questioned and tested. The overly sensitive Generation Snowflake, and its no-platforming approach to suppress academic freedom, must not be allowed to prevent counter opinions being aired, otherwise democracy is threatened. Rigid, prescriptive attitudes and thinking must be resisted so that free expression can continue and flourish.

In the past, our laissez-faire approach to debate has generated a disproportionate number of citizens with enquiring minds who defied convention. Disagreements (e.g., Newton and Darwin) with the Establishment have been an expression of healthy and vibrant liberty. Opposing views about religion and politics are an integral part of a democratic society. For many years, religion was seen to be an obstruction to scientific progress. Eventually, as religious influence waned, scientific truth began to prevail over religious dogma.

EU Remainers cry that our departure from the EU will be hurtful since it will mean less scientific collaboration. Collaborative efforts—where novel thinkers come together, seeding each other with ideas and sparking innovation—is clearly desirable and beneficial. However, big discoveries in the past have normally been made by individuals or small groups. In general. bigger groups with a more corporate approach to research tend to stifle curiosity. An exception would be cost-prohibitive projects such as CERN's Hadron Collider where international collaboration and a multi-disciplinary approach has been essential to its success.

Opponents to Brexit claim that barriers to research collaboration will be erected when we depart from the European Union. History tells us that being an island nation has encouraged us to explore, not only the world (e.g., Cook, Drake, Raleigh etc.) but what makes the world, the planets, and other galaxies (e.g., I. Newton, S.W. Hawking). For four centuries, Britain has been a science and technology superpower.

All scientists want to learn something about the riddles of nature and the world in which we live. The main characteristics of the scientists detailed in this publication are their shear brilliance and total domination of their subject. Those now deceased were so rare that they numbered one in several hundred million of the world's citizens. Yes, they accustomed themselves with all relevant material on their subject, and they aired their hypotheses and tested their theories with fellow scientists, both informally and at learned societies and international symposia, but they were single-minded in their research and invariably wanted precedence of publication. Scientific institutions are normally challenging rivals. As is the case in most sectors of society, competition is the spur for major innovation.

None of the scientists and engineers mentioned herein were requiring of a 'nanny state'. On the contrary, they were self-motivated, self-possessed individuals with a zest for knowledge and practical application in their field. The way of science is cluttered with discarded theories. A combination of passion and perseverance helped these men and women power through inevitable disappointments, uncertainty and disheartening failures before success was finally achieved. Some of them were brilliant communicators (e.g., Davy and Faraday) whilst others developed into gifted teachers (e.g., Dalton and JJ Thomson).

1.4   Technological Advances; The Victorian Era and Empire

For almost two centuries, ending in about 1875, most of the technological advances in the world were invented in Britain, or, put to large-scale use here. We ushered in the first Industrial Revolution—designing and building innovative machines for factories and transport. Great Britain was particularly transformed during the Victorian era when ambitious, brilliant engineers devised amazing inventions which revolutionized our lives and laid the foundations for the modern world in which we live. After Napoleon Bonaparte was defeated, at the Battle of Waterloo, in 1815, Britain focussed on industrial expansion and avoided any costly European entanglements until WW1, in 1914.

For a long period, Britain was pre-eminent in science; the world's powerhouse, driven by brilliant technologies, creating world-beating products that were exported around the globe, eventually accruing socially desirable benefits, including universal education and health care. By the mid-nineteenth century, a quarter of international trade passed through British ports and more than a third of the world's industrial output of traded commodities poured out of British mines, mills, factories and workshops. Britain's output per worker was higher than any other country.

In the nineteenth century, Britain was at its zenith of power, its sphere of interest extending across the world. At one point, we ruled the biggest empire ever seen. Presently, Britain is suffering a an increasingly troubled relationship with its past. Sadly, like most colonial nations, Britain has had its shameful moments, but these should be judged in the context of the international standards of the age. History is ridden with moral complexities. Our ancestors had different perspectives and understandings of right and wrong. Whilst not air-brushing our past, neither should we denigrate the whole of our history—the British Empire was not inherently wicked.

Almost everyone harbours a racial bias but how can white people today readily correct any complicity with, or investment in, any racism of our forefathers? Should a company with colonial or racist roots in the nineteenth century, but behaving scrupulously since then, be punished for what happened two centuries ago? Should a coincidental association with a tainted relative be held against our ancestors or us? We must build up bonds of social trust and mutual respect with our contemporaries in diverse communities.

Selective amnesia about our past is to be avoided and a binary debate about Empire and race is not helpful. Of course, it would be beneficial to us all to better understand the moral and political norms of those historical times so we can educate our children to do better. A non-sectarian society is what most of us desire. Some want to purge themselves of inherited guilt but what are the solutions—more hate; entrenched divisiveness; endless recriminations? Whilst self-flagellation may make some white people feel better about themselves, it is a poor substitute for constructive activism in today's world.

Conscious of our impressive history, we have memorialised our exceptional citizens in the form of statues, building and street names. Segments in our society, now query our legacy as a colonial power, generating the notion of hereditary guilt, attacking Britishness and Britain's contribution to the World. Suffering from post-colonial guilt syndrome, they are critical of sentimental jingoism and Empire nostalgia and attack our national institutions, public monuments, historical artefacts, flag and heritage. They deface sacred memorials and want to silence the lyrics of patriotic songs. Such citizens have an abiding sense of imperial guilt gnawing away at part of their national conscience. Many of them have an anti-capitalist agenda.

While generally promoting pride in Britain as a great nation, we must recognise that, deplorably, it has also committed several horrific crimes. Naturally, whilst celebrating our historic achievements and glories, we should also feel shame at past disgraces but remember that we cannot fully atone for the historic failings of our forebearers because we did not commit them. It is a peculiarly contemporary, but short-sighted presumption, that our forefathers knew what we know now, and had the same values that we have today. Standards evolve according to political whim and the national mood.

Few of our citizens have lived perfect lives, and no atrocity should be sanitized. It will not be possible to teach the lessons of reprehensible historical events, and chart a better future, if we deliberately expunge those happenings. We cannot 'unwrite' history, but we could rewrite it, contextualising it to give a more balanced view of our colonial past. Even great men and women can make great mistakes, but their towering accomplishments invariably outweigh their failings. For those notable achievers with a chequered history, the contentious aspects of their legacy, as well as their accomplishments should be recorded.

Rather than ruminating excessively on past shortcomings, we should concentrate on solving current and future problems, using our intellectual, economic, political and cultural influence to solve them. Where there are polarised views, we should unite behind plans that bridge societal divides. With shared endeavour, we can build a genuine multiracial meritocracy where constructive difference of opinion is encouraged, whilst hateful division is acknowledged to be destructive, and should be avoided. Scientific and engineering inputs can contribute to achieving a better society.

2

"Revolutions"—Scientific, Agricultural and Industrial

The problems of the world cannot be solved by sceptics or cynics whose horizons are limited by the obvious realities. We need men (and women) who can dream of things that never were and ask, 'Why not?'. (John Fitzgerald Kennedy. Address before the Irish Parliament, Dublin, 28 June 1963)

2.1 Revolutionary Change

A revolution is a profound turning point in history. It might refer to a period of radical colonial, social or political upheaval that normally occurs over a relatively short period of time.

Examples of colonial, social or political revolutions of note include the American Revolution (colonial; 1775–1783), the French Revolution (social; 1787–1799) and Russia's October Revolution in 1917 (political). England's Glorious (or Bloodless) Revolution in 1688 was neither a revolution nor a coup d'état. However, coupled with the Bill of Rights 1689, it permanently established a constitutional monarchy by which parliament controlled the monarch, met frequently and made the laws of the land.

There are other types of revolution that take place over decades, even centuries and examples of these will follow.

© The Author(s), under exclusive license to Springer Nature Switzerland AG 2022
J. Bailey, *Inventive Geniuses Who Changed the World,*
https://doi.org/10.1007/978-3-030-81381-9_2

2.2   The World's Scientific Revolution

The Scientific Revolution represented a period of drastic change in scientific study during the sixteenth and seventeenth centuries. Science became an autonomous discipline, distinct from both philosophy and technology. Developments in mathematics, physics, astronomy, biology and chemistry transformed society's view of the world around us. No longer did people simply theorize about how the world worked. Instead, they used individual experience, abstract reasoning and scientific exploration to gain actual knowledge.

They called for a more in-depth method of scientific inquiry and the development of exhaustive proof. They applied an 'inductive approach' to science, whereby evidence came first, and hypotheses followed. They conducted controlled, methodical, experimental procedures, espousing the view that observation, experimentation, and reliable measurement would further scientific-understanding and reveal Nature's secrets, thereby unlocking the mysteries of matter. Quantitative results were subject to validation and verification.

The Enlightenment (aka Age of Reason) was an intellectual and cultural movement in the eighteenth century that emphasized reason over superstition, and science over blind faith.

Britons have long been explorers who believe in human progress and scientific advancement. The Scientific Revolution in Great Britain prompted higher learning and communication to permeate all social classes which in turn benefitted the scientific discipline. The burgeoning interest in the understanding of nature prompted the use of science rather than religion to explain natural phenomena.

Learned societies, dedicated to the advancement of science, fuelled the spread of knowledge. Those such as the London-based Royal Society (founded in 1660) and the Royal Institution (1799), as well as the Royal Society of Edinburgh (1807) received royal patronage. In Birmingham, scientists, engineers and industrial luminaries met monthly at the Lunar Society (1766). Consequently, science became embedded in the State, its institutions and culture. In the coming years it would help Britain rise in international prominence.

Basic or pure science seeks to expand knowledge but is not focussed on developing a product or a service of immediate public or commercial value. On the other hand, applied science (or technology) uses scientific knowledge to solve a defined, real-world problem or exploit an opportunity.

Whilst science and technology in the UK has a long history, its standing reached its zenith in the second half of the seventeenth century. This marked the emergence of modern science when British scientists were constantly making discoveries about how nature works and how we might harness it to enhance human power over it.

After his *annus mirabilis*, in 1666, the theorist, Isaac Newton, emerged as the figurehead for British scientific achievement, setting science on its modern course, replacing abstract ideas. The age of light replaced the age of darkness. The Scientific Revolution was truly an era of scientific enlightenment.

2.3 Britain's 2nd Agricultural Revolution as a Precursor to Its Industrial Revolution

Britain's so-called 2nd agricultural revolution was not a single event, but a series of episodes during which significant and substantive changes were made to land usage, farming methods and agrarian technology. During this period, the British agricultural sector advanced more rapidly than it did in any other European nation.

In one hundred years, between 1670 and 1770, agricultural output grew faster than the population. Thereafter, productivity remained amongst the highest in the World, until about 1870. The increase in food supply contributed to a rapid growth in population in Great Britain, virtually doubling from 5.5 million in 1700, to 10.5 million in 1801.

The increase in agricultural output can be traced to 9 or more inter-related factors:

(1) Perhaps the most important factor, in the overall mix, was the enclosure of land. A series of Acts of Parliament authorized the enclosure of open fields and medieval common land, creating legal property rights and ending traditional grazing and crop-growing rights. The controversial legal processes for this privatization, particularly in the eighteenth century, led to small holdings being consolidated to create larger fields and farms where the property-owning farmer was free to adopt better farming practices.

This resulted in a major reduction in small landholders compared to the Continent, bringing more land into effective agricultural use. It helped the larger landowner redistribute land in his favour but disadvantaged the landless. In these farming communities, deep inequalities developed. But the shift to large-scale commercial farming did mean that the land was managed more efficiently and economically than it was under traditional subsistence farming.

(2) The more productive enclosed farms meant fewer agricultural workers were needed to farm the land. Many left rural areas and moved to the cities in search of work in the emerging factories built during the Industrial Revolution which started in about 1760. By 1850, only 22% of the nation's workforce worked in agriculture—the smallest proportion for any country in the World.

(3) Increased availability of farmland by utilising previously barren land and reclaiming fen land.

(4) The planting of new crops such as (i) turnips which could be left in the soil over winter, reducing the area of fallow land and (ii) nitrogen-fixing legumes and clover. The latter is excellent for pasture.

(5) The "little ice age" had ended, so a climate more favourable to crop growth resumed.

(6) Better animal husbandry, boosting livestock size and quality via (a) regulated selective breeding by mating two animals with particularly desirable characteristics and (b) the ability of ruminant animals to forage on turnips during the winter. Animal waste was used as fertiliser, returning nutrients to the soil. This, together with nitrogen-fixing clover and legumes helped improve soil health and sustained cereal yields.

(7) Improved crop yields by, for example, adopting the Norfolk 4-course crop rotation system to restore soil fertility and reduce the risk of pests associated with a monoculture,

(8) Development of farm machinery including:

   (a) The Dutch plough with improved shape and, in 1730, Joseph Foljambe's Rotherham swing plough with iron fittings and an absence of depth-wheel. These ploughs could be pulled with fewer oxen or horses.

   (b) Attempts to supplant "broadcast" sowing by hand—Jethro Tull's 1701 wheeled seed drill gave the correct spacing and drill depth but was too fragile to be widely accepted. It was not until the 1800s that mechanical drilling became widely accepted in England. Two Suffolk drill makers (J. J. Smyth of Peasenhall and R. Garrett of Leiston) helped to popularise it.

   (c) In 1786, Scottish millwright, Andrew Meikle invented the first successful threshing machine which separated seed from husk and stalk.

(d) A hay-tossing machine was invented in the early 1800s by R. Salmon of Woburn, reducing the labour needed to turn and dry hay.

(9) The success of regional markets was aided by the expansion of Macadam roads, inland waterways and railways (after 1825).

Market-oriented agriculture developed with farmers becoming free-market capitalists. Except for the Corn Laws (1815–1846) the lack of internal tariffs, custom barriers and feudal tolls made Britain the largest coherent domestic food market in Europe, differentiating it from other nations which had regional markets and took protective measures against foreign competition. In fact, free trade in all goods improved the prosperity of almost every man, woman and child in Victorian Britain.

The agricultural advances proved to be a major turning point in Britain's history. They transformed the countryside and created sufficient food supplies for its citizens to be better nourished and healthier, so the growing population far exceeded earlier peaks. Infant mortalities fell.

Agricultural productivity increased to such a degree that, even after satisfying the needs of an expanding population, many agricultural workers found themselves redundant. As farming became less labour intensive, unemployed agricultural workers and their families were forced, by their circumstances, to migrate to look for wage labour in cities or overseas.

From the 1870s, domestic food production increasingly gave way to food imports, as the population of Great Britain increased to 37 million by 1901. Open to cheaper imports, there followed a period of agricultural depression until 1914.

2.4   Birthplace of the First Industrial Revolution—Why did it Occur? Why did Occur When it did? Why did it Occur in Great Britain Before Other Nations?

In England, before 1760, work was mainly conducted by villagers in, or near, their own homes. Jobs in agriculture were seasonal and there was much agrarian hardship. Britain's damp climate is ideal for raising sheep. This led to a long history of producing textiles like wool, linen and cotton. Textile weavers and other artisans worked at home or in small workshops. Other small-scale industries included metal production in the midlands and coal mining in the northeast. For ordinary families, life was a struggle with a constant battle against famine and a wicked landlord or employer.

The first Industrial Revolution took place, in England, with a long sequence of events, starting in about 1760. It was more of a gradual evolution than a revolution. It is associated with a relatively peaceful social revolution. Within two generations—roughly between 1760 and 1840—the customary way of achieving production changed. Although it had no clearly defined beginning or end, it is an epochal period in history when the pace of change accelerated due to the coupling of technology with industry.

The Industrial Revolution marked a period of radical development that transformed largely rural, agrarian societies into industrialized urban ones. Its birthplace was England. The question sometimes asked is, "What was different about the economy of England in 1760 compared to Florence in 1300, China in 500, Rome at the time of Christ, or Athens at the time of Plato?".

Firstly, the surplus of agricultural labour coincided with the need for labourers in the new emerging industries that were being established during the Industrial Revolution. As the nation had more people working in factories and mills, so there were more people to purchase the goods they manufactured.

Goods that had once been painstakingly crafted by hand started to be produced in mass quantities, by a machine, in a factory. The Industrial Revolution is associated with new chemical manufacturing (e.g., bleaching agents for cotton, and fertilisers for the land); iron and steel mass production processes; improved efficiency of waterpower; increasing use of steam power from static and mobile steam engines; the development of machine tools and the rise of the factory system.

Pre-industrial society was very static and often cruel. Even then, there was child labour, dirty living conditions and long working hours. The Industrial Revolution in England brought about a relatively peaceful social revolution. By 1820, it was normal to bring workers into a factory situation, where they were overseen.

By our modern standards, the industrial towns where they lived were unenviable slums and working conditions were harsh. But to many people who had come from a cottage in the country to an urban terraced house, it was liberating. The lives of some of the poor were uplifted by soap; cotton clothing; coal in an iron range; earthenware pottery in their kitchen; an iron bedstead; glass in the window frames, and a better choice of food.

Necessary improvements to sanitation and the health of the urban poor were triggered by several acts of parliament and implemented by civil engineers such as Newlands and Bazalgette. Florence Nightingale improved nursing conditions, Jenner immunised the population against smallpox, whilst Lister pioneered antiseptic surgery.

The North of England was the heart of the Industrial Revolution, becoming the powerhouse of the British economy. At their peak, some centres of production represented the most advanced economies in the entire globe. The Industrial Revolution witnessed the triumph of the new middle class of industrialist and businessmen over the landed class of gentry and nobility. Producers in towns organised themselves into guilds to represent the interests of their specific trade.

The people who made the revolution possible were, in the main, practical men, with little formal education. By their achievements, they attained intellectual equality with classically educated contemporaries. Grammar schools had come to prominence in the sixteenth century. Located in towns, they taught classical subjects. At the time, the Universities of Oxford and Cambridge, took little interest in modern science or engineering subjects and were closed to those who did not conform to the Church of England. It was the Dissenting Academies that devoted more attention to the teaching of mathematics and scientific disciplines.

2.5 Why Was Britain So Receptive and Responsive to the Application of Science, Technology and Engineering in Industry?

In the distant past, Great Britain was perceived as a little, isolated, irrelevant, sea-faring island on the margins of continental Europe. Now, the United Kingdom of Great Britain and Northern Island comprises a quilt of four constituent countries, a family of nations, and a nation of families. At its height, and for over a century, the UK and its predecessor designations presided over the biggest empire in history and was the foremost global power.

As an island separated from mainland Europe, it was not ravaged by military and financial plunder during the many European wars in the seventeenth to nineteenth centuries. After the Napoleonic Wars (1803–1815) it possessed an enormous mercantile fleet and the most powerful navy able to protect sea lanes and links to distant trading posts. For years, it was chiefly interested in trans-continental commerce, dominating world trade. By 1913, the British Empire reigned over 412 million people or 23% of the World's population. but its industrial pre-eminence declined after WW1.

Great Britain became a constitutional monarchy with well-respected common law and property rights. Common law is rooted in centuries of English history and consists of rules and other doctrines developed gradually by the judges of the Royal courts. It emphasises the centrality of the judge and the idea that law is founded in the distillation and continual restatement of legal doctrine through the decision of the courts. Much of international commerce prefers the common law system because it is pragmatic, relatively quick, more responsive and better suited to fleet-of-foot businesses. In contrast, the approach by the EU and others is weighed down by civil law systems derived from Roman law but latterly based on 19th century codifications.

Relying on its mercantilism and imperialism, Britain's inter-continental trade links unlocked mass markets, providing economies of scale for its manufacturers.

Commitment to science and its application would eventually form the basis for success in commerce and industry, driving the economy and national prosperity. Its long tradition of technological ingenuity and scientific achievement resulted in, for example, static and mobile steam engines for a multitude of industrial sectors. Engineers were the heroes of their time, making their steam engines ever more versatile, powerful and efficient.

In textile mills—where much of the productivity growth originally occurred—key developments included the flying shuttle (John Kay 1733), spinning jenny (James Hargreaves 1764), water-powered spinning frame (Richard Arkwright 1769), spinning mule (Samuel Crompton 1779) and high speed, cast iron, power looms (Richard Roberts 1822). In Cromford, Derbyshire, R Arkwright built the prototype of the modern factory, including the first factory housing development.

Textile machinery and steam engines were coveted by Continental governments to promote their infant industries. Machine tools were exported legally or clandestinely for the autonomous development and emancipation of Continental industries.

Starting in the sixteenth century, deforestation led to a scarcity of wood for both lumbar and fuel. Shortage of wood was relieved by the exploitation of coal for which the UK had ample reserves which were relatively easy to access. The country's transition to coal as its principal energy source was virtually complete by the end of the seventeenth century. In the early 1700s, Britain had the cheapest energy in the World. Coal production rose from 2.5 to 10 mt in 1800.

The mining and distribution of coal would set in motion some of the dynamics that led to Britain's Industrial Revolution. The coal-fired steam engine was, in many respects, the most important enabling technology of the time. During the Steam Age, the steam engine turned the wheels of mechanised factory production. No longer did the manufacturer have to locate his factory near sources of waterpower. Instead, large enterprises began to consolidate in rapidly growing industrial cities.

As well as land ownership, the notion developed that ideas and innovation should be afforded better protection. It is generally considered that the concept of patents and patent law started in Venice, in 1474, mainly in the field of glass making, to give legal protection against infringers. The granting of intellectual property rights spread to England, but it was further formalised by devolving its supervision to independent courts. In this way, seventeen years of statutory protection could be obtained.

Under the developing patent system, James Watt, for example, was able to reveal the detailed workings of his varied steam engines, whilst enjoying a state of monopoly in their production and sale.

The shortage of wood necessitated a switch from wood charcoal to coke (a product of coal) in the iron smelting process. This would trigger developments culminating in the mass production of affordable steel, helped by the availability of cheap iron ore.

The Railway Age accelerated when G. Stevenson used steel to construct the rugged rail track for the 1st inter-city railroad upon which his son's steam locomotive, the Rocket, would travel. As well as iron, novel construction materials such as concrete and steel were used by Brunel to build a tunnel under the Thames, magnificent railway bridges and ocean-going passenger liners.

Industrialists located domestic supplies of other key raw materials such as lead, copper, tin, limestone, pottery clay, as well as freely available waterpower. Rubber and cotton were imported from the dominions.

The Industrial Revolution was an exciting and productive time of intense research and development in practically every field of scientific exploration, spurred on by a universal entrepreneurial spirit. With an abundance of capital for investment, Britain had the pre-requisites of commercially successful innovation, and the social needs and resources to sustain and institutionalise a process of economic expansion, based on technological advances. In a world of commercial opportunities, bankers and industrial financiers rose to a new prominence. In 1773, the London Stock Exchange was established, albeit unregulated at the time.

In 1776, the Scottish economist, Adam Smith, wrote his seminal work, "The Wealth of Nations". He defined real wealth as the output of the land and manufacturing, together with the labour of the society to generate goods which command value-in-exchange. He argued that (i) the division of labour in the economy results in a web of mutual interdependency that promotes stability and prosperity through the market mechanism and (ii) the means of production and distribution should be privately owned, unfettered by regulation.

Economic progress in Britain was assisted by a Protestant work ethic and confidence in the rule of law. There was investment in learning. By international standards, the British people were relatively well educated, with literacy rates of 60% amongst males, with some highly skilled. These are attributes essential for a highly technical revolution to unfold.

Its stock of knowledge was increasing rapidly. There was an increasing professionalism in science and engineering. Technology started to revolutionise the world, particularly when it substituted capital and energy for labour. Mechanical engineers played a vital part in the increasing mechanisation of manufacturing industry and transport.

In academia, background scientists such as Hooke, Boyle, Black and Joule were studying the relationship between heat, energy and mechanical work. Faraday was working on the links between electricity and magnetism, laying the foundations for an industry based on electric motors, alternators and generators. Clark Maxwell was investigating electromagnetism, preparing the world for revolutionary technologies of the future.

Concurrent with the increasing output of agricultural products and manufactured goods arose the need for more efficient means of delivering these products to market. Construction of an extensive network of canals started in about 1773, to be followed by railways after 1825.

The economies of the World would benefit from the knowledge expansion begun in Britain. Industrial revolutions eventually followed in Europe and America but not immediately in Asia or the Middle East, despite these regions previously having empires and a strong scientific footing. British industry blossomed in the technology-intensive sectors such as textiles, railways, steam navigation, telegraphy, telephony, ceramics, sanitary wear, steel and cement. In 1912, the textile industry of Great Britain reached its peak, producing about 8 billion yards of cloth. All these industries relied on the energy released from coal which had held the solar energy captured over millions of years by fossilised forests.

In the twentieth century, however, different geological stores of photosynthetic energy were exploited, namely oil and natural gas. At the start of WW1, in 1914, the industrial heartlands changed, with foreign markets setting up their own manufacturing industries. The golden age of British industry came to an end.

In the late nineteenth and early twentieth century, there was a 2nd Industrial Revolution. Countries such as America and Germany were quicker to develop industries based on petroleum, such as the internal combustion engine and organic chemistry (e.g., dyestuffs and synthetic polymers). They also majored in other industrial sectors such as electrical engineering, optical instruments, photography and cryogenics.

Now in the twenty-first century, our love affair with petroleum, the internal combustion engine and plastic is coming to an end. New industries are emerging based on electricity as the powerhouse with a new focus on saving the planet from man's neglect.

3

The Steam Age—Evolution of Steam Engines and the 1st Steam Locomotive

That works very well in practice, but how does it work in theory? (Various scientific, economic, and political commentators)

3.1 Précis. The Age of Steam Power

Britons were world pioneers of the Steam Age. Thomas Savery built the first operational steam pump with no moving parts. Thomas Newcomen designed an atmospheric steam pump. James Watt was proposed by some commentators to be the "Father of the Industrial Revolution" because he designed super-efficient beam, rotary, and double acting steam engines for a host of emerging industries. Richard Trevithick built high pressure steam engines in both stationary and mobile formats, one of the latter becoming the first steam locomotive to operate on cast iron rails. His prototype was the progenitor of railway engines. Robert Stephenson designed and built the Rocket steam locomotive which became the template for steam locomotives across the world for the next century.

By the late seventeenth century, a national shortage of wood had created a demand for coal as a substitute fuel. Whilst Britain possessed huge reserves of coal, some of its extraction required deep mining techniques. However, the curse of coal mines was their tendency to fill with water. Likewise, the flooding of copper and tin mines, particularly in Cornwall, was of great

© The Author(s), under exclusive license to Springer Nature
Switzerland AG 2022
J. Bailey, *Inventive Geniuses Who Changed the World*,
https://doi.org/10.1007/978-3-030-81381-9_3

concern. Two Devonshire men—Savery and Newcomen—used their engineering skills to build static steam engines that could be used to pump out unwanted water from mines.

The application of steam pumps in coal mines led to a colossal expansion of the coal industry, at relatively low cost, because the coal fines that were usually waste material could be used to generate steam.

The invention of the steam engine was crucial to the industrialization of modern civilisation. Before them, we relied on power generated by wind, water, humans, and animals. Steam engines rank amongst the greatest inventions of all times. Their artificial source of power facilitated the exploitation of our mineral wealth.

They relied on the burning of coal to produce heat to vaporise water into steam. The subsequent condensation of steam in a confined space created a vacuum which facilitated either a siphoning process or the movement of a piston.

The steam engine was a complex invention that underwent a process of incremental development which, over the years, incorporated many important innovations. These resulted from an increased understanding of 'atmospheric pressure' and the nature of 'vacuum', as well as novel engineering improvements.

Later, steam engines were used to hoist coal and mining machinery leading to an abundant supply of low-cost coal. At the time, long-distance freight was carried by road or canal. The fastest way to move people between urban centres was by horseback. Engineers and industrialist would use steam engines to power trains, steamboats and various machines present in manufacturing sites across the industrial world. The steam engine, in various formats was one of the most successful inventions of all times.

As they got smaller, steam engines could be set up wherever mechanical power was needed. They powered Great Britain to prominence as the first industrialized nation in the World. Britain then emerged as the most powerful trading nation in the world.

Over a period of 131 years, between 1698 and 1829, six British engineers and entrepreneurs were largely responsible for the development of the steam engines and steam locomotives that contributed so significantly to the first Industrial Revolution. These great British men were Thomas Savery, Thomas Newcomen, James Watt, Richard Trevithick, George, and Robert Stephenson.

During this period of advancement, several excellent, British background scientists were working in related fields. These include Robert Hooke, Robert Boyle, Joseph Black (theory of latent heat {'hidden' heat and energy in steam}) and James Joule who determined the exact rate of exchange at which mechanical work is converted to heat.

However, the inventions that these six engineers assembled appear to have been stimulated, not by the application of scientific knowledge and theories, but by their familiarity with on-site technical operations, their ingenuity, personal engineering skills, practical expertise, craftsmanship, repeated improvement trials, as well as a stroke of luck. Learning resulted from attempting new things.

3.2 Thomas Savery (C. 1650–1715)—1st Generation Steam Engine with No Moving Parts

Thomas Savery was born in a manor house near Modbury in Devon. He was a member of a well-known Devonshire family. He was well educated and became a military engineer.

In 1698, Savery invented and patented the first, crude steam engine. It had no piston and no moving parts. A coal-fired steam boiler was connected to a cylinder that was double-valved to (i) a suction pipe and (ii) a discharge pipe. Once the cylinder was filled with steam, cold water was introduced, causing the steam to condense, creating a partial vacuum, and drawing water up the suction pipe. The steam supply valve was then opened, and the boiler pressure forced the water up the discharge pipe for ground-level disposal.

Such a 'pump' could be used to drain water from mines and was more effective than using horses hitched to a turnstile. It was known as the "miners' friend" or an "engine to raise water by fire". It continuously converted heat from fire into useful work. However, it could only raise water 20 feet and required steam pressure so high that the soldered joints often gave way.

Savery 'protected' his design with a broad, 14-year patent that covered "all vessels or engines for raising water or occasioning motion to any sort of millworks by impellent force of fire". An Act of Parliament extended the protection for a further 21 years.

Because Savery was a prolific inventor, he was invited to become a fellow of the Royal Society in 1706.

3.3   Thomas Newcomen (1663–1729) Atmospheric, 2nd Generation Steam Engine and Pump Shaft

Thomas Newcomen was born in Dartmouth, Devon, about 15 miles from Savery's birthplace. Newcomen grew up to be a blacksmith and then an iron-monger, specialising in designing and manufacturing tools for the mining industry. Some of his customers were tin miners who faced difficulties with the flooding of their mines, as they became deeper.

In 1712, Newcomen side-stepped the problems with Savery's high pressure steam and used, instead, the power of atmospheric pressure. He invented an engine that operated by injecting water into a cylinder containing steam. The steam condensed to create a partial vacuum, whereby a piston returned to its original position due to atmospheric pressure acting on its upper surface. Through a rocking beam, this power stroke raised a pump rod hanging within the mine shaft, and so lifted water out of the mine.

He made alternating steam admission and water injection self-acting. The 'working stroke' of the engine could be repeated indefinitely, providing the source of steam could be maintained. In this way, he created a stereotypical pumping engine which, with refinements, lasted for 200 years.

To achieve his objective, Newcomen replaced Savery's simple steam condensation cylinder with a cylinder containing a piston and added a lever to transfer the force of the piston to a pump shaft. By utilising the sequential expansion and condensation of steam, the engine transformed heat energy into mechanical motion. When a piston is made to move, first one way, and then the other, it is called a "reciprocating" action (Fig. 3.1).

Casting the cylinder and getting the piston to fit tightly was pushing the limits of technology at the time. Newcomen deliberately made the piston slightly narrower than the cylinder and used a ring of wet leather or rope to make a tight seal. Such a feature meant that a Newcomen engine was not prone to bursting, unlike a Savery engine.

His 'atmospheric steam engine' is the world's oldest known practical steam engine for pumping water, used widely for the drainage of mines and raising water to power waterwheels.

Unfortunately, because of the wide scope of Savery's patent, Newcomen was required to build his engines under licence from Savery. The first Newcomen engine was installed at the Coneygree Coal Works near Dudley Castle, in Staffordshire, in 1712. It operated at 12 strokes per minute and could raise 10 gallons of water up to 156 feet on each stroke.

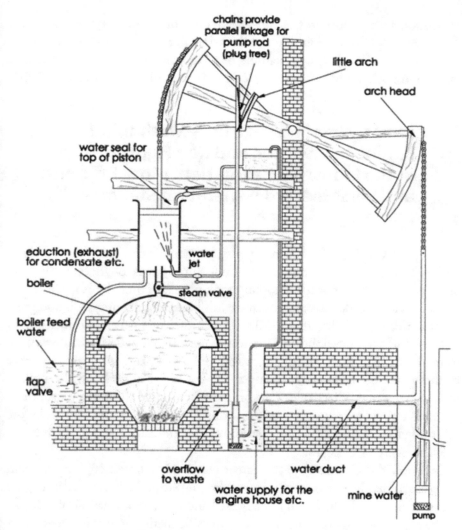

Fig. 3.1 Newcomen's atmospheric steam engine

Thereafter, this basic engine was manufactured, virtually unchanged for about 75 years and over 2,000 were built. Because of its voracious appetite for coal and its low efficiency, it was uneconomic for Cornish miners. Instead, it was confined mainly to collieries where coal for steam generation was cheap.

Due to the extensive royalties paid to Savery, Newcomen failed to gain financially from his engineering breakthrough, and he died, not an affluent man, in 1729.

One of his original engines—built in 1725—can be seen at the Newcomen Engine Museum (TQ6 9YY) in Dartmouth, working, albeit with hydraulics, for demonstration purposes. A full-scale replica of the Coneygree engine can be seen, 1.5 km from the location of the original, at the Black Country Living Museum (DY1 4SQ). A late example of a Newcomen engine can be seen at the London Science Museum (SW7 2DD). It was built by F Ashover in 1791.

3.4   James Watt (1736–1819)—Mechanical Engineer, Inventor, and Key Figure in the Industrial Revolution. More Efficient and Versatile, 3rd Generation Steam Engines

James Watt was born in Greenock, Renfrewshire, Scotland, then a fishing port, where his father—a carpenter—built ships. James' boyhood health was often poor and much of his schooling took place at home, by his mother.

He had an aptitude for mathematics; exhibited engineering skills and great manual dexterity. With these skills he became an instrument maker and repairer, but then developed an interest in steam engine design and went on to improve steam engine technology. A popular story is that he became aware of the power of steam by watching it lift the lid of a kettle on the boil. A pressure cooker is, effectively, a piston making a single stroke, in a cylinder.

James trained for a year as a scientific instrument maker in London, before returning to Glasgow, repairing instruments for the astronomy department at the University of Glasgow.

In 1763, the University gave Watt a model Newcomen engine to repair. He realised that heat was wasted by condensing steam inside the cylinder, the substantial mass of which had to be heated and cooled for every double stroke of the engine. Recognising its short-comings, he made two key improvements to the Newcomen engine—these were: separate vacuum and condensing chambers (in 1765) and rotary motion, in 1781. These two modifications were, possibly, the two most significant improvements to steam engines.

In the first advance, a vacuum was formed in a jacketed steam cylinder by releasing its steam content into a separate, water-cooled, chamber, where the used steam condensed. By keeping the insulated steam cylinder permanently hot, this eliminated the need to alternatively heat and cool the working cylinder, thereby significantly improving operational efficiency and speed. Moreover, fuel consumption was reduced to one third that of a Newcomen engine., making steam engines more affordable. In 1769, aged 33, Watt patented his new engine, obtaining patent protection for 14 years. The key features of his engine are shown (Fig. 3.2).

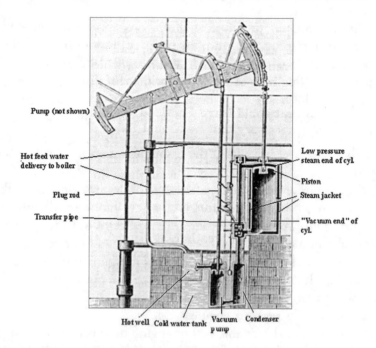

Fig. 3.2 Watt's beam engine. *Source* https://courses.lumenlearning.com/boundless-worldhistory/chapter/steam-power/. Robert H. Thurston, History of the Growth of the Steam engine, D. Appleton & Co 1878 License: CC BY-SA: Attribution-ShareAlike. https://creativecommons.org/licenses/by-sa/4.0/legalcode https://commons.wikimedia.org/wiki/File:Watt_steam_pumping_engine.JPG

Watt exhausted his development funds before his engine was perfected. Matthew Boulton, who owned an engineering works in Birmingham, persuaded Watt to seek an Act of Parliament to extend his patent by a further 16 years, from 1783, to 1799. Boulton had access to craftsmen with the precision-boring and instrument-making skills that Watt's vision required for steam-tight engines.

In 1774, Boulton and Watt formed a partnership and would become the most successful engineering company in the country, meeting the considerable demand for steam engines. The partnership was a perfect combination of Watt's scientific and engineering ingenuity, with Boulton's factory and commercial skills.

When Watt's engines were installed in mines, to extract water, the annual fee was based on one third of the fuel saving realized. News of these super-efficient engines soon spread. They became the engine of choice across the nation, and, with developmental modifications, they grew into the key power source for the surge in industrial output that was about to begin.

Their 'beam' engines produced a simple, reciprocating, up-and-down action, ideal for pumping, but with few other applications. Foreseeing business opportunities in corn, malt, and cotton mills, Boulton urged Watt to adapt the reciprocating action to a rotary motion for driving mill machinery.

This required the beam to be both pushed upwards and pulled downwards. The rigid chain linkage on an up-and-down engine would have to be replaced. In 1781, Watt created a 'parallel motion', by equipping his modified engine with a fly-wheel drive from the beam, and a set of so called 'sun and planet gears'.

It has the curious property of imparting a complete rotation of the flywheel for each one-way stroke of the piston. The rotary motion is doubly effective since the piston gives power strokes in both directions of its movement.

The introduction of rotary engines enabled engines to drive wheels. It was ideal for power looms facilitating the industrial mechanisation of weaving and cotton spinning. It could drive furnace bellows and power other mechanical devices. Soon paper, flour, and iron mills, as well as distilleries, canals and waterworks were utilising Watt engines.

In 1782, he patented a double-acting engine in which the piston pushed as well as pulled. Condensation of steam, both above and below the piston was allowed. This was achieved by directing steam from the boiler into the piston cylinder. The pressure of steam pushes the piston to one side, moving the piston rod. When the piston reaches the end of the stroke, the side valve shifts the steam to the other side of the piston, forcing it back. In this way, almost twice as much power was generated.

Watt charged his customers a premium for using his machines. In 1871, to describe the efficiency of his engines, he coined the term "horsepower". He compared each machine to a work horse, having made an arbitrary calculation that a draft horse, at the surface of a mine shaft, could lift a 550-lb weight, a height of one foot, in a second (equivalent to 33,000 foot-pounds of work per minute). He claimed, therefore, a 20 horsepower engine could lift the equivalent of 11,000 lb. through a height of one foot, in one second. Hence, "horsepower" is an arbitrary unit of measure, where one Imperial horsepower (550 fps) is equivalent to 746 watts, whilst a 'metric horsepower, is 735.5 watts.

Objects with mechanical energy (either potential by virtue of their position or kinetic by virtue of their motion) can do work and impart energy on other objects. Almost every transfer of energy that takes place in the real world does so with an efficiency below 100% and results in some thermal energy. Energy transfer also occurs when there is a temperature difference between bodies and heating of one of them occurs. Work and heat are not forms of energy but the means by which energy is transferred.

Relevant to the work of Watt and Faraday are three types of unit to measure energy. Concentrating on Imperial and cgs units, these are (i) mechanical units including foot-pound-force (fpf), horsepower-hour and joules (ii) thermal units such as British thermal units (btu) and calories and (iii) electrical units, watt-seconds. The units may be different, but they are equivalent.

Starting his research in the 1840s, J.P. Joule performed classical experiments on the relationship between work expended and the heat produced. Heat and work are two different ways a system can interact with its surroundings and have its internal energy transformed. Joule's Mechanical Equivalent of Heat (aka Joule's constant, J) is the number of units of work that must be done on a system to produce a single unit of heat. He demonstrated quantitatively that work and heat are interconvertible, and 4.186 J equal one calorie.

In 1785, both Watt and Boulton were elected Fellows of the Royal Society of London. Watt's improvements to the steam engine transformed our world from an agrarian society to one based on engineering and technology. Between 1775 and 1800, they produced 451 engines, of which 268 were rotative. They had transformed the prototype from a simple pump into a sophisticated prime mover which could be applied to a wide variety of industrial processes.

Watt had made steam power, fuelled by an expanding coal industry, a practical reality that radically improved our core industries, leading to an unprecedented increase in British trade. To that extent, he is one of the fathers of the Industrial Revolution. As a result, Britain was able to exploit huge and rapidly expanding international markets. Birmingham, in 1791, was possibly the biggest manufacturing town in the World. Its products were in demand all over the globe.

For almost 200 years, coal-fuelled steam was the outstanding source of power for industrial and transport systems. However, in the twentieth century, large-scale generation of electricity dislodged it from its supremacy. Michael Faraday would have apart to play in this displacement (see Sect. 10.3.7).

Watt died a wealthy man, in 1819. Over 60 years later, in 1882, the "watt"—a unit of measurement of power of electrical and mechanical energy was named in his honour. In the International System of Units, one watt is equal to one joule of work performed per second.

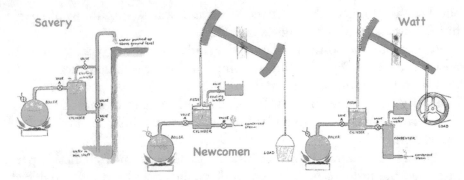

Fig. 3.3 Advances in steam engines. Savery → Newcomen → Watt

One of the Watt-Boulton pumps, built in 1786, for the Barclay & Perkins Brewery (to both grind barley and pump water) can be seen at the National Museum of Scotland, in Edinburgh (EH1 1JF). It was in service for 87 years.

Watt's attic workshop is preserved in the Energy Hall at the Science Museum in London (SW7 2DD). Also, in the Museum is a rotative beam engine, built in 1788, and used at Boulton's Soho factory, where it drove 43 metal polishing (or lapping) machines for 70 years.

Advances to the basic Savery 'suction pump', via Newcomen's 'atmospheric beam pump', to Watt's permanently hot steam cylinder, separated from a condensing chamber, are shown schematically (Fig. 3.3).

3.5 Richard Trevithick (1771–1833)—High-Pressure, 4th Generation Steam Engine and 'Father' of the Steam Locomotive

Richard Trevithick was born in a mining district of Illogan in Cornwall. His father worked in numerous Cornish copper mines, rising to managerial positions at four mines, including the Stray Park copper and tin mine in Camborne. His father's occupation provided the initial arenas for his son's technical ingenuity.

Richard's schoolmaster at Camborne described him as a slow, disobedient, obstinate child. However, he had an intuitive ability to solve complex engineering problems. On leaving school, he worked under his father, overseeing the potential application, and deployment of, steam engines invented by James Watt. He started to introduce fuel-saving innovations to Watt's engines.

In 1797, Trevithick married and worked as an engineer at the Ding Dong mine near Penzance. The engines designed by Newcomen and Watt relied on a vacuum, created by condensing steam, to drive a piston. Trevithick worked on building and modifying steam engines to avoid paying royalties due to Watt on his 'separate condenser' patent.

In 1798 Trevithick was engineer at the Dolcoath mine and was joined by his cousin, Andrew Vivian, who became Manager. Richard built a stationary engine. Exhaust steam was vented through a vertical pipe or chimney, straight to the atmosphere, thus obviating the need for a condenser, and avoiding any infringement of Watt's patent.

Trevithick's breakthrough came in 1799, when he made the first model of a high-pressure steam engine which utilised the pressure of steam expansion (then called 'strong steam'). This eliminated the need for a condenser and featured a smaller but stronger cylinder and, hence, a more compact and lighter engine. Trevithick's engine operated at 50–60 lb per square inch (psi) compared to Watt's mediocre 14.7 psi. This meant that the piston was impelled with up to four times the force. Without these radical changes, locomotion, at the turn of the century, may not have been possible. Although the Trevithick engine was more expensive to construct, it was more fuel efficient—a vital consideration in Cornwall, where coal was imported from Wales.

Static steam engines were used to pull wagons horizontally over relatively short distances. It was obvious that it would be beneficial to have self-propelled, mobile, steam engines that could move anywhere. The drive for steam locomotives began (from Latin 'loco moveri'—move by change of position).

Once Watt's patent expired in 1800, Trevithick's compact, high-pressure version of Watt's engines paved the way for steam cars and locomotives.

His engine was small enough to fit in his "Puffing Devil" or "Puffer" carriage/car (which puffed steam into the atmosphere as opposed to the noiseless condensing engines). In 1801, Trevithick first full-sized road locomotive was demonstrated to the public in his hometown of Camborne. Four days later, the boiler overheated and was irreparably damaged.

In 1802, Trevithick and Vivian were granted a patent (No. 2599) for high pressure, non-condensing engines, including an application thereof for driving carriages. To prove his ideas, he built a stationary engine at the works of the Coalbrookdale Company in Shropshire. He chose this company because he believed that their ironmasters were best able to construct a boiler capable of withstanding such high pressures. The engine ran at 40 piston strokes per minute, with an unprecedented boiler pressure of 145 psi. To assess the work done, it forced water to a measured height.

Trevithick realized that the orientation of the cylinder had little influence on its wear. Therefore, another development was the change from vertical, to horizontal cylinders. In 1803, a second road locomotive was constructed at Camborne. It had a horizontal cylinder which improved its steadiness in motion. Having wheels of larger diameter enabled it to pass more easily over the poor road surfaces of the age. The Steam Carriage was transported to London for trials but attracted no commercial interest.

In 1803, one of Trevithick's large cast iron boilers, in Greenwich, exploded killing four men. Although it was caused by negligence of the boilerman, Watt and Boulton endeavoured to capitalise on the disaster, condemning Trevithick's machines as inherently unsafe. Trevithick responded by incorporating a safety valve, an early type of fusible plug, and a steam gauge which became standard features on new boilers.

In 1803, Samuel Homfray, part-owner of the Penydarren ironworks, near Merthyr Tydfil, in South Wales, invited Trevithick to his site. Homfray was interested in the high-pressure engines that the Cornishman had developed and installed in his road locomotives. Homfray encouraged Trevithick to consider the possibility of converting his compact road engine into a railway locomotive.

In 1804, Trevithick designed the first steam locomotive engine to operate on cast iron rails and to be independent of animal power. It resulted in the first demonstration of the use of steam power for the transportation of men and goods on smooth rails which offered less resistance than a road surface. Ultimately, the world's first working locomotive pulled an impressive load of 4 wagons, carrying 10 tons of iron and 70 men, the nine miles between Penydarren and Abercynon, at five miles per hour. Here, was the loading point for the Glamorganshire Canal.

Effectively, the chemical energy from coal created steam which drove the piston and powered mechanical energy which was transmitted through a coordinated system of crosshead, flywheel, crank, and cogwheels to generate the kinetic energy of forward moving wheels.

The locomotive was of primitive design and some of the cast iron plates on the tramroad broke repeatedly due to the additional weight of the locomotive. Because of continual derailments, the trams returned to being pulled by horses and the engine was used in a static role.

After 1806, Trevithick concentrated on improvements to his boiler design and this led to the so-called Cornish engine, clones of which spread far and wide across the world. Had he concentrated on stationary engines, then he may have become more prosperous. Instead, his career began a downward trajectory.

In a letter to Davies Gilbert (*aka* Giddy who funded the road locomotive*)*, he wrote *"However much I may be straitened in pecuniary circumstances, the great honour of being a useful subject can never be taken from me, which to me far exceeds riches"*.

Trevithick was a man of great stature and prodigious physical strength who enjoyed wrestling. However, he had no financial or business sense. He died penniless in 1833. His workmates had to club together to pay for his burial. He was buried in the now closed, St, Edmund's burial ground, in Dartford. A stained-glass window in his memory was presented to Westminster Abbey by the President of the Institute of Civil Engineers. It is now in the nave of the north-west tower chapel.

Trevithick's 3 horsepower, high pressure, stationary engine, built by Hazledine & Co. in Bridgnorth, in 1806, is on display at London's Science Museum (SW7 2DD). A full-scale conjectural reconstruction of the Penydarren locomotive is on display at the National Waterfront Museum in Swansea (SA1 3RD). A working model of Trevithick's locomotive can be seen at the Ironbridge Gorge Museum (TF8 7DQ). It also featured on a £2 coin (Fig. 3.4).

Trevithick was the first engineer to employ high pressure steam generated within a cylindrical boiler. He used the engine's exhaust to draw the fire and he coupled a locomotive's driving wheels, all of which would become

Fig. 3.4 Trevithick's steam locomotive featured on a £2 coin

common practice in later steam locomotive design. He proved that the steam locomotive was a viable proposition and inspired others, including George Stephenson and his son, Robert. Economy in the conveyancing of coal became a key driver.

4

Advances in Forms of Transport—Steam Locomotives, Cycle Tyres, Oceanic Liners, and Jet Aircraft. Transport Infrastructure—Canals, Roads, and Commercial Railways

Nil desperandum (never despair) has always been my motto—we may succeed yet–perseverantia. (I K Brunel entry in his diary, 17 December 1831, on hearing funds to revive the Thames Tunnel project were not forthcoming)

The United Kingdom may only be a small island nation, but we have nurtured some highly imaginative and ingenious citizens responsible for some exceptional global firsts, leading to seismic shifts in our understanding of the world and how we have exploited it. In particular, the Industrial Revolution changed Britain and the rest of the world for ever. It confirmed Great Britain as a global superpower.

Manufacturing was transformed from cottage industries into urban industrial phenomena, driving economic growth. Public transport became available to rich and poor, reshaping our country and improving the lives of many Britons. Because of these changes, society experienced a 'time-travel' compression. Life's constant struggle for survival was gradually replaced by some material comforts and leisure.

4.1 Précis. The Canal Age, the Railway Age, Oceanic Travel and the Jet Age

After the Roman invasion of a country called Albion in AD 43, it became Provincia Britannia. The occupying Romans set about building several

© The Author(s), under exclusive license to Springer Nature Switzerland AG 2022
J. Bailey, *Inventive Geniuses Who Changed the World*, https://doi.org/10.1007/978-3-030-81381-9_4

thousands of miles of paved roads and a few navigable canals, primarily for the transportation of troops and military supplies across their newly conquered territory. This network subsequently provided vital infrastructure for commerce, trade and the transportation of goods. Surprisingly, the framework for internal communication by land or water lacked substantive development until the mid-seventeenth century.

In the eighteenth century, road and water travel were revolutionised by the turnpike and canal systems that made the Industrial Revolution possible. The modern age is characterised by speed of physical communication and the ability to traverse long distances with increasing ease by road, water, rail and air.

Railways—a British invention—gave the Industrial Revolution its impetus and staying power, and transformed the lives of millions, especially when they were adopted internationally. Ocean and air travel made it possible to visit distant continents, whilst pedal bicycles and motor cars democratised travel locally.

British inventors and engineers involved with transport and its infrastructure helped to facilitate the Canal Age, the Industrial Revolution, the Railway Age, oceanic travel, and the Jet Age.

James Brindley was semi-literate, but a natural genius—wise without formal education. Self-taught, he had uncommon talents regarding the application of mechanical principles. He became a civil engineer and canal-builder, executing works new to Great Britain. He combined technical brilliance with the beautifying transformation of our rural landscape, to begin the construction of a network of inter-connecting, man-made waterways. Their pathways were strategically chosen, originally passing though sleepy villages and towns that would then develop into industrial powerhouses. They became transport arteries across the country, linking sources of raw materials to industrial heartlands where they were converted into useful items utilised by citizens in major British cities and beyond.

John McAdam developed new methods of road construction which, with minor refinements, are still in use today.

George Stephenson was an unschooled miner's son; a rags-to-riches success story, learning his engineering skills empirically by trial and error, thereby mastering the laws by which steam engines worked. He and his son, Robert Stephenson, were the cornerstones of railway infrastructure. They promoted the Railway Age, making key refinements to steam locomotives and super-intended construction of vast swathes of the railway network, including the world's first inter-city line. First came the Stockton and Darlington Railway which was the prototype modern railway offering a passenger service, as well as freight transport. Alongside the growing railway network, Cooke and Wheatstone were involved with a telegraphic signalling method for British railways.

Isambard Brunel played his role in railway construction, tunnels, railway stations, docks, novel bridge design, as well as creating futuristic oceanic passenger liners with engineering innovations adopted decades later. His largest ship was used to lay the first successful trans-oceanic telegraphic cable. Brunel is one of the titans of engineering, combining an artistic flair with engineering prowess to create some Victorian edifices celebrated for their fine architectural features and functionality.

Almost every road vehicle on the planet is cushioned from the road by air entrapped in pneumatic tyres. Their development came in two stages, both resulting from the inventiveness of two Scotsmen, namely, R.W. Thomson and John B. Dunlop. The first pneumatic inflatable tyre was invented to improve the speed and comfort of horse-drawn carriages, whilst the second was developed for cycles.

Dunlop was a veterinarian. He was a most unlikely inventor, being neither an engineer nor cyclist, just an inveterate tinkerer. He is credited with realizing that vulcanised rubber could withstand the wear and tear of being a tyre, whilst retaining its resilience.

Coupling the 'safety bicycle' with the pneumatic tyre meant that people became less dependent on horses and horse-drawn carriages and had the personal means to travel beyond their local communities. It was particularly liberating for women. The timely arrival of his pneumatic tyre was critical to the rapid uptake of the motor car, providing both traction and a cushioning effect, giving a more comfortable ride.

Not wishing to be totally isolated on an island state, we invented faster and safer forms of oceanic passenger travel (Brunel) and speedier transport through the skies. Frank Whittle's 1928 thesis and 1930 patent revolutionized aviation in the way military and civilian aircraft are propelled via the jet engine. His name is immortalised in the annals of aviation history. Like others reported in this manuscript, he had an innate desire to search into the unknown.

Whittle invented the turbojet method of aircraft propulsion. He became the Father of the Jet Age, leaving an impressive jet engine lineage which radically changed the speed at which we travel around the globe.

Whittle's story is one of triumph over adversity. With his steely determination, he fought an epic battle against social and academic obstacles, officialdom, lack of funding, entrenched technical opinions and discouraging opposition. Having engineering brilliance, he developed an unrivalled grasp of the fundamentals of thermodynamics and aerodynamics, leaving a profound mark. His jet engine was to aviation, what Stephenson's Rocket was to the railways.

Whittle filed his first patent for jet-propelled aircraft in 1930. He built and ground-tested the first liquid-fuelled, turbojet, bench engine in April 1937. His designs were extremely radical. The Air Ministry, however, were wedded to piston engines and dismissed his plans as impracticable. Their slow realization of the importance of his invention allowed Germany to seize the initiative in jet development during WW2. Germany won the race to develop the first jet-powered flight in August 1939. Britain's Gloster Meteor was the first Allied jet fighter to fly a combat mission in late July 1944. The German Me-262 jet fighter entered squadron service in June 1944 but was not active operationally, with the Luftwaffe, until early October.

Whittle's jet engines were masterpieces of simplicity in design and construction and, collectively, may be described as the most significant mechanical invention in the twentieth century, transforming the aviation industry.

After WW2, most turbojet manufacturers based their engines on Whittle's blueprints and there followed another short, but golden era of British engineering and industrial prowess. With their great power and compactness, his engines were at the forefront of aviation development for many years. His turbojet technology was applied to commercial passenger airliners and transport aircraft, allowing them to fly higher, faster, and further than ever before. They changed our lives by making the World "a smaller place". The Jet Age had begun, and Britain was at its forefront.

At the end of WW2, the aircraft industry was our biggest industry employing about one million people, sustaining about thirty separate companies, some household names. British aero-engineers and test pilots were highly prized. Until 1966, when foreign aircraft were allowed, the annual Farnborough Air show was a pageant of British ingenuity and innovation, demonstrating superlative engineering.

Notable civil aeroplanes with jet engines include the Vickers' Viscount (world's 1st turboprop with Rolls Royce Dart engines); de Havilland's Comet (1st pressurised commercial jet airliner with Ghost turbojet engines); Vickers/BAC's VC10 with rear-mounted, Rolls Royce Conway turbofan engines for long-haul and short take-off. Until 2015, only the supersonic Concord had travelled across the Atlantic faster than the subsonic, VC10.

4.2 James Brindley (1716–1772)—Canal Engineer and Builder. The 'Canal Age'

4.2.1 Background and Early Life

James was born in relatively humble circumstances in Tunstead, Derbyshire. His family were originally from an area of Leek, in Staffordshire. When James was about 8 years old, they returned there when they inherited a farm from Quaker relatives. Aged 17, he was apprenticed to a millwright and wheelwright in Sutton, near Macclesfield. Despite being semi-literate, he showed exceptional engineering ability and skills. The millwrights were the engineers of the coming age.

By nineteen, he was not only repairing machinery, but making innovative improvements and inventing new gadgets. In 1742, he moved to Leek and set up his own business. As a millwright, he rebuilt the water mill at Leek (now the site of the Brindley Mill/Museum, ST13 8FA). He improved the performance of the water wheel as a machine, making it a multi-purpose machine for new industries. For instance, in 1757, he used it to improve the grinding of flints which were used in the rising pottery industry.

This early career focus on water mills was his entry to the control of flowing water. Although self-taught, Brindley became highly regarded in the mining and water working industries as an intuitive engineer and problem solver. Today, he would be known as a civil engineer. His reputation brought him to the attention of the 3rd Duke of Bridgewater who was looking for a way to reduce the cost of transporting coal from his collieries, in Worsley, to Manchester. The city was a centre for textile-manufacturing and required coal to power steam engines. In the wintertime, eighteenth century roads were particularly wretched—often flooded and impassable. The Duke had a man-made, 'narrow canal' or "navigation" in mind.

4.2.2 Navigable, Inland Waterways, Including Fossdyke and Sankey Brook. French Artificial Canals

The 11-mile Fossdyke canal connects the tidal River Trent, at Torksey, across the flat Lincolnshire fens, to the River Witham at Lincoln. It was probably built around AD 120 by the Romans who wanted to connect the important settlement of Nottingham (on the Trent) with the strategic fortress and inland port of Lindum Colonia (Lincoln). When the Roman army departed, the works decayed, and the channel silted up. The cut was scoured in 1121, during the reign of Henry I and used for a period by the Normans.

In 1566, the Exeter ship canal was the first to be built since Roman times. Further works on Fossdyke took place in 1671–2, after which tolls were charged for the coal, wool and other commodities that were borne along its length. A reliable channel was not fully reinstated until 1744, after which it was routinely maintained.

Before 1700, most British, navigable, inland waterways had been exploited by aristocratic landowners to carry agricultural products. However, rivers carried navigable challenges due to tides and currents and their course was often sub-optimal. In 1704, after the canalisation of two rivers, the V-shaped Aire and Calder Navigation linked Leeds to Wakefield, via Castleford. It is noteworthy that, over a century later, in 1821, Leeds was linked to the North Sea via the Humber estuary, and it was then possible to transport textile goods to the port of Goole, with corn, from Lincolnshire, travelling in the opposite direction.

The Sankey Brook waterway which opened in 1757, was promoted as a river navigation scheme, but an entirely new channel was cut, converting the brook to an artificial canal. It was eight miles long, rising 60 ft through eight locks. It carried coal, from a region that was to become St. Helens, to the River Mersey and on to Liverpool.

It was the French who had pioneered large scale artificial canals for water-borne transport. In 1757, the 3rd Duke of Bridgewater decided to visit the famous 'Canal Royal en Languedoc', opened in 1681 and renamed the 'Canal du Midi' after the French Revolution, in 1789. This visit, together with activities at nearby Sankey Brook convinced him that a canal would be beneficial to his business, and he was willing to finance it.

4.2.3 The Bridgewater Canal and Brindley's Input to the Golden Age of Canals. Deaths of Brindley (1772) and the Duke of Bridgewater (1803)

The 23-year-old Duke of Bridgewater, his land agent, John Gilbert, and Brindley looked at ways of constructing a canal and improving drainage of the Duke's mines, on his estate, at Worsley. The finally agreed plan was to construct an extensive series of subterranean canals in the mine; to connect them to a surface canal, thereby draining the mining seams of unwanted water and facilitating direct delivery of coal to canal boats waiting at the head of the canal. Worsley would become the birthplace of British canals.

Brindley pioneered the principle of 'contouring'—keeping the canal level by taking circuitous routes around hills, to avoid the expensive digging of cuttings. Brindley also improved the art of 'puddling'—compacting a deep layer of a semi-solid mixture of clay or heavy loam with gravel and water to make the canal water-tight. His 'puddle linings' became the standard water-proofing method until the twentieth century. He made his canals "narrow" to minimise the amount of water required to fill them. (N.B. At a meeting of canal company proprietors, in 1769, the size of a narrow boat was agreed to be 74′6″ long and 7 feet wide).

The original 10-mile (16 km) Bridgewater Canal was to cross the Mersey and Irwell Navigation at Barton, continuing to the edge of Manchester. The route required no locks but did involve the construction of an aqueduct to take one canal over another. The aqueduct comprised three sandstone arches and was 38 feet above the river, requiring embankments. No aqueduct of this scale had ever been attempted in England, and it represented a new level of engineering expertise.

In the 1890s, the construction of the Manchester Ship Canal necessitated the construction of a new aqueduct at Barton to allow larger vessels to pass underneath. Therefore, in 1894, Brindley's stone aqueduct was replaced by E. L. William's steel 'swing' aqueduct—the only one in the world. It rotates through 90 degrees, carrying an 800-ton tank of water, allowing the Bridge-water canal to continue operating without the need to empty the enormous water trough.

The first boat passed over the first Barton aqueduct in 1761, travelling as far as Stretford. The original boats on the canal were horse-drawn. By 1762, the price of coal, in Manchester, fell by 50%.

The section of the Bridgewater Canal to its terminus, at Castlefield, was completed in 1765, giving a level link of water from the Duke's mines in Worsley to the centre of Manchester. Arguably, this was the first truly man-made, still-water canal in Britain which did not follow, wholly, or in part, an existing brook, river or tributary. In Manchester, it was eventually connected with the Rochdale Canal and the Manchester Ship Canal (see Fig. 4.1).

In 1762, Royal assent was given for a new extension to the Bridgewater Canal, from a junction at Waters Meeting, near Stretford, to the Mersey tideway at Runcorn Gap, opposite Widnes, thus forging a link to the Port of Liverpool.

The Runcorn basin is 83 ft above the Mersey, so a flight of ten locks was built to connect the two. The Bridgewater canal between Manchester and Liverpool was completed in 1776, four years after the death of Brindley, from pneumonia, at his home, Turnhurst Hall. The final Runcorn section of the

Bridgewater Canal is now cut off, at a point adjacent to the A533 road to the Silver Jubilee Bridge.

In 1795, the final phase of the Bridgewater Canal began. This resulted in a north-easterly extension from Worsley to Leigh, to connect with the Wigan branch of the Leeds and Liverpool Canal. This would provide navigation through Lancashire and Cheshire. It gave a y-shaped canal, 41 miles (66 km) in length, completed between 1759 and 1800, three years before the canal visionary and father of Britain's industrial navigation network, the 3rd Duke of Bridgewater, died.

4.2.4   Brindley's Input to Other Canal Projects

Because of his success with the Bridgewater Canal, Brindley became a key figure in the pioneering phase of canal building, leading to its golden age.

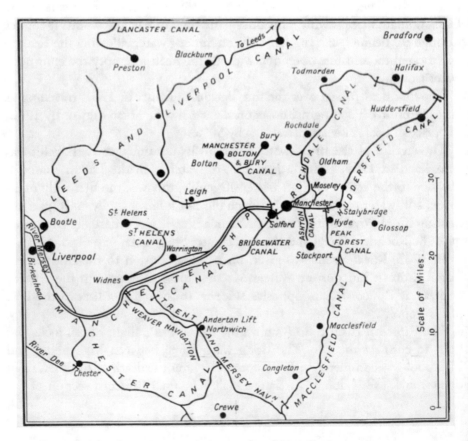

Fig. 4.1 Network of canals around Liverpool and Manchester

He played an essential role in determining where, and how, national canals should be constructed. His vision was to link, via the Midlands, the four great rivers of England, namely the Trent, Mersey, Thames, and Severn to the five important English ports of Liverpool, Hull, Manchester, Bristol, and London—three inland, and two coastal. In this way, cargo could be exchanged with seagoing ships for import and export.

This target was his "Grand Cross" scheme. The first link between the rivers Trent and Mersey became known as the Great Trunk and the eventual intention was to join Hull (on the east coast) with Liverpool (on the west). The ultimate joining of the four rivers was not completed until 1793, twenty-one years after Brindley's death.

He surveyed, gave advice on, and/or supervised the construction of numerous canal projects, few of which were completed before his untimely death, from pneumonia, in 1772. During his career, he laid out over 360 miles (580 km) of canals. He undertook all his works with no written calculations or drawings. Some of his notebooks, written phonetically, are held at the National Waterways Museum, Ellesmere Port in Cheshire (CH65 4FW). This was a fitting place to deposit them since the canal port at Ellesmere was the busiest in the country, through which many of the wealth generating goods of England were trans-shipped.

4.2.5 Trent and Mersey Canal (T&M). Wedgewood Pottery

One of his other big projects was the Trent and Mersey (T&M) Canal, which is 93.5 miles long, linking the R. Trent (at the mouth of the River Derwent, in Derbyshire) to the R. Mersey at Runcorn. His friend, the industrialist, Josiah Wedgewood, was the main backer for this project since he wanted a mode of transport for his pottery that was safe and smooth, minimizing breakage of his ceramics from the Potteries to the coastal port on the Mersey—the Atlantic gateway. It also enabled china clay to be brought from Cornwall, right to the door of his Etruria factory and village, in a district of Stoke-on-Trent, opened in 1769. Because of its whiteness and purity, china clay is used to manufacture bone china and porcelain. Locally sourced Staffordshire clays had a variety of colours and found numerous applications.

The Trent and Mersey Canal was connected to Brindley's first canal, the Bridgewater, at Preston Brook, south east of Runcorn, in Cheshire. Starting in 1766, it took eleven years to build the Trent and Mersey Canal, being completed in 1777, five years after his death. It included 70 locks, 160 aqueducts, and 5 tunnels. The twenty-six Cheshire locks raise the T&M about 250 feet, from the Cheshire plain, at Wheelock (Lock 66), to its summit level at Kidsgrove (Lock 41), on the edge of the Potteries. Because some of the locks are not in easy walking distance, this section got the name 'Heartbreak Hill'.

Brindley's greatest feat on the T&M was his project for the construction of a 2,800 yd (2,633 m) tunnel cutting beneath Harecastle Hill. It was 9 ft wide and 12 ft high—the first major canal tunnel. For a while, it was the longest man-made tunnel on Earth. Boatmen negotiated it by lying on their backs and pushing against the wall with their feet ('legging it'). In 1914, the Brindley tunnel at Harecastle was closed due to severe subsidence but a parallel sister tunnel is still navigable.

4.2.6   Canal Mania

There was a period of frantic canal-building between 1790 and 1810 which became known as 'canal mania'. The promotion of canals across the country and its effect on Britain's economy was dramatic. The artificial waterways became the major transport arteries and veins of the Industrial Revolution. In the nineteenth century, 18,000 families lived on working boats. These hard-working, ordinary men and women, as well as others across the nation helped build its prosperity. They served to transform manufacturing from cottage industries into urban industrial phenomena, driving economic growth. They shaped our landscape and our lives.

Britain's most important canals were sponsored, not only by aristocrats and bankers, but also by coalmine owners, textile manufacturers and pottery barons hoping to open new markets for their products. Stone, iron and copper ores, lime, sand, beer, salt, agricultural produce, and manure, which earlier could not support the cost of transport, were moved on canal boats.

Fly boats, on regular schedules, moved day and night, carrying merchandise and even passengers. The barges were made, not to carry luxuries necessarily, but common things that people buy by the pennyworth, such as pots and pans, bales of cloth and ribbon. These things had been manufactured in villages that were growing into towns, away from the capital, London.

4.2.7   Canals in the Nineteenth Century. Competition from Railways. The Manchester Ship Canal

These transport arteries kick-started the Industrial Revolution. They reached their peak of popularity in the 1840s, after which the railways often became a more attractive mode of transport for many, but not all, products. By 1850, about 4,800 miles of inland waterways had been constructed in Britain.

Particularly noteworthy is the Manchester Ship Canal. The Lancashire cotton mills had long depended on raw cotton to be shipped from Virginia to Liverpool, from where it was carried, initially by horse and cart to Manchester. Stephenson had partially bridged the transport 'gap' between Liverpool and Manchester by initiating the first inter-city railway line in 1830 (see Sect. 4.4.4).

New generations of machine, such as the Lancashire semi-automatic weaving loom, would take productivity for woven cloth to a new level. Manchester became the first 'industrial city'. In 1871, one third of global cotton fabrics were made in Manchester. Cotton mill owners realized that transport costs for their raw material and finished product could be reduced if a channel, suitable for ocean-going cargo ships, were built, from the Irish Sea, at Liverpool, to Manchester.

The herculean task of constructing the largest ship canal ever seen was overseen by civil engineer, Thomas A. Walker. Manchester is 36 miles inland, and 60 feet (18 m) above sea level. The so-called 'Big Ditch' is 120 feet wide and now, 28 feet (8.5 m) deep (equalling the Suez Canal). Starting at Eastham, a series of five stepped locks conveyed the cotton-carrying vessels to the port of Manchester which became the third busiest in Britain. It rendered Manchester and Salford a hub of wealth-creation and helped to make Britain the richest country in the world. At its peak, in 1958, the Ship Canal carried over 18 m tonnes of cargo.

A typical narrow boat could carry up to 35 tons of coal. In the winter, it might be stationary due to frozen water. In a hot summer, water levels could fall to a point where boats were grounded. The winners of the transporting race would be those who could convey materials most cheaply, quickly, and reliably throughout the year.

4.2.8 Canals in the Twentieth and Twenty-First Centuries

In the eighteenth and nineteenth centuries, the canal network advanced the Industrial Revolution but then became a victim of it. For about 80 years, the railways would take over domination of transport and travel. Competition from road transport, after WW1, caused further decline of the canal system.

Canals, inland waterways, and railways were nationalised in 1948. In the 1960s, about 2,000 of the 4,800 miles of canals were navigable. Commercial canal traffic ended after the 1960's. The remaining 2,000 miles of navigable canals now have become cruising waterways for leisure purposes and managed by the Canal and River Trust.

4.2.9 Brindley's Legacy

Brindley was a self-made man from a humble background. The genius of his canal construction techniques have sealed his place in history and contributed to the Industrial Revolution, particularly in north Staffordshire, the Potteries, Birmingham, the Black Country and Manchester—regions and cities which became hotbeds of manufacturing and homes for other industries. His uncommon talents regarding the outstanding application of engineering principles are commemorated by his peers in the form of a portrait which hangs in the Institute of Civil Engineering.

4.3   John Louden McAdam (1756–1836)—Road Builder

John was born in Ayr, Scotland to parents of a minor branch of Ayrshire nobility. He was the youngest of ten children, and the second boy born to Susannah and James McAdam. Susannah was the niece of the 7th Earl of Dundonald, and James was the Baron of Waterhead.

In 1763, his father co-founded the Bank of Ayr. James soon became a man of consequence. In 1762, John narrowly escaped a premature death when the family home—Lagwyne Castle on the outskirts of Carsphairn—burnt down. The family moved to Blairquhan Castle (also known as Whitefoord Castle) near Straiton.

John attended the parish school at Maybole. Even as a boy, he became interested in roads and roadmaking. It is said that, whilst still at school, he superintended the construction of a model section of road between Maybole and the neighbouring village of Kirkoswald.

Due to his lifestyle and mismanagement, James' financial empire collapsed, and he had to sell the Waterhead estate. In 1770, when John was 14 years of age, his father died. John's upbringing was entrusted to the care of his uncle in America. His uncle was a prosperous merchant, and John settled down in New York. During the American War of Independence, the McAdam's family were loyal to the British Crown and John served in the British reserves. He was a British government contractor engaged in 'war prizes' (for captured enemy warships and sailors) from which he made a small fortune, in a short time.

When the thirteen colonies became an independent nation, in 1783, McAdam and his family were no longer welcome in America and they returned to Scotland. Before his departure, the US government expropriated most of his assets. However, he did have sufficient deposits in European banks to enable him to purchase a small estate at Sauchrie near Maybole.

Here, he became magistrate, Deputy-Lieutenant for the County of Ayrshire, and, in 1787, Trustee of the Ayrshire Turnpike. During his travels he became perturbed by the lamentable state of the roads which were made of gravel, offering little resistance to the wheels of heavy carriages, wagons, and carts. The roads were easily ploughed up into fresh ruts and potholes.

At his own expense, he spent much time experimenting with new methods of road building. To his Sauchrie estate, he laid down the first stretch of 'macadamised' road.

In 1798, Britain entered the Second Military Coalition against the French who were under the leadership of Napoleon Bonaparte. McAdam was appointed agent for providing logistical support to the English navy in all ports of the west of England. He left the County of Ayr with a network of good roads and moved to Falmouth. At his new home, and again at his own expense, he continued his experiments to improve the methods for roadmaking.

He dispensed with the Roman approach to road building which consisted of a surface or 'metalling' of inter-meshed paving stones on a multi-layered base. His research findings suggested that drainage was key, and a road should be raised slightly above the level of the surrounding land, be above the water table, and have a slight (convex) camber so that drainage ditches on both sides of the road could take away rainwater.

Previous innovations by Trésaguet and Telford required a structured foundation layer. McAdam asserted that heavy foundations of rock-on-rock were unnecessary since the native soil alone should be enough to support the weight of road and traffic on it.

The road itself should be built from layers of stone, broken into angular shaped fragments and carefully size graded. The lower 20 cm layer of the road was restricted to stone fragments smaller than 7.5 cm; the upper 5 cm layer comprised stone fragments of about 2 cm. No stone was to weigh more than 6 oz. Rather than using a roller to compact the stone, the passage of traffic would cause the small stones of different angles to nestle together to give a solid surface, protective of lower levels.

Its longevity would depend on the durability of the stone from which it was constructed. He recommended crushed granite, Whinstone or pebbles of Shropshire or Staffordshire stone for the base layer. Materials that could imbibe water and be affected by frost were to be avoided.

McAdam predicted that such a road would be almost impervious to the weathering action of wind and rain and would be hard and unyielding to traffic. His overall approach resulted in a smoother surface and carriage ride, whilst being cheaper to build.

In 1801, when he was 45, he was appointed to the Bristol Turnpike Trust, so he moved to Bristol in 1802. In 1816, the year after the Battle of Waterloo, he was appointed Surveyor-General of the Bristol roads and it was then that he put his ideas into major practice.

From the beginning of the eighteenth century to the Highways act of 1878, many British roads were the responsibility, not of central Government, but of private enterprise. So-called turnpike trusts levied tolls and this income provided both good roads and good profits. At their peak, there were 8,000 toll gates in Britain.

By 1818, McAdam was acting as consulting surveyor for thirty-four road trusts. In 1819, he published his "Practical Essay on the Scientific Repair and Preservation of Roads". This was followed, in 1820, by his "Present State of Road-making". In 1823, a committee of the House of Commons was set up to inquire into the feasibility of applying his new system of road-making throughout the country. In 1827, he became General Surveyor of Metropolitan Roads for the whole of Great Britain.

From his mid- to later years, McAdam had spent a vast amount of time and his own money on travel, inspection, survey, and experimentation, all for the public good, demonstrating philanthropy at the highest level. The government wished to reward him for his great service with a knighthood, but he declined. It appears he was overlooked for nomination as a fellow of the Royal Society of London or Edinburgh. He died a relatively poor man and was buried in Moffat cemetery, Scotland.

Macadam roads were satisfactory in the days of horse-drawn carriages. However, long after McAdam's death, in 1836, the advent of the motor car, at the end of the nineteenth century, highlighted serious shortcomings with Macadam roads. Firstly, the jagged surface caused punctures of pneumatic tyres. Fast moving cars create an area of low pressure at the road surface which sucks dust from the surface, creating dust clouds and damage to the integrity of the road.

In 1902, in Monaco, Ernest Guglielminetti demonstrated that treatment of a dusty road surface with a mixture of tar, gravel and sand, cured the problem. Also, in 1902, Edgar Purnell Hooley patented a mixture of coal tar and ironworks slag as 'tarmac' which also offered improved durability. The first Tarmac surfaced road in the world was built, in Nottingham, in 1903.

Macadamisation was the greatest advancement of road construction since Roman times, being an effective and economic method of road making, which was widely adopted in the USA and across Europe. After minor refinements to include tar, slag, and other components to bind the surface and fill the interstices, it is still in use today.

Transport development in general was one of the factors that made the Industrial Revolution possible. It facilitated rapid and efficient transportation of goods and passengers throughout the Kingdom, reducing costs and forming an integrated free market.

4.4 George Stephenson (1781–1848) and Robert Stephenson FRS, HFRSE, DCL (1803–1859)—Civil and Mechanical Engineers. Refinements to Steam Locomotives and Pioneers of Steam Railways. The 'Railway Age' and the Electromagnetic Telegraph

4.4.1 Their Early Years

George Stephenson was the son of illiterate, working-class parents. His father was employed in the coal industry, following work, from one active pit to the next. George was born in a miner's cottage in Wylam, Northumberland, He lived next door to the Wylam horse-drawn waggonway which connected the colliery to the River Tyne.

This mining area of Northumberland was the birthplace of three additional pioneering railway engineers—George and Timothy Hackworth were born in Wylam, whilst William Hedley was raised in nearby Newburn.

In 1804, the first commercial railway was built at the Penydarren ironworks, in Wales, by R Trevithick (see Sect. 3.5) to carry workmen and goods to the canal at Abercynon. This was achieved with a moving steam engine, independent of the pulling power of a horse.

Coincidently, in 1805, whilst seeking to replace Wylam's horse-drawn system of transport, Trevithick's assistant, John Steel, built a steam locomotive, in Gateshead, to Trevithick's design but it proved far too heavy for the wooden waggonway.

Later, in 1813, William Hedley—resident engineer at the Wylam colliery and his assistant Timothy Hackworth—built a steam locomotive, 'Puffing Billy'. Hedley solved the problem of slippage of iron wheels on iron rails by adopting two sets of driving wheels. The travelling engine hauled coal 5-miles to the River Tyne, travelling sedately at 5 mph, but often breaking down. The locomotive can be seen at London's Science Museum (SW7 2DD). Their second locomotive, 'Wylam Dilly', remained in service until 1862.

From an early age, George Stephenson was fascinated by machines. His parents were too poor to send their son to school and he was illiterate until the age of 18. George entered the coal industry in his early teens, moving from one colliery to the next, undertaking a variety of jobs.

For a time, George's father was a fireman at the Water Row colliery in Newburn, a few miles from Wylam. Between 1798 and 1801, George worked with his father at Water Row colliery. Because of his youthful ability, he was appointed engineman, in charge of a new pumping engine at the colliery. At 18-years of age, he realized the importance of education, and used some of his earnings to study at night-school. Writing, however, proved persistently difficult for him.

George married in 1802 and Robert, his only son, was born, in 1803, in Willington Quay, Northumberland. A year later, the family moved 15 miles to Killingworth, near Newcastle, home to several coal mines. George worked as brakes man in the Killingworth West Moor colliery.

In 1805, Robert's mother, George's wife, died from tuberculosis when Robert was only three years old. Robert was taught, firstly at the local school in Long Benton, after which he attended the John Bruce Academy, 6-miles away in Percy Street, Newcastle, between 1815 and 1818. It was a commercial school preparing boys for trade and business and Robert attended as a day pupil. At the age of 15, he began an apprenticeship with an engineer who was viewer (manager) at Killingworth colliery. Robert would prove to be an engineer with talents equal to, or greater than, those of his father.

In 1812, having successfully corrected faults with a newly installed atmospheric pumping engine, George was promoted to "engine-wright" at all Killingworth collieries, working on stationary steam engines.

His knowledge was essentially practical, learning empirically through trial and error. He could quickly assess the virtues and defects of contemporary machines and find solutions to problems that baffled engineers with more theoretical knowledge. By this process, he mastered the laws by which steam engines worked, thereby becoming an expert in steam-driven machinery. Thereafter, both George and son Robert were to play pivotal roles in the development of commercial railways.

4.4.2 George Stephenson's 1st Steam Locomotive for Freight Haulage on Privately Owned Rail Tracks

George convinced the Killingworth mine manger to experiment with steam locomotion. By 1814, George developed the first fully effective steam engine locomotive (called Blücher in honour of a Prussian general who was an ally of Britain in the Napoleonic wars and would later assist the Duke of Wellington in defeating Napoleon, at the Battle of Waterloo). It was used to haul coal on the Killingworth waggonway. It had greater pulling power than previous locomotives. It was the first locomotive to have single-flanged wheels and was able to haul up to 30t of coal, on 8 wagons, at 4 mph, up a gradient of 1 in 450.

By 1818, the waggonway had been entirely re-laid with cast-iron edge-rails developed and patented by George. He built sixteen engines at Killingworth. A replica of one of his early 'travelling engines' can be seen at the Stephenson Railway Museum (NE29 8DX).

4.4.3 George and Robert Stephenson's First Steam Locomotive for Freight and Paying Passengers. The Stockton and Darlington Railway (S&DR). Standard Gauge Rail Track

Durham coal was transported from inland mines to the River Tees for loading onto ships for shipment to new markets. As was the case in Northumberland, the journeys from the local mines were undertaken in wooden wagons, running on wooden tracks, the wagons pulled by horses. During the Napoleonic Wars, horse fodder became more expensive, whilst coal became more plentiful and cheaper. As the demand for coal increased, a more efficient method for its transportation was sought.

In 1821, Parliament granted permission for the Stockton-on-Tees to Darlington railway (S&DR). Darlington banker, Jonathan Backhouse and Quaker mill owner, Edward Pease, were willing to invest in new transport infrastructure and were persuaded by George Stephenson to invest in steam locomotion, rather than a horse-drawn railway system or canal.

A new act of parliament permitted the use of "a loco-motive or moveable engine". This was the genesis of modern railways, the crucible in which the modern railway was forged. There were no established blueprints for stations, sidings, cuttings or embankments, railway bridges, accommodation bridges for farmers or signalling system.

In 1821, Robert terminated his apprenticeship to help his father conduct a survey of land over which the Stockton to Darlington railway (S&DR) would travel.

George had ignored advice that his son should gain a university degree. However, upon completion of the survey, Robert was permitted to spend 6-months at Edinburgh University, attending lectures in mathematics, chemistry, and geology.

In 1823, George and his son, Robert, opened the world's first purpose-built locomotive works on Forth Banks, Newcastle-upon-Tyne. The first locomotive engine produced by the new company was initially called 'Active' and then became 'Locomotion Number One'. The Forth Street Works built about 3,000 locomotives by the end of the century, supplying railways at home and abroad.

The Stockton and Darlington 26-mile, single track line was officially opened in 1825. It was the first public railway in the world to use steam traction to operate both freight (from 1825) and passenger services (routinely from 1833). On its maiden journey, Locomotion Number '1' pulled 36 wagons, taking two hours to cover the 9-mile journey. As well as a single covered coach (named "Experiment") for dignitaries, it pulled twenty-one primitive, uncovered wagons for ordinary passengers, together with freight wagons.

The S&DR was an instant success—being popular with local people, and profitable too. By 1828, it was carrying 50,000 tons of coal plus 40,000 people. Although much of it was closed after 38 years, parts of it are still in active use.

Its main purpose was to haul coal from coal mines nestled around Shildon. At the 'Brusselton Incline', coal wagons were transferred from a wagonway reliant on the pulling power of a stationary steam engine, to a railway track and the hauling power of Locomotion No. 1. The wagons travelled via Darlington, to the port at Stockton-on-Tees, where the coal was transferred to sea-going boats.

At Darlington, the wagons traversed the Skerne Bridge, over the River Skerne. It is the oldest railway bridge still in use today. It was designed by Durham architect, Ignatius Bonomi, the so-called first railway architect. The bridge, Locomotion No.1 and Robert Stephenson's Rocket were immortalised by appearing on the Bank of England 1990's £5 note. As integral parts of the embryonic public railway system, they represent technological and engineering achievements allowing the Industrial Revolution to flourish.

The S&DR track gauge was required to accommodate wagons used on the older, horse-drawn waggonways serving coal mines. This requirement appears to be the main reason that 4 feet 8.5 inches was subsequently adopted as 'standard gauge'.

Shildon is described as the "cradle of the railways" since it was the first place in the world to witness a steam engine hauling a train of passengers on a public railway. The S&DR owned the track and was responsible for its maintenance and upkeep. It was an open access, public railway, allowing private contractors to use the track for a fee. Some companies—such as the Union—even carried passengers in horse-drawn railway coaches.

The design of Locomotion No. 1 combined, and built on, the incremental improvements that engine-wright, George, had incorporated in his Killingworth locomotives. It had a centre-flue boiler (boiler pressure 50 psi); a pair of vertical cylinders and a wheel configuration of 0-4-0 to spread the weight. It may have been the first locomotive to use coupling rods to connect its driving wheels together. Since cast iron rails exhibited excessive brittleness, S&DR used wrought iron, malleable rails.

Eventually, the steam locomotive was able to travel at 15 mph. It served for several years for hauling freight before the boiler exploded, killing the driver who had inactivated the safety valve. Rebuilt, it can be seen at the Head of Steam/Darlington Railway Centre and Museum (DL3 6ST). A working replica can be viewed at the Beamish Museum in County Durham (DH9 0RG).

4.4.4 The Liverpool and Manchester Railway (L&MR). The Rocket Travelling Engine Built by Robert Stephenson & Co.

The commercial success of the Stockton and Darlington line would change business opinions about modes of transport. The S&DR was the template for railway finance, management, operation and engineering. In the 1820s, the Industrial Revolution was forging ahead. The first textile mill had been built in Manchester, in 1781, by Richard Arkwright. Merchants and manufacturers shipping goods had a choice between canals and roads. The mills of Manchester (later nicknamed 'Cottonopolis') were producing so much cotton that a faster, more efficient mode of transport of goods to, and raw materials from, the port of Liverpool was needed. To service this need, the Liverpool and Manchester Railway Company (L&MR) was founded, in 1823. Director Henry Booth was one of the most vigorous proponents of the railroad.

After three surveys, the Liverpool to Manchester railway was finally approved in 1826. Construction of the 35 miles (56 km) Liverpool to Manchester line began. It took 4.5 years to complete. George Stephenson was employed to oversee the project, being responsible for the track, bridges, tunnel, platforms and rolling stock etc.

Because of the limited power of early steam engines, George had to keep the railway as flat as possible. Therefore, the new line required many bridges, viaducts, and tunnels. He had major obstacles to overcome including (i) passage over the treacherous peat bog at Chat Moss; (ii) resistance from canal operators; (iii) objections by landowners, including the Earl of Sefton who was opposed to a railway line passing near his ancestral home at Croxteth Hall and (iv) initial lack of support from Parliament.

Major achievements were the Wapping tunnel; a two-mile long cutting through rocks at Olive Mount; sixty-four bridges; a 9-arch viaduct at Sankey Brook; and a railway track floating on the bog at Chat Moss. In 1831, as an exercise in corporate advertising, to depict these achievements, the L&MR employed Rudolf Ackermann—one of the most distinguished lithographic publishers of the day—to commission a series of on-the-spot drawings, from architect Thomas T Bury. The aquatint of *"The View of the Railway across Chat Moss" engraved* by Henry Pyall, after T. T. Bury is particularly noteworthy.

George realized that individual railway lines would eventually be connected, so he insisted on a common gauge. For the track, the Stephenson gauge of 4 feet 8.5 inches (1435 mm) was accepted. This is the distance between the inside edges of the rails. It became the "global standard-gauge", being adopted by 55% of the world's railway tracks.

As the track was being laid, the L&MR directors quarrelled about what should pull the carriages and wagons. Therefore, in 1829, they organised performance trials to find the best locomotive engine to operate on the line being built between the two cities. Rainhill was to become an international historic civil and mechanical engineering landmark, denoting that one of the most significant developments in transport history took place there.

The prize for winning the so-called Rainhill Trials was £500, over and above the cost price. Performance parameters were strength, power, efficiency, and reliability, rather than speed alone. Of the five entrants, the self-propelled steam locomotive, Rocket, was declared the winner.

Early steam locomotives were ponderous and could readily explode. George's 25-year-old son, Robert, probably with advice from his father, designed and built the Rocket which won the Rainhill Trials. Help in design also came from L&MR Director, Henry Booth, who suggested the adoption of many narrow-bore fire tubes, rather than one simple large flue, a concept originated from French inventor, Marc Seguin.

Rocket (1829) and one of its successors, Planet (1830) with inside cylinders, were watershed machines that set the standard for locomotive design and became the template for steam locomotives for the next 150 years. For instance, they had a separate double-walled firebox; a 25-tube boiler, and blast pipe exhaust for improved draught. Water was converted to steam by radiant heat from glowing coke, as well as convection from the hot exhaust gas. Steaming and boiler efficiency were optimised, so giving enhanced heat transfer and mechanical energy.

Pistons—set close to horizontal—were connected directly to the driving wheels having 'guiding' flanges. The wheel arrangement was 0-2-2 with a single pair of driving wheels. Rocket proved to be exceptionally reliable. It has been heralded as an iconic object and a potent symbol of Britain's industrial heritage. Although many of its features were not new, when combined, the individual elements became part of virtually every locomotive built during the reign of steam on the world's railways.

Unless on loan, Rocket can be seen at the Science Museum in London (SW7 2DD). In late 2019, it will be on long-term display at the National Railway Museum in York (YO26 4XJ).

4.4.5 Speed of Travel *vis-à-vis* Horses; Railway Mania; the Flying Scotsman and Mallard; Changes to Society and GMT

The Liverpool to Manchester railway was the first inter-city network to use only coal and water to generate enough steam power for hauling both passengers and freight. So profound was the inaugural journey that the prime minister at the time—the Duke of Wellington—was one of the passengers on one of the eight trains travelling from Liverpool to Manchester.

It represented a successful 'marriage' between two innovations: the steam locomotive and the robust, stable track on which it ran. A hard metal wheel on a hard metal rail suffers less frictional resistance than a wheel on a muddy track. The weight of the locomotive was distributed through more axles. Wrought iron rails on proper supports were more robust. Overall, it enabled man to travel faster than a horse.

Such an "iron horse" was able to haul rolling stock at a maximum speed of 29 mph, twice the speed of previous steam locomotives and faster than anything else made by man. Although steam locomotives had been around for 25 years, since Trevithick's concept in 1804 (see Sect. 3.5), they had never achieved such feats of speed and power.

The railways telescoped distance and time. In his estimate of the benefits likely to accrue from the railways, Henry Booth said:

Speed—despatch—distance are still relative terms, but their meaning has been totally changed within a few months: what was quick, is now slow—what was distant, is now near; and this change in our ideas will not be limited to the environs of Liverpool and Manchester—it will pervade society at large.

Historically, people had used horses to travel quickly. The fastest recorded speed for a galloping horse and rider over 1.5 miles (2,414 m) is 38 mph (61 km/hr). A carriage horse can trot at 10–15 mph for 2–3 h with short walking rests. Over long distances, train speeds well over these velocities were ultimately achieved.

Two notable locomotives could travel at 100 mph or more. The Flying Scotsman and Mallard were designed by Nigel Cresley (axle plan oo⊖⊖⊖o) and built in the Doncaster Railway Works for the London and North Eastern Railway (LNER).

Class A3 LNER numbered 4472, The Flying Scotsman, built in 1923, was the first steam locomotive to be officially authenticated as reaching 100 mph (161 km/hr) in 1934.

Four years later, in 1938, Class A4 LNER numbered 4468, named Mallard, built that same year, became the holder of the world speed record for a steam locomotive. It achieved a verifiable speed of 126 mph (202 km/hr) for a short time. It is on display at the National Railway Museum, in York (YO26 4XJ).

Having linked the two great commercial centres—Manchester and Liverpool—L&MR provided a scheduled service of fast passenger trains. Passenger travel was more profitable than freight transport. Investors saw great potential and, after the economic depression between 1840 and 1843, Britain became gripped in 'Railway Mania'. Steam would dominate the railways for over a century.

Steam locomotives revolutionized travel/transport, allowing greater volumes of people and goods to be moved at unprecedented speeds. It opened extensive travel to both rich and poor. Businessmen in Bristol saw the virtues of rail transport and appointed I. Brunel to construct the Great Western Railway between Bristol and London (see Sect. 4.5.4) which was opened in stages between 1838 and 1841.

The frenzy of railway construction that started in Great Britain then spread throughout the world, beginning the Railway Age. In 1845, across Britain, a spider's web of 2,440 miles of railway were open and 30 million passengers carried. Britain was being transformed. Connectivity was being created to all corners of the kingdom. Heavy goods, previously shipped by sea around coastal ports, could be transported inland, by rail.

In 1820, about 75% of the population lived in the countryside, mostly within twenty miles of where they were born. Then a social revolution occurred. Thirty years later, in 1850, about 50% of the population lived in new cities, 8% in the countryside and the rest in small towns and villages.

By 1850, most cities and towns in Great Britain had a rail connection. Rail-track and passenger numbers increased exponentially between 1830 and 1915 when there were 19,000 miles of track and 1.5 billion passengers carried, together with huge quantities of freight. A mass transport system was in being and the railway system was a crucial part of the British economy, having a profound effect on the whole population. Taking the increasing population into account, the number of journeys per head of British population increased at an unprecedented rate. On average, there were 0.65 railway journeys per head of population in 1841, 20 in 1881, and 32 in 1911.

Within a single generation the railways brought about a major change to our nation, transforming how we lived, where we lived and worked, what we ate and how and where we spent our leisure time. Railways steamed us to the modern age, transforming landscape both physically and culturally.

These "iron roads" transformed the British landscape, carving their tracks through nineteenth century society. They created novelty and excitement, as well as fear and anxiety about the speed and direction of change. They became the symbol of progress, bringing opportunity and prosperity. Victorians saw railway locomotives and the railway network on which they travelled as the pinnacle of their engineering achievements.

Whilst the L&MR was the first inter-city railway to use only coal and water to generate sufficient steam power for hauling both passengers and freight, other 'firsts' included a double track; a signalling system; fully time-tabled travel and the first to carry mail. To facilitate time tabling across all railway networks, unified time keeping was necessary, so Greenwich Mean Time (GMT) was adopted.

4.4.6 Telegraphic Signalling

From the beginning, the railway system needed a method of communication that was faster than the train itself. To manage rail traffic, signal boxes and railway stations had to warned of approaching trains.

Volta's battery and Faraday's work (see Sect. 10.3.7) on electro-magnetism in the 1830's, led Cooke and Wheatstone to develop the multi-wire, electro-magnetic, telegraphic, signalling method which they patented in 1837. This involved the movement of a conducting coil over a magnet, in one location, inducing an electric current that is transmitted to another location, where it affects a galvanometer, thereby transmitting messages in code.

The original method depended on five lines of wires to trigger five magnetic needles arranged to point at different letters at the receiving station. By introducing the idea of a code, this was reduced to two needles in 1838. The telegraph spread all over the British railway network. By 1852, 4,000 miles of telegraph had been installed. Eventually, it became a nationalized communication service.

4.4.7   Their Distinctions, Decorations, and Legacy

George Stephenson was a self-made civil and mechanical engineer who became a successful businessman. He was constantly innovating, improving his engines and tracks. He was the guiding spirit of the Railway Age. In 1847, he became the first President of the Institute of Mechanical Engineers (IMechE). He retired to Tapton House, near Chesterfield, where he died in 1848.

Robert was a born leader, respected by all with whom he had dealings. In 1833, Robert was appointed chief engineer of the London Birmingham Railway. From 1838, to the end of his life, Robert Stephenson was engaged on over 160 railway projects, not only in Great Britain, but all over the world. As recognition of the latter, he received European decorations from Belgium, France, and Norway.

It is said that, by 1850, Robert had been involved with the construction of about one third of Britain's railway system. He believed that railway development should be comprehensive, benefitting the whole population, particularly poor people, not just the privileged few.

In 1847, he was elected Member of Parliament for Whitby, a seat he held until his death. In 1849, he was elected Fellow of the Royal Society (FRS) but, in August 1850, he declined a knighthood, from Queen Victoria, in recognition for civil engineering works that were nearing completion.

He was regarded as one of the finest engineers of the nineteenth century. He served as President of both the Institute of Mechanical Engineers (1849–53) and the Institute of Civil Engineers (1855–7). In 1855, he was made an honorary fellow of the Royal Society of Edinburgh (HFRSE). In 1857, Oxford University awarded him an honorary DCL degree (Doctor of Civil Law).

After a life of severe mental and bodily exertion, Robert died in 1859, aged 55, four weeks after the passing of his friend, Isambard Brunel. Robert's remains are interred at Westminster Abbey, alongside Thomas Telford. An obituary, in the Engineer, on 21 October 1859, recorded the following:

> He battered down no towns nor dwellings of men, to their impoverishment, but only threw down the old where it impeded progress of the new. He used powder, not to slay men, but to perforate the hills for their free passage through them. He wasted no iron in great guns or cutting weapons, but worked it into tools and rails, and locomotive engines – all reproductive implements. He levelled hills and filled up valleys and made the rough places smooth.

By 1887, rail travel was safer, more reliable, and more comfortable for all social classes. In that year, Robert Louis Stevenson used a sprightly rhythm, in a children's poem, to evoke the movement in a steam train, as well as the visual images seen. In an extract from his poem, "From a Railway Carriage", he wrote:

> Faster than fairies, faster than witches,
> Bridges and houses, hedges and ditches,
> And charging along like troops in a battle,
> All through the meadows the horses and cattle,
> All of the sights of the hill and the plain
> Fly as thick as driving rain,
> And ever again in the wink of an eye,
> Painted stations whistle by.

4.5 Isambard Kingdom Brunel FRS (1806–1859)—Mechanical and Civil Engineer; Marine Technologist

4.5.1 Family Background and Education. His Father's Influence

Isambard Kingdom was born in Portsea, a short distance from the Portsmouth dockyard. His father—Sir Marc Isambard Brunel - was born in Normandy. 'Isambard' is an Old German name meaning 'glittering iron'. Marc's parents expected their second son to join the Roman Catholic priesthood, but he resisted, instead receiving an education in science and mathematics. Despite his Catholic routes, Marc was an agnostic, possibly an atheist. He served in the French navy but, because of his royalist sympathies, he felt compelled to escape the French Revolution (1789–1799). In 1793, he fled to America, where he became Chief Engineer in New York City.

Isambard Kingdom Brunel's mother was Sophie Kingdom, an orphan who had met Marc, in France when she was working there as a governess. Sophie—accused of being an English spy—was arrested and held for 18 months. After the execution of Robespierre in 1795, she was released, fled France and returned to England.

Whilst in America, Marc perfected a mechanical means for producing wooden pulley blocks, as used in the rigging of sailing ships. In 1799, Marc travelled to England and persuaded the British Admiralty, with the world's biggest navy, to adopt his machines. He was tasked with installing them at the Portsmouth Royal Dockyard, where they continued in service for nearly half a century. The Portsmouth Block Mills represent a key episode in the Industrial Revolution, being the first factory in the world to run on steam, use all-metal machine tools, and pioneer mass production.

Two years after Isambard's birth, the family moved from Portsmouth to Lindsey House, Chelsea. He soon exhibited an aptitude for mathematics His father groomed him to follow in his footsteps, teaching him arithmetic, geometry and scale drawing, enabling Isambard to draw a freehand perfect circle, and make detailed technical drawings of mechanical subjects. By the age of 8-years, he had command of the basics of engineering. His first school was a progressive one, in Hove, near Brighton. Whilst there, he undertook a survey of the town.

Isambard was a fluent French speaker. When he was fourteen, his father sent him to the College of Caen, followed by the prestigious secondary school, Lycée Henri-Quatre, in Paris. The academic training was followed by a period of apprenticeship to Louis Breguet, maker of watches and scientific instruments. This academic and practical education had a breadth and quality unavailable in England. He spent two years in France, before returning to England, as apprentice to his father.

In 1818, Marc invented and patented the Miners' Cage or 'tunnelling shield', a technological advance that made it possible to tunnel reasonably safely through water-bearing strata. The device was a reinforced shield of cast iron in which thirty-six miners could work in separate compartments (12 long × 3 compartments high), digging at the tunnel face. Periodically, the shield would be driven forward by large jacks, and the exposed tunnel behind would be lined with bricks.

It is claimed that Marc's inspiration came from his naval background and the shipworm, *Teredo novalis*, which has its head protected by a hard shell, whilst it bores through ships' timbers.

Marc had a diverse family business—making army boots, knitting stockings, printing, as well as sawing and bending timber for navy ships. Unfortunately, he faced a series of setbacks: his sawmill burnt down; the demand for army boots evaporated and his bankers, Sykes & Co. became insolvent, so no one would honour his cheques. In 1821, because of their financial difficulties, Marc and Sophie spent thirteen weeks in Southwark debtors' prison. This was Sophie's second incarceration. The family shaming would leave an indelible mark on Isambard. Luckily, after prompting from the Duke of Wellington, the Treasury paid Marc a grant of £5,000 and he was released.

4.5.2 The Thames' Tunnel

In 1822, Isambard started working for his father. His first notable achievement was the part he played in planning and supervising the construction of the first ever substantive tunnel under any navigable river. The Thames' Tunnel would link the docks, at Rotherhithe, to the warehouses, at Wapping.

Construction started in 1825 with Marc Brunel as Chief Engineer. In 1827, Isambard, aged twenty-one, was appointed Resident Engineer, when the incumbent became ill. He attained experience in brickwork and the use of cements. Using the 'tunnelling shield' workers had to dig through clay, mud, and quicksand. It was extremely dangerous for all concerned. Because of seepage of heavily polluted river water, air quality was poor. Sometimes the partly built tunnel collapsed, and it was deluged with water from the river above. On one such occasion, Isambard rescued one man and, in 1828, Brunel himself nearly drowned but was rescued by a contractor. Brunel worked long hours and developed a reputation for 'leading from the front'.

There were so many accidents (including six fatalities) and cost over-runs that the Thames' Tunnel Company went bankrupt. The tunnel was walled off and abandoned for seven years. By 1834, Marc Brunel had succeeded in raising adequate funds to restart construction and the tunnelling was completed in 1843, an engineering feat for which Marc was knighted in 1841. The tunnel is now part of the London Overground system.

The original engine house, in Rotherhithe, is now the Brunel museum (SE16 4LF). It commemorates Isambard's first and last key projects—firstly the Thames' Tunnel, as the birthplace of the London Underground tube network, and lastly, the Great Eastern steamship, with the hallmarks of a modern ocean liner.

4.5.3 The Clifton Suspension Bridge

After his near-drowning, Brunel, aged 23, was sent to Brislington, on the edge of Bristol, to recuperate. It was here, in 1829, that he heard about a competition to find someone to design a new bridge over the River Avon, at Avon Gorge. Brunel's first submission was aesthetically pleasing and fitted in with the geological features of the gorge. However, it was rejected by competition-judge, Thomas Telford, who deemed Brunel's span dimension to be too ambitious, putting the bridge at risk from crosswinds.

To combat the wind, Brunel's solution was to use short suspension rods to bring the chains down to deck level. He slightly reduced the free suspended span to 194 m by building the Leigh Woods' Tower proud of the cliff face. Brunel's second submission was successful. Two distinguished mathematicians had verified his calculations were mathematically exact. This acceptance showcased Brunel as an engineering radical.

In 1831, riding high on the iron boom, work started on a suspension bridge which would be lighter and stronger over that distance of 194 m. Unfortunately, work was stopped after a few days due to the riots in Bristol, a city where the wealthy had a vote, but the poor wanted change. It resumed in 1836 but, sadly, as with the Thames Tunnel project, money eventually ran out. In 1843, with only the two Egyptian-style towers built, the project was abandoned.

When Brunel died in 1859, the bridge was incomplete. As a memorial to him, work resumed in 1862. John Hawkshaw and William Barlow made the following modifications to Brunel's original design—the roadway was widened, and the suspension chains increased from two to three on each side. The bridge was completed in 1864 and is now a Grade I listed structure. Its length is 702 feet; 214 m centre to centre of piers. Its height is 245 feet (76 m) above high water at deck level.

When it was opened, in 1864, it was the longest, single span, chain bridge in Queen Victoria's Empire—the ultimate symbol of status and power. It is still an iconic structure and an engineering masterpiece, reflective of Brunel's brilliant mathematical mind. It is a rare example of a man-made structure enhancing its natural setting. Tours are arranged from the Visitor Centre which can be found on the Leigh Woods side of the bridge (BS8 3PA).

4.5.4 The Great Western Railway (GWR) from Paddington Station, London, to Brunel's Original Terminus at Bristol Temple Meads on Broad Gauge Track

The story of the Great Western Railway is one of the most interesting and romantic in the history of the railways. It forms part of the story of England and Englishmen in the nineteenth and twentieth centuries. It established Brunel's international status as an engineering genius.

The opening of the Stockton-on-Tees to Darlington railway line, in 1825, became a world-wide sensation (see Sect. 4.4.3). In response, the merchants of Bristol wanted to improve communication with London. They formed a committee to advance an ambitious proposal of laying a railway to London.

In 1833, with his Bristol connection, Brunel, aged twenty-seven, was chosen to superintend the construction of the Great Western Railway. At the time, this was the biggest civil engineering project in the world. When completed, in 1841, it would be the longest (118 miles) railway on Earth and reduce travel time from London to Bristol from two days to five hours.

Within 9-weeks, Brunel had completed a thorough survey that would set the standard for a well-built railway, minimising gradients and curves. As a friend of George Stephenson, Brunel had travelled on the Manchester to Liverpool railway line. Brunel felt that he could make a train journey more comfortable by keeping track gradients to a minimum and using a wider gauge.

Parliamentary approval of the railway was eventually granted in 1835 and work began immediately. A large part of the construction was contracted out. There were no mechanical excavators. All excavation was done manually. Itinerant navvies, many of whom had helped build the network of canals in earlier years, began their task. As well as their physical strength, their technical knowledge of the character of rocks and soils, together with their experiences in excavating tunnels, erecting embankments, and making cuttings was of enormous value. A good navvy could excavate, with pickaxe and shovel, twenty tonnes of earth per day, removed by wheelbarrow.

These construction gangs lived alongside the route of their labours. They had a reputation for heavy drinking, fighting and lawlessness. In fact, navvies were the unsung heroes of both canal and railway construction. In the years between 1850 and 1859, a weekly journal conducted by Charles Dickens, called "Household Words" was published. H. J. Brown worked alongside navvies and, on 21 June 1856, wrote an article entitled "Navvies as they used to be" (Volume xiii, pages 543–50). It described the pitiful working and living conditions they experienced.

Even with the most amenable terrain chosen for the London to Bristol route, there were still some major obstacles to overcome and creative solutions to be found. These included:

1. Crossing the Brent Valley. This was achieved with the Wharncliffe Viaduct—the first of Brunel's many viaducts. It was the first railway viaduct to be constructed with hollow piers.
2. Crossing the River Thames via the Maidenhead Bridge. Since the Thames was a navigable river by barges, the Thames Navigation Commissioners would allow only one bridge pier to be positioned in the river. To satisfy them, Brunel designed a brick bridge with two graceful 128' (39 m), elliptical arches. He boasted that it was the flattest, yet widest brick arch in the world.
3. High ground between Twyford and Reading. This necessitated a 2-mile cutting, sometimes 60 feet in depth.
4. The hill at Box, in Wiltshire. This required a tunnel 1.8 mile (2.9 km) long, together with six ventilation shafts, 28 ft (8.5 m) in diameter, varying from 70 to 300 ft deep. Excavating through solid Bath Stone in the eastern section was a formidable task. Each week, a ton of gunpowder was consumed. Between 1,500 and 4,000 men toiled 24 h each day, working from both ends of the tunnel. More than 100 navvies lost their lives in the process. Construction started in 1836 and the tunnel was opened in 1841. At the time, it was the longest railway tunnel in existence.

Artist, J. M. W. Turner was particularly interested in industry and technology. He encountered both Faraday and Brunel at new Somerset House, home of both the Royal Society and the Royal Academy of Arts. Turner's famous painting, "*Rain Steam and Speed—the Great Western Railway*" debuted at the Royal Academy, in 1844. It featured the GWR's broad gauge locomotive, "Fire Fly" (designed by Northumberland engineer, Daniel Gooch), speeding over Brunel's Maidenhead Bridge. Passengers, some in open carriages, travelled at average speeds of 50 mph (80 km/h) which were unprecedented in the 1840s.

The painting normally resides at London's National Gallery (WC2N 5DN). It symbolizes the march of technology, of renewal and reform. It juxtaposes "new railway bridge and old road bridge; speeding train and drifting boat; steam-driven locomotive and horse-drawn plough; linear track and meandering river".

Completion of the GWR, in 1941, was transformative, generating numerous new opportunities for its pioneering Chief Engineer, Brunel.

As travel became faster, the need for time synchronisation arose. GWR adopted 'Railway Time' in 1840. 'London Time' followed, in 1847, with the legal adoption of Greenwich Mean Time (GMT), in 1880.

To aid comfort, and to raise the carrying capacity of the rolling stock, Brunel recommended a broad gauge of 7' 0.25" (2,140 mm) as opposed to the narrow gauge of 4' 8.5" which is now the adopted standard (see Sect. 4.4.3). The clash between broad and narrow gauges intensified in 1845. The broad gauge aggregated 274 miles; the narrow, over 1,900 miles. In 1846, the Gauge Regulation Act declared that any gauge other than 4' 8.5" could only be used under special authority.

Despite Brunel's fastidiousness about gradients and gauge, reaction to his railway was mixed. Detractors said the track was uneven and bumpy and the passenger carriages were badly sprung. Shortcomings were overcome and the line extended to many other destinations.

4.5.5 Extension of the Great Western Railway (GWR): Bristol and Exeter Railway and the South Devon Railway. Brunel's "Atmospheric" Railway and the Prince Albert Bridge

Beyond Bristol, lay the important agricultural centre of Exeter and the harbour at Plymouth. Accordingly, GWR became associated with the Bristol and Exeter Railway, as well as the South Devon Railway. To take a reasonably straight line to Plymouth, the southern skirts of Dartmoor had to be crossed and this meant traversing some steep gradients, up to 2.5%.

Blinded by his own brilliance, Brunel made a major blunder in 1845. He became aware of a new means of motive power and visited Dalkey in Ireland to see a short, but steep, section of railway driven by "atmospheric power". The system was based on the principle of Newcomen's atmospheric steam engine (see Sect. 3.3).

A 20" metal pipe was laid between the rails. A continuous slot along the top of the pipe allowed an arm connected to a piston within the leading carriage to travel along the pipe as the carriage moved. At intervals of about 3 miles, steam-driven pumps ahead of the train evacuated air in the pipework. Atmospheric pressure acting on the piston provided the motive force, pushing the train forward, dispensing with the need for a locomotive. A hinged, leather flap maintained the vacuum.

Brunel's new railway was opened, in 1847, with great fanfare and potential. In early 1848, a full timetable was introduced, covering a 20-mile, level section between Exeter and Newton Abbey.

After less than one year, however, Brunel recommended that atmospheric power be abandoned, and the line reverted to locomotive working. The pipeline failed because of a fatal flaw with the leather flaps which lost their flexibility in winter. It was a financial disaster and Brunel's reputation suffered hugely.

Another noteworthy feature of the Cornwall—Devon network, was the railway bridge he built across the River Tamar, a tidal estuary. The bridge had to be high enough for naval ships to pass at high tide, the Admiralty demanding a minimum headway of 100 feet.

At Saltash, between Plymouth and Truro, the turbulent River Tamar is 1,100 ft wide. Here, he built the bold and beautiful Royal Albert Bridge. It was Brunel's final and, perhaps, one of his greatest, civil engineering achievements, opening travel between London and Truro.

Since there were no securing points to attach tension chains, Brunel designed a bridge with two self-supporting, lenticular-trusses, each with a profile resembling a convex lens. Each truss comprised a wrought iron, tubular, parabolic arch with sets of suspension chains hanging on each side of the tube, in a catenary curve. The load on the bridge deck is transferred to the two ends of the truss which sit on solid abutments.

The overall result was a bridge with two long spans, each of 455 ft, together with seventeen short, approach spans: ten on the Cornwall side and seven on the Devon side. As regards the construction of the central pier in the middle of the Tamar, Brunel copied a form of 'diving bell' used by his father during the construction of the Thames Tunnel.

A 90 ft cylinder was floated out and sunk in the middle of the Tamar. Water was pumped out and the cylinder sealed at the top. Men passed through air locks to work inside the cylinder—pressurized with compressed air to prevent water ingress. They excavated 12 ft of mud and 3 ft of rock before an adequate bedrock was located. They then built a masonry pier to a height above water level. Some of them suffered from the compression sickness (the 'bends').

Over a 40-week period, the lenticular-trusses were floated, in turn, on pontoons, in to position and then raised incrementally, by hydraulic jacks, into their final position, 100 ft (31 m) above the Tamar. As the trusses were lifted, supporting piers were built beneath them (Fig. 4.2).

Brunel was too ill to attend the official opening ceremony, in 1859, and six weeks later, he died, aged 57.

Fig. 4.2 The Royal Albert Bridge over the River Tamar in 1858. *Source* https://commons.wikimedia.org/wiki/File:Royal_albert_bridge_hist.jpg. By unknown author—unknown source, public domain

4.5.6 From Bristol to New York via Paddle Steamer, PS Great Western. Integrated First Class Travel from London to New York via a Single, Seamless Ticket

Brunel next extended his vision of travel, from London to Bristol via railway, to New York, across the Atlantic Ocean, via steam ship instead of sailing ship. His dream coincided with the desires of Bristol businessmen wanting to trade with New York and compete with the Port of Liverpool. For the first time in his life, he worked harmoniously with a multi-disciplinary team who formed the Construction Committee for the Great Western Steam Ship Company. Other members were T Gruppy (marine engineer and businessman), Royal Navy Captain Claxton and W Patterson (ship builder).

A standard sailing ship would have taken 28 days or more to make the crossing. Brunel aimed to halve that time. Steam power was already dominating transport on land. The problem with steam-powered travel across the ocean appeared to be the lack of adequate space to carry sufficient coal for the journey. Brunel realized that a substantial increase in the size of a ship was necessary.

His calculations were based on a simplistic "cube versus square" rule. He argued that fuel capacity correlates with the volume of the hull—a cubic measure. However, fuel consumption correlates with the surface area of hull pushing through the water—a square measure. He reasoned, therefore, that doubling the size of a ship would create eight times more coal-carrying capacity but would require only four times as much coal.

The first vessel he built was the paddle steamer PS Great Western, an oak-hulled ship. She was 72 m (236 ft) long and 10.7 m wide (18 m with both paddle wheels) and had a gross registered tonnage of 1,340 GRT and displacement of 2,300 ton. Four masts with auxiliary sails were fitted as a precautionary measure.

No longer dependent on favourable trade winds, she sailed to New York, in 1838, the journey taking 15 days, travelling at an average speed of 8.66 knots, arriving with 203t of coal unburned.

A smaller, passenger steamer—the Sirius, owned by a competitor company—sailed from Cobh (formerly Queenstown) in County Cork and arrived a day before the PS Great Western, making the Sirius the first, scheduled, passenger ship to cross the Atlantic under steam power alone. The Sirius travelled at an average speed of 8.03 knots, took three days longer, and consumed more coal than the PS Great Western.

For two years, the PS Great Western was the largest steamship in the world, making 45 Atlantic crossings in 8 years of service. She held the record for the fastest transatlantic crossing, twice in each direction, and was the Concorde of her day. She was broken up in 1858, the year SS Great Eastern was launched.

4.5.7 SS Great Britain. Iron Hull, Steam Propulsion and Screw Propeller

In 1838, Brunel, aged 32, was a restless and relentless workaholic. He had a boundless appetite for innovation. Building on what he had previously learnt, he yearned to build something bolder. Not an inventor as such, he took other peoples' practical ideas, combined them together and made them more efficient, becoming a pioneer of marine technology.

The industrial-scale production of wrought iron lowered the price of iron plate at a time when timber was becoming scarcer and more expensive. This favoured the construction of steamships with iron hulls. The latter are not subject to woodworm or dry rot.

In 1839, the keel was laid for the screw ship (SS) Great Britain which, at the time, was the largest ship ever built. Iron plate is stronger and lighter than timber, so internal space could be gained, meaning more passengers and freight could be carried.

His friend and fellow engineer, Thomas Guppy, persuaded Brunel to abandon the partially constructed paddles in favour of screw propulsion. The reasons were multi-fold: screw propulsion is lighter and occupies less space; is cheaper and is always fully submerged, resulting in higher efficiency. It lowers the ship's centre of gravity, making it more stable in choppy seas. The elimination of the paddle boxes reduces the resistance through water and improves manoeuvrability at docking.

Propeller technology was embryonic. It had never been applied effectively in a commercial environment and certainly not for trans-continental vessels. Aided by his ability to draw a circle freehand, Brunel designed curved blades with different pitches. By cutting and shaping, 8 different propeller designs were tested. Thirty-two trials took place on various ships, including SS Archimedes. Performance data were meticulously recorded to determine the optimal thrust and efficiency.

They initially chose a 16' diameter, 6-bladed screw propeller, with a curved design. Even the most modern screws, resulting from computer-aided design, are only marginally more efficient than Brunel's design.

Propellers need a higher speed of rotation than does a paddle shaft. For the first time, use was made of inverted tooth, 'silent pitch chains' to turn the propeller shaft. To reduce the noise from the engine room, wooden teeth were used in the sprocket wheels, teak in the upper and lignum vitae in the lower.

The ship was so big that it could not pass through the Bristol dock gates, so at high spring tide, the gates, as well as some brickwork and masonry, were removed. Launched in 1843, as a lavish ocean liner, it made 67 oceanic crossings in 8 years.

The distance from her home port of Bristol to New York, was 3,354 miles and the journey took 14 days. She returned with perishable goods grown in America, creating new markets in Britain, and changing British diets. As a result, Anglo-American economies became more inter-connected.

SS Great Britain enjoyed 90 years of working life, serving as an emigrant ship to Australia; a cargo ship; a troop carrier in the Crimean War; a supply ship in WW1 and a floating warehouse.

Arguably, it is the most innovative ship ever built. She was the first vessel with an iron hull, steam propulsion and screw propeller to cross any ocean. Many of her seminal features were adopted for subsequent naval and merchant ships.

On 19 July 1970, she was towed under Brunel's Clifton Suspension Bridge, on her way to Bristol, 127 years after she emerged. She now sits in dry dock, as a maritime museum, next door to the Brunel Institute (BS1 6TY). She is listed in the National Historic Fleet subgroup of the National Register of Historic Vessels. A full-scale model of the 1,000 horsepower V-engine that propelled her—designed by Isambard's father, Marc Brunel—is also on display.

4.5.8 PSS Great Eastern

Fast, efficient means of communication were at the heart of the complex structure of international trade developed by the Victorians and upon which their wealth creation depended. It was the successful development of railways and steamships, together with the electric telegraph, that gave them their unique place in world history.

Brunel's sketch book was haunted with gigantic ships. In 1850, lots of refuelling stations were required to serve ocean travel in a global empire. The Eastern Steam Navigation Company was formed in 1851, with the goal of exploiting the increase in trade and emigration to India, China, and Australia. Brunel was asked to design a ship capable of sailing to Australia without refuelling on route. He set a personal goal of making the whole journey without re-coaling.

By 1854, construction of the paddle and screw steamship PSS Great Eastern had begun at the Millwall dock, on the Thames, in London. It was built side-on to the river as it was too long for a conventional launch. Originally called Leviathan, the vessel was to be six times the size of SS Great Britain. It was 692 feet (211 m) long; 118 feet wide and had a gross registered tonnage of 18,915 GRT. Its size necessitated several propulsion systems: it had a 4-bladed screw propeller and paddle wheels on both sides, as well as six sailing masts. It was the largest moving object ever built. It displaced 32,160 ton.

The Great Eastern was the first ship to be built on an industrial scale, and the first modern ocean liner, comprising 30,000 iron plates, each held in place by 100 rivets giving a total of three million rivets. In those days, there were no welders; 400 riveters worked 12-h days.

The PSS Great Eastern was built in partnership with John Scott Russell, a naval architect. He performed experiments to find the optimum shape of a hull to minimise resistance when sailing. Using his 'wave-line' theory, the traditional bluff bow of the day was replaced with a sleek bow with hollow lines which, arguably, displaced water more easily. This style was later adopted by the sailing clippers.

Some of the other engineering innovations advocated by Brunel and Russell are still in use today. For instance, the PSS Great Eastern was built like a bridge beam, with an innovative cellular construction.

It was compartmentalised, having high, water-tight, upright walls (bulkheads) to stop propagation of any flood water to other parts of the vessel. No doors were cut through the bulkheads below the lower decks. There were two 107 m (350 ft) × 18 m (60 ft) high, longitudinal (fore to aft) bulkheads on either side of the engine room. These were coupled with fifteen transverse (starboard to port) bulkheads; nine were 18 m high and extended to the upper deck, whilst six extended either to the iron lower deck or the water line. A steel roof separated the boiler rooms from the coal bunkers, giving a total of fifty watertight compartments.

For the PSS Great Eastern, adoption of a double-skinned hull was an additional safety feature in case the outer hull was pierced. The Great Eastern was virtually unsinkable. This was tested in 1862 when the ship rubbed against the 'North East Rips' at Montauk Point, near Long Island, New York. It opened a gash (25 m long × 2.7 m wide) in the ship's outer hull.

Unlike the Great Eastern which had a double hull, the RMS Titanic, launched in 1911, was only double bottomed. Some of the transverse bulkheads in the Titanic were only 3 m above the waterline, to E-deck, to give passengers unencumbered spaces on this deck and above. The nominally watertight compartments did not have a "roof" section, so were only watertight horizontally. After collision with the iceberg, the forward six compartments flooded, causing the bow to dip and the ship to sink rapidly. Taller bulkheads may have slowed down the sinking process, giving more time for passenger evacuation.

In the case of the Great Eastern, longitudinal (fore to aft) stiffeners offered strength against buffeting waves. Crosswise, complete, and partial bulk heads created strength and stability when going over waves.

Against Brunel's wishes, people bought tickets to watch the launch of the ship in 1858 but its release was embarrassingly unsuccessful, and it took three attempts, over three months, to get her into the water. On its maiden voyage, an explosion in the engine room (caused by negligence rather than poor design) caused serious damage. Brunel's personal investment in the Great Eastern was a folly which affected his finances and health and may have shortened his life.

Although a technological triumph, as a passenger ship, the PSS Great Eastern was a commercial failure. Its owners lost a lucrative mail contract to the P&O line. When vast coal deposits were found in Australia, the concerns about refuelling faded. Increasingly a white elephant, it came out of passenger service in 1863. It was too big to pass through the Suez Canal, opened in 1869. Although an engineering triumph, it suffered many vicissitudes and was known as the 'unlucky' ship because it was plagued by mishaps. However, some of the safety features insisted upon by Brunel, in the 1850s, are mandatory in today's ship construction.

4.5.9   PSS Great Eastern and the Transatlantic Electric Telegraph

In 1856, the Atlantic Telegraph Company was formed with the sole purpose of laying and operating the first Atlantic telegraph cable. Their ambition was partially realized in 1857–8 but the telegraph stopped working after a few weeks. In 1865, a higher-specification cable weighing 980 kg for each kilometre was laid (see Sect. 7.4.2). Because of the enormous weight of the cable, there was only one ship in the entire British navy capable of carrying it. Brunel's Great Eastern, with its huge size and powerful engines made her eminently suited to the task.

The durable telegraph cable reduced communication time between North America and Europe from 10 days (by ship) to several hours. The electric telegraph was the forerunner of the internet and World Wide Web (see Sect. 15.4.5).

In 1889, the PSS Great Eastern was still the largest ship afloat, but in that year, it was scrapped.

Appendix A, Tables A.1 and A.2 summarize the key features of Brunel's three Great ships.

4.5.10   Brunel's Personal Accidents; Fatal Illness; Character and Career Over-View; Absence of Honours and Decorations, and Final Resting Place

Brunel was dogged by accidents throughout his working life and by serious illness in his fifties. During the Thames Tunnel project, he was almost drowned. Twice he was nearly killed during the construction of the GWR, and he fell from a burning ladder on the PS Great Western and was severely injured.

During the building of the PSS Great Eastern, when his health was failing, he was sent overseas for recuperation. He was suffering from kidney disease. At the age of 53, he had a stroke on board the PSS Great Eastern. Five days after the ship was damaged by an explosion on its maiden journey, he died.

His funeral was a relatively quiet affair, attended by family, friends and professional acquaintances, including his friend, Robert Stephenson. Isambard is buried, in the family vault, with a modest, white marble block headstone, in Kensal Green Cemetery, in the Royal Borough of Kensington and Chelsea, four miles from his childhood home. Historic England have assigned it a Grade II listing (Entry number 1265238).

In contrast, when Robert Stephenson died four weeks later, there was a national outpouring of grief and his body was committed to Westminster Abbey. Queen Victoria gave permission for the funeral cortege to pass through Hyde Park—an honour previously reserved for royalty. Nine years after the death of Isambard, a memorial window was erected in the Abbey.

The Brunel family vault is surrounded by railways—the Bakerloo Line and London Overground on one side, and the West Coast Main Line on the other. Underneath will be the bored tunnel route for the high-speed train, HS2. A few miles away will be the HS2/Crossrail interchange station at Old Oak Common, making Kensal Green one of the best-connected neighbourhoods in the national railway network.

Brunel was the son of a French immigrant, born into a world constructed of brick, timber and stone. The world changed and iron and steam became the technologies of choice. Isambard was adept at finding new uses for these technologies, to achieve engineering marvels of unprecedented ambition and scale.

He was an engineering polymath who, alongside George and Robert Stephenson, transformed Victorian transportation and Britain's transport infrastructure. However, opinions about Brunel are mixed because of his conflicting characteristics. On the one hand he was a genius with a restless intellect and artistic flair; a creative innovator; driven to achieve; an uncompromising workaholic; persuader of politicians, investors and public; a perfectionist but audacious.

On the other, he was gnawed with self-doubt; prone to accidents; controlling and reluctant to delegate—a dictatorial micromanager; a hard taskmaster; often brusque with his employees; uncompromising and quick to argue; obsessed with detail and over-confident, with a cavalier approach.

At just over five feet tall, Isambard was self-conscious about his stature and favoured stovepipe hats to make himself appear taller. His forename, Isambard, has Germanic origin and means 'iron bright'—quite appropriate for Robert Howlett's iconic photograph where Brunel stands in front of the launching chains for the Great Eastern (Fig. 4.3).

Notwithstanding the above, he is one of the nineteenth century engineering titans, arguably Britain's finest engineer and key figure in the Industrial Revolution, bringing about major changes in a short time. His influence in engineering is unparalleled, shifting the general perception about what engineers could do and how they could change the world, adopting a more scientific and systematic approach to their projects. In 1830, he was elected Fellow of the Royal Society—rather surprising since the Society was somewhat dismissive of the scientific efforts of engineers. In a BBC poll in 2002, he was voted the second greatest Briton.

He and Robert Stephenson were the most celebrated and sought-after engineers in the country and overseas. Isambard was involved simultaneously with hundreds of projects, a visionary who pushed back the boundaries of engineering by his innovative approach. He is associated with some of the great edifices of Victorian engineering. With his artistic flair, his concepts were

Fig. 4.3 Isambard Kingdom Brunel, 1857. *Source* By Robert Howlett—Robert Howlett (1831–1858), public domain, https://commons.wikimedia.org/wiki/File:Rob ert_Howlett_(Isambard_Kingdom_Brunel_Standing_Before_the_Launching_Chains_ of_the_Great_Eastern),_The_Metropolitan_Museum_of_Art.jpg

tasteful in design, bold and original, sometimes pioneering new constructive materials.

Few, if any, active in the nineteenth century have left as great a legacy as Brunel. In Great Britain, he designed and built railways (25 routes covering 1,400 miles), 125 bridge designs, viaducts, tunnels, railway stations, piers, docks, and a railway village in Swindon. Many of his structures are still in constant use, 150 years later. He conquered the oceans with ingenuity and passion. He powered passenger travel across oceans via three ships of ground-breaking design and construction. It meant that people could travel more cheaply, comfortably and trade in a new way.

Unlike his father, Isambard was not an inventor but an inspiring inno-vator whose life-time achievements eclipsed those of his father. Whilst his father was knighted, in 1841, by Queen Victoria (probably at the prompting of Prince Albert) for his work on the Thames Tunnel, Isambard was never so honoured—receiving no Royal decorations despite having achieved so much more nationally, in his short life. His motivation may not have been aristo-cratic acknowledgement, preferring the acclaim of discerning observers who praised his constructions for their fine architectural features and functionality.

4.6 John Boyd Dunlop (1840–1921)—Inventor of the 1st Commercial Pneumatic, Inflatable Tyre. The Chemistry of Rubber. History of Pneumatic Tyres for Motor Cars. Rise and Fall of the Companies Named After J.B. Dunlop

4.6.1 Natural Rubber and Its Vulcanisation

India/natural rubber/caoutchouc is an organic substance, harvested from the latex sap of the rubber tree, *Hevea brasiliensis*. Joseph Priestly discovered that it erases (rubs out) black lead pencil marks and gave it the English name, "rubber". This tree is native to Amazonia and the term India rubber refers to the material's origin in the West Indies. Mesoamerican tribes used the mate-rial for centuries. In 1735/6, the French scientist and adventurer, Charles de la Condamine discovered it in Peru. The French word 'caoutchouc' may refer to a Caribbean word, 'cauchuc', meaning "juice of a tree".

Milky latex sap is produced in special, secretory plant cells called laticifers. Natural latex is secreted in response to physical damage, acting as the first line of defence against herbivores and pathogens. The milky rubber secretion comprises about 70% water. The polymeric chains of the rubber molecules are intertwined with one another in a random way. This spatial configuration of cis-1,4-polyisoprene gives rubber its elastic character. Because there are few cross-links between molecules, natural rubber is a thermoplastic, that is, it loses its elasticity in cold weather and becomes sticky in hot.

The challenge was to convert this unstable substance from the tropics to something more useful. In 1823, Scottish chemist, Charles Macintosh sandwiched natural rubber between outer and inner layers of fabric, making a raincoat which exploited the water-repelling of natural rubber, whist circumventing its stickiness.

The tackiness of the untreated rubber can be eliminated chemically, whilst retaining pliability, by the process of 'vulcanisation' which involves the heating of rubber with sulphur. It is named after Vulcan, the Roman god of fire. The sulphur molecules unwind and combine with the *cis*-double bonds of the rubber molecules, forming bridges of sulphur chains from one rubber molecule to another. Vulcanisation with a small amount of sulphur leads to a soft product used in rubber bands.

In 1839, American, Charles Goodyear, started experimenting with the vulcanisation of rubber. In 1842, Englishman, Thomas Hancock, associate of the raincoat firm, Charles Macintosh, came into possession of vulcanised rubber produced by Goodyear. Unbeknown to Goodyear, Hancock developed his own process and took out a UK patent, in November 1843, for the vulcanisation of rubber using sulphur. Goodyear's USA patent 3633 followed in June 1844.

4.6.2   Compound Fillers; Colourants for Tyres; Toxicological and Environmental Concerns

When vulcanised rubber is used to make tyres, compounding it with a filler such as carbon black, increases tyre strength and wear resistance and gives vulcanised rubber its distinctive black colour for automobile tyres.

When carbon black was used in the tread (for improved tread life) and white zinc oxide was added to the rest of the tyre mix, white side-walled (WSW) tyres resulted.

There are now toxicological concerns about workers in tyre factories being exposed to materials in the micrometre size range. Moreover, whenever a vehicle brakes, accelerates or turns, particles of rubber are shed from its tyres. These contribute to airborne and microplastic pollution.

4.6.3 Polymers Associated with Natural and Synthetic Rubber. The Use of Petrochemicals to Manufacture Synthetic Rubbers

Polymerisation is the joining together of many small molecules to make exceptionally large molecules. For example, simple alkenes such as ethylene ($CH_2 = CH_2$) will undergo polymerisation to give polyethylene ($-CH_2-CH_2-$)n.

Likewise, a conjugated diene such as 1,3-butadiene will undergo free-radical polymerisation to form polybutadiene:

$$CH_2 = CH - CH = CH_2 \rightarrow (-CH_2 - CH = CH - CH_2)_n$$

When the hydrogen atom at carbon-2 of butadiene is replaced by a methyl group, the 5-cabon alkene, isoprene, results. The isoprene unit is one of nature's favourite natural building blocks. It occurs, not only in natural rubber, but in a wide variety of compounds isolated from plant and animal sources. For instance, the terpenes contain multiples of this 5-carbon skeleton. For example, menthol, camphor and limonene contain 10 carbon atoms; vitamin A has 20, whilst β-carotene, has 40.

Natural rubber, gutta-percha and chicle (in natural chewing gum) are derived from the milky latex sap of specific plants. 'Natural polymerisation' of C_5-H_8 isoprene sub-units gives poly-terpenes with cis-linkages (e.g., *rubber*); *trans*-links (e.g., gutta-percha) or both *cis*- and *trans*-links (chicle). The trans isomer, gutta-percha, from the Malaysian sapodilla tree, contains fewer branched poly-terpene chains than natural rubber, so it is more stable, not so elastic, is an insulator and resistant to seawater. For these reasons, it was used to coat cables for the trans-Atlantic telegraph (see Sect. 7.4.2).

Natural rubber is a *cis*-polyisoprene with the methyl groups on the same side of the double bond:

$$nCH_2 = \underset{\underset{CH_3}{|}}{C} - CH = CH_2 \rightarrow \left[-CH_2 - \underset{\underset{CH_3}{|}}{C} = CH - CH_2 - \right]_n -$$

Isoprene
[2 – methly butadiene]

Natural rubber
[Poly isoprene]

If the hydrogen atom at carbon-2 of butadiene is replaced with a chlorine atom, chloroprene results. This can be polymerised to polychloroprene (Neoprene, Duprene) which was the first successful rubber substitute in the USA.

Another way to modify the properties of a polymer is through the process of 'copolymerisation'. Here, two or more, unsaturated compounds polymerise together. A particularly important copolymer is the one between butadiene and styrene (C_6H_5–CH = CH_2) to give styrene butadiene rubber (SBR). This was the most important synthetic rubber developed to replace natural rubber, the latter being unavailable to the allies during WW2.

Another polymerisation of commercial importance is the acid catalysed polymerisation of isobutylene, $CH_2 = C(CH_3)_2$ with a little isoprene (1–3%). This yields butyl rubber or isobutylene isoprene rubber (IIR), used to make inner tubes, as well as the inner lining of tubeless tyres.

4.6.4   Background to John Boyd Dunlop and the Development of the Company Named After Him

John was born on a farm in Dreghorn, Ayrshire, Scotland. He developed a love of animals and studied veterinary science at the Royal (W. Dick) School of Veterinary Studies, in Edinburgh. He qualified, in 1859, aged only nineteen. After gaining his veterinary diploma, together with an honorary fellowship of the Royal Veterinary Society of Edinburgh, he set up a veterinary practice in the city, before moving to Downpatrick, 21 miles from Belfast, in 1867. Here, he joined his brother's practice and then moved to Belfast to set up his own.

Dunlop's veterinary work involved extensive travelling, often in carriages with wooden or iron-rimmed wheels, over roads that were very uneven. In 1887, he noticed his rather delicate son, John junior, was struggling to ride his trike over cobbled stones. To make his ride more comfortable, Dunlop wrapped the solid rubber tyres of his son's trike with rubber sheets, bonded with liquid rubber. He inflated the composite wheel/tyre with a football pump for a cushioning effect and thereby created the first practical pneumatic tyre for cycles.

Another story is that Dunlop modified the tyres for his invalid mother's wheelchair. He took an air-tube of rubber, covered it with cloth and then with rubber again. Dunlop gifted a pneumatic cycle tyre (made from Arbroath sailcloth and India rubber) to the National Museum of Scotland, in Edinburgh (EH1 1JF; museum reference T.1910.27).

In July 1888, he was granted patent 10,607 for an improvement in tyres of wheels for bicycles, tricycles, or other road cars. In March 1889, patent 4116 included a means for securing tyres to the wheel rims. Air was compressed in an elastic inner tube provided with a crude non-return valve for inflation. To reduce the number of punctures suffered, it was surrounded by a non-stretchable casing of tough canvas; the latter being covered with an India-rubber tread to take the friction of the road.

It encouraged a shift from the 'high-wheel', penny-farthing bicycles to so-called 'safety bicycles' with a 'diamond' frame, two equally sized wheels, and chain-driven rear wheel.

By deforming around the surface of the road, a pneumatic tyre has less rolling resistance than a solid rubber tyre, so less energy is lost to the road as the wheel rotates. Within months of Dunlop's patent, competing inventors jockeyed for position in the burgeoning pneumatic tyre market.

In 1888, Robert William Edlin built the first bicycle for Dunlop in his Gloucester Street premises, in Belfast. This was followed by racing bicycles in 1889. Edlin had to modify the frame and fork shapes to accommodate the fatter tyres.

Willie Hume was the first cyclist to recognise the supremacy of Dunlop's pneumatic tyre. Hume was captain of the Belfast Cruisers Cycling Club. In 1889, he sensationally won all four races at the Queen's College Sports, and all but one race at an event in Liverpool.

Edlin assumed he had a long-term deal with Dunlop to produce more cycles, but this was not to be, and Dunlop switched his allegiance to partners in Dublin. William Bowden—a Dublin cycle agent—persuaded Dunlop and William Harvey du Cros to join him in the formation of a public listed company to exploit the patent.

Du Cros was a financier; an enthusiastic cyclist; cycle sport's trainer for his 6 sons, and President of the Irish Cyclists' Association. Two of his sons were beaten in a race by W. Hume using Dunlop's pneumatic tyres. This fired du Cros' interest in the tyres.

In 1889, Dunlop's patents became invested in the Pneumatic Tyre and Booth Cycle Agency. Its prospectus said that the Agency had been formed to take over the British and Irish interests of Dunlop's patents, as well as the cycle business of Messrs Booth Brothers, Dublin and of the cycle works of Messrs Edlin and Company in Belfast. Du Cros was Joint Managing Director of the

Company but kept all financial arrangements under his control. He became the main advocate for the pneumatic tyre and would go on to popularise it.

4.6.5   Robert W. Thomson's Patent

Unfortunately, in 1890, Dunlop's patent 10,607, was officially declared invalid because of the belated discovery of prior art by fellow Scot, Robert William Thomson.

The latter had been granted British patent 10,990, in 1845, for a rubberized fabric tube (bladder) inflated with air. In the absence of rubber of adequate durability, the inner tube of rubberised canvas was enclosed in two overlapping hemispheres of leather, rivetted at the top join, and bolted to the wheel hub at the bottom join. Patents for Thomson's 'pneumatic tyre'/'aerial wheel' followed in France, in 1846, and in America, in 1847.

Such pneumatic tyres were meant for horse-drawn carriages, to lessen the power required to pull them, rendering their motion easier and diminishing the noise they made in motion. Thomson tyres were manufactured at the Macintosh Works in Manchester. Successful tests were performed in Hyde Park, London, in 1847. However, the aerial wheel was never commercially successful because of the poor state of roads and the high cost of the rubber for the bladder. The adage that, "few patentees grow wealthy" applied to Thomson. Forty years would pass before the technology available met the public demand for pneumatic tyres.

It was accepted, then, that Thomson was the rightful inventor of pneumatic tyres, albeit by a different and more expensive method than Dunlop's. The latter had effectively 're-invented' the pneumatic tyre and was able to utilise vulcanised rubber of more consistent quality.

4.6.6   Monopoly from Intellectual Property Protection—Patent Acquisition and Extension. Increasing Popularity of Cycling but Dunlop's Retirement

In 1890, because Thomson's patent had expired, and Dunlop's patent had been rescinded, anyone was free to produce a pneumatic tyre. To protect his enterprise from widespread competition, Dunlop started acquiring specific design patents spawned by his tyre's introduction, thereby extending the Company's sphere of influence from tyres to the whole wheel, including the rim and cycle valve. This assured the company's prosperity.

Chief among them was the timely acquisition of the Charles Kingston Welch patent (14,563) which was granted in September 1890. This was for an easily demountable, wired-on pneumatic tyre, thereby securing, for Dunlop, the simplest, most practical form of pneumatic tyre. It conferred lucrative patent protection until 1904. The 'well-base rim' was first used in competition in the Dublin to Limerick cycle race in 1892. Eventually, it would be used on motorcycles, motor cars, and even aeroplanes.

Dunlop's original tyre had a structural casing that was awkwardly glued around the outside of the metal wheel rim. In contrast, the Welsh 2-part tyre allowed the outer casing to be easily removed from the inner tube to facilitate its repair, when punctured. An inextensible wire hoop was moulded into each edge of the tyre's casing, the diameter of which was carefully matched to the rim. The tyre was manipulated on to the rim and held fast when the inner tube was inflated. The resulting Dunlop-Welch combination was widely applied in the UK cycle industry but was not universally adopted in the emerging motor car industry until the Bartlett beaded edge tyre became obsolete in the mid-1920s.

The 'wired-on' approach was not the only way of securing a tyre to a rim. In 1890, five weeks after Welch's patent, American W E Bartlett, working for the North British Rubber Company, in Edinburgh, took out the Bartlett patent (16,783). Bartlett's "beaded edge" tyre had some similarities with the Welch design, as well as the Gormully & Jeffrey "clincher" model in America.

Hard vulcanised rubber beads running along the two edges of the tyre were slotted ("clinched" in a tight grip to fasten securely) into a flat based rim with turned over edges. When the inner tube was inflated, it held the ridged edges of the outer carcass to the groves of the rim. For the original model, the edges of the rim were only slightly inclined inwards. In later versions, the rim was turned inwards in the form of a hook.

At the start of the twentieth century, when cars were fitted with Bartlett pneumatic tyres, his design was problematic with fast cornering cars because the bead slid laterally from its seating, causing tyre dislodgement.

In 1893, because of local complaints about fumes, Dunlop's manufacturing site was relocated from Lincoln Place, Dublin, to Coventry which became the heart of the British cycling industry. Factories were also established in the USA, France, and Japan. The Pneumatic Tyre and Booth Cycle Agency became the Pneumatic Tyre Company.

In the mid- to late 1890s, bicycles had a far-reaching social impact. Unsurprisingly, as bicycles became cheaper, their popularity increased. More people had their own means of personal travel, reducing their dependence on horses and horse-drawn carriages. It gave them liberation, enjoyment, and freedom to travel beyond their local community. With the advent of the drop-frame lady's safety bicycle, society cycling became even more popular since ladies could retain their 'feminine identity'.

In 1896, financier Ernest T. Hooley, of dubious principles, acquired the Pneumatic Tyre Co. for £3 m and then floated it as a public company, making a profit of £2 m. Aged 56, Dunlop resigned from the Pneumatic Tyre Co. and retired to Ballsbridge, an affluent neighbourhood of Dublin. Dunlop transferred his patent rights, and, in return, was given 3,000 shares together with £500 cash, equivalent to 20% of the capital. He died at his home in 1921, aged 81. In 2005, he was inducted into the American Automotive Hall of Fame.

In 1897, the North British Rubber Company sold Bartlett's patent rights to the Dunlop Pneumatic Tyre Company for a reputed price of £973,000. The Dunlop company also acquired the Clincher patent's rights from the Palmer Cord Tyre Co., as well as the Westwood tubular-edged cycle rim patent (2102 of 1890), and the Woods 2-way cycle valve, allowing for easy inflation, and controlled deflation when required. Effectively, Dunlop held a monopoly on the 2-part tyre-tube combination. Through the rest of the decade, the Dunlop Company prospered by defending these extensive intellectual properties in the courts.

4.6.7 Rise and Fall of the Dunlop Company in Its Different Guises

In the late 1890s, the Company began to acquire its own rubber mills to process natural rubber. In 1910, to ensure a steady supply of raw materials, the Company acquired rubber estates in Malaysia and, by 1926, it was the largest single landowner in the British Empire. It had become a fully integrated company, sourcing and processing its own raw materials to manufacture finished products.

In 1900, it manufactured its first tyres for motor cars. Early pneumatic tyres comprised an outer casing designed to embrace, reinforce, and protect an inflated air-tight inner tube, and to provide essential wear resistance and grip to the road. There were further demands imposed on automobile tyres. For instance, each wheel and tyre assembly functioned to isolate the structure of a fast-moving vehicle from the vertical and longitudinal vibrations suffered by the wheels as they traversed the road.

In 1912 the Dunlop Pneumatic Tyre Co. passed its activities to the Dunlop Rubber Co. It acquired 200 acres in Birmingham for its headquarters and a new production site. Construction at Fort Dunlop started in 1916.

In 1924, the Dunlop Rubber Co. began to diversify, entering the sports market (golf balls/Slazenger), raincoat manufacture (via Charles Macintosh) and aircraft tyres. In 1920, 90% of its turnover was in tyre production. This fell to 72% by 1928. By 1930, the Dunlop Company was the eighth largest public company in Britain.

In 1941, Japan invaded Malaysia and the Dutch East Indies which meant that Japanese had control of 95% of the world's natural rubber. Natural rubber was a commodity of vast economic and military importance. The quest to develop a range of synthetic rubbers was expedited in the USA.

In WW2, the Dunlop Company produced the bulk of rubber tyres and boots for Britain's war effort.

Concomitant with the weakening of the British car industry in the 1960s, Dunlop went into decline, accelerated by mis-timed mergers and poor management decisions.

In 1984, Dunlop's European and US tyre business was acquired by Japan's Sumitomo Group—its former subsidiary. In 1999, America's Goodyear and Sumitomo formed a global alliance that saw Goodyear take control of Dunlop. How pleased Charles Goodyear would have been.

In 2014, Dunlop ceased production of racing car and motorcycle tyres at the Fort Dunlop site. This heralded the death of the British tyre industry for road vehicles. Only the Dunlop Aircraft Tyre Co. continues to operate in Birmingham.

According to the International Rubber Study Group, global production of natural rubber, in 2018, was 13.9 million metric tonnes, about 70% of which was used for tyre production. Some purchasers of natural rubber have agreed to source it only from trees grown on low quality, degraded land, forsaking the need to clear tropical forests to plant them.

The 2018 global production of synthetic rubber from hydrocarbons derived from crude oil was 15.2mt, about 60% of which was used in the tyre industry.

Tyres evolved side-by-side with the automobile. Key milestones in the history of rubber tyres are shown in Appendix B.

4.7 Frank Whittle OM, KBE, CB, FRS, FRAeS (1907–1996)—Aeronautical Engineer; Inventor of the Turbojet Method of Aircraft Propulsion; Father of the Jet Age

4.7.1 Early Life, Family Circumstances and Education

Frank was born in a terraced house in Newcombe Road, Earlsdon, Coventry, Warwickshire, the eldest of three siblings. His father, Moses, was a skilled and inventive mechanic who worked as a foreman in a factory manufacturing machine tools. Frank was raised in strict Wesleyan tradition, in a household where inventiveness was encouraged. At an early age, he became familiar with the drawing board and other implements of the draughtsman trade.

Frank's father bought him a clockwork model of a Blériot aircraft, and this sowed a seed. Frank became obsessed with aeroplanes. In 1916, a WWI aeroplane—built at the local Standard Works in Coventry—force-landed near his home. When it took off, it nearly decapitated him.

When Frank was nine years of age, the family moved to 9 Victoria Road, Royal Leamington Spa because his father had purchased the Leamington Valve and Piston Company. Frank cultivated his engineering skills by helping in his father's workshop. Frank earnt pocket money by drilling slots in valve stems and doing lathe work. He developed an understanding of the metal alloys designed to withstand the high temperatures encountered in a cylinder head.

In 1917–18, he attended the Milverton Primary School, in Rugby Road, from where he won a scholarship to the School for Boys at Binswood Hall which later became a grammar school, named Leamington College. He was nicknamed 'grub' because of his small stature.

He had an intent interest in scientific and engineering subjects. He obtained a copy of Aurel Stodola's "Steam and Gas Turbine" and learnt about the dynamics and functions of turbines. In the School Certificate Examination, he obtained six credits, no distinctions and failed to matriculate.

Sadly, in the post-war period, his father's business faltered; the family was made homeless and for a time, they lived above the workshop, before moving to a poorer district in Leamington.

His passion for engineering and flying continued to develop. Frank was determined to become a pilot but his first attempt to join the Royal Air Force (RAF) as a pilot at the Hatton camp, went unrewarded because of his diminutive stature. After a vigorous fitness and bodybuilding dietary programme, he grew three inches, boosted his build, and passed the medical. In September 1923, aged 16, he reported to the Aircraft Apprentice School at RAF Cranwell, in Lincolnshire. He trained as a fitter/rigger. The RAF was to provide a most intensive and comprehensive training course.

He underwent three years of training as an aircraft mechanic, learning about the theory of aircraft engines and gaining practical experience in RAF engineering workshops. Throughout his apprenticeship, he maintained an interest in the Model Aircraft Society, making model planes. This exempted him from attendance at organised team sports which he disliked. It also gave him a priceless opportunity to reveal his growing depth of knowledge to the RAF hierarchy at Cranwell, many of whom were also model aircraft enthusiasts.

In 1926, impressed by his academic and practical skills, he was selected for officer and pilot training at the RAF College, also at Cranwell. Unfortunately, he had a strong dislike for the strict discipline and coarse barrack-room behaviour. He did not quite fit with the Cranwellian, team-spirited, public-school fraternity. Eventually, he would show that he was a better leader than follower. From working class boy, through to cadet and pilot, he learnt to become both an officer and a gentleman.

Upon graduation, in 1928, he was posted to 111 fighter squadron, at Hornchurch, as a Pilot Officer. Over-confident in aerobatics, he was inclined to perform to the gallery. He was successful at crazy-flying displays. He was reprimanded for the unofficial ones and congratulated for those which he had been commanded to perform.

To cool his dare-devil ardour, he was transferred to an instructor training course at the Central Flying Training School, RAF Wittering. He became an Instructor at No. 2 Flying Training School, RAF Digby. This was followed by two years as a Test Pilot, at the Marine Aircraft Experimental Establishment, RAF Felixstowe.

4.7.2 Theory of the Sub-sonic Turbojet Engine. Concept Development of the Whittle Unit, WU

A requirement of the officer training course at Cranwell was that each officer-cadet had to produce a thesis for graduation. Aged 21, Whittle wrote a thesis entitled, "Future Developments in Aircraft Design". In 1928, the maximum speed of a RAF aeroplane was under 150 mph, at a service ceiling of 20,000 feet.

Frank reasoned that, to achieve longer ranges and higher speeds, planes would have to fly at higher altitudes, where air resistance is lower. He was thinking of 500 mph, in the stratosphere (above 33,000 ft or 10 km), where air density is one quarter that at sea level, thereby reducing resistance in proportion to speed. However, at high altitudes, the air is too thin to properly engage propellers and the oxygen content is insufficient to run piston engines efficiently.

Whittle considered ways by which propeller and piston engine combinations could be improved upon, or even superseded. He considered rocket propulsion or gas-turbine-driven propellers as possible ways forward. He quickly dismissed rocket propulsion as unsuitable.

Initially, he described a conventional, reciprocating, piston engine driving a low-pressure fan instead of an external propeller. The engine and fan would be enclosed within a hollow nacelle or fuselage. The fan would inhale air through the hole in the front. It would be heated, in part, by the exhaust and other waste heat from the engine. However, by the introduction of burning fuel, the flowing air would be given a boost to its velocity, before expulsion through a nozzle at the rear, producing a propulsive jet. This powerful pushing force is called 'thrust'.

At the time, he had not considered piston-less, propeller-less, gas turbines for jet propulsion. The idea of using a gas turbine to get the effect of a rocket suddenly occurred to him, at RAF Wittering, in October 1929. He mused with the idea of substituting a windmill-like, spinning wheel (or turbine) for the piston engine. Hitherto, a gas turbine was regarded as a machine for supplying 'shaft-power'. Whittle envisaged it as a means of providing propulsion for aircraft by utilising a **jet** of exhaust **gases** to make the **turbine** rotate. Hence the names 'gas turbine' or 'jet engine' appeared.

A turbine could be used to extract some power from the exhaust to drive a compressor, like those used for superchargers. The thrust left over from the exhaust would propel the aircraft. Fifteen months after his 1928 thesis, he had stumbled on the answer for which he had been searching.

Put simply, the jet engine sucks in air at the front and propels exhaust gases out, at high speed, at the back. The engine is forced forwards as the exhaust gases stream backward. (Compare an inflated balloon: when the contained air is released backwards, the balloon flies forward). It would require a new type of gas turbine—one which produced a propelling jet instead of driving a piston. It would also need a compressor to have a much higher pressure-ratio. Such a turbine would have to deal with much larger amounts of energy for a given size and weight.

In an internal combustion engine, the processes of fuel intake, compression, combustion, and expansion all take place in one compartment—the cylinder. The intermittent reciprocating motion of pistons can be converted to rotary motion by means of cranks. In turn, they can drive a propeller which facilitates propulsion in a piston-driven aeroplane.

In a jet engine, there are three major linked components, namely a rotating gas compressor; a downstream turbine on the same axle and one or more combustion chambers between the two. The compressor of air is found in the 'cold', front end of a jet engine. Early models were composed of a single impeller, or a series of alternating stationary and spinning blades that suck air through an inlet and compress some of the incoming air, raising its pressure, before passing it to the combustion chamber.

In an **axial** compressor, the flow of air is substantially unidirectional, that is parallel to the axis of the shaft. A **centrifugal** air compressor is a scaled-up version of a fan unit in a vacuum cleaner, running at very much higher speeds. They were better understood in the context of engine superchargers. A **centrifugal** compressor turns the air flow from the axial direction to the radial, via curved vanes on the impeller, and then back to the axial direction. Air enters near the axis of the rotor and is accelerated to supersonic speed as the vanes on the rotor fling it to the circumference of the rotating wheel.

In Whittle's first practical design, he chose a single-sided centrifugal compressor. Its diameter was 19″ and, theoretically, was capable of breathing 1,560 pounds of air per minute, giving a pressure of over four atmospheres in the combustion chamber. To double the air flow, a symmetrically double-sided centrifugal compressor was later adopted.

In the combustion chamber, energy is added by spraying fuel into the high-pressure air from the compressor and igniting it. The higher the temperature in the combustion phase, the greater the expansion of the gases. However, the targeted temperature had to be limited to a level that could be safely tolerated by the materials used in the turbine and exhaust components.

Whittle aimed initially to burn 168 Imperial gallons of fuel per hour, in a compartment measuring 6 ft^3 (the size of a suitcase), achieving a combustion intensity several times greater than the prevailing state of art in boiler technology. The burning mixture can reach a temperature of about 900 °C.

The superheated gases, now at even higher pressure, rush from the combustion chamber, through a turbine. Blades on the turbine disc deflect the high-speed gases flowing through them, thus generating torque. The motion of the turbine drives the compressor at the front, and any pre-compressor, inlet fans, via a central shaft. Whittle's original design anticipated a shaft rotating at 17,750 rpm, giving a tip-speed of 1,470 feet sec^{-1} for the compressor impeller, and 1,250 feet sec^{-1} for the turbine.

The original 16.5″ turbine operated like an extremely powerful windmill, generating 3,010 hp; static thrust of 1,389 pounds of force (lbf). The hot exhaust gases exited through a tapered exhaust nozzle. Just enough energy was extracted from the exhaust gases to drive the compressor. All the rest provided the energy of the propulsive jet. The various forces react either forward or rearward, and the amount by which the forward forces exceed the rearward forces is the rated thrust of the engine.

The turbojet is a simple heat machine, converting most of the thermal energy into propulsive force by the kinetic energy of hot gases escaping at high speed from the tail pipe. For an aeroplane travelling at 600 mph, the air leaving the exhaust will be travelling at about 1,300 mph—over twice the speed of the cold air entering the engine at the front.

In more recent jet engine variants, such as a turbofan engine, a fan mounted at the front sends some of the inhaled air around the outside of the engine, by-passing the combustion chamber. 'Low bypass' turbofans send most of the air through the core of the engine, whilst 'high bypass' turbofans send more air around it. The by-passed air cools the engine and minimises noise.

Whittle reasoned that a jet turbine essentially has a single moving part—a rotating shaft attached to a compressor at one end, driven by a turbine at the other. In the absence of the reciprocating parts of a piston engine, rotational speeds and power output of a jet turbine could be increased exponentially. The faster and higher the aircraft flew, the better would a jet engine work. Nowadays, efficiency is maximised at supersonic speeds, at which conventional propellers fail to function properly. This sets a jet engine apart from a reciprocating internal combustion engine.

Moreover, because air intake, compression, combustion, and exhaust take place continuously and simultaneously, a jet engine produces power seamlessly throughout its running. This is unlike a reciprocating piston engine which only generates power during a fraction of its cycle.

In 1687, Isaac Newton published his laws of motion (see Sect. 6.3.5.1). The third law states that "For every action, there is an equal and opposite reaction". The action (namely, the force of exhaust gas) is equal and opposite to the reaction (the force of the plane moving forward). This process is a vivid and practical illustration of the third law.

According to Newton's second law of motion, "The resultant force exerted on a body is directly proportional to the rate of change of linear momentum of that body". Thus, the force on the gases expelled from the rear of the aeroplane is equal to their rate of change of momentum and is equal to, and opposite from, the force on the plane.

As the aeroplane moves through air, its thrust is countered by its drag. Conservation of linear momentum occurs when the increased backward momentum of the expelled gases exactly balances the increase in forward momentum of the frontal air that is pushed out of the way by the forward-moving aircraft.

4.7.3 Air Ministry Response to Whittle's Patent. Competition from Piston Engine Propulsion

Whittle was convinced that his blueprint for a gas turbine engine was a viable way of producing the necessary jet thrust to propel an aircraft. He approached the Air Ministry with his design sketches and calculations. Ministry officials were advised by their consultant—Dr A Griffiths, at the Royal Aircraft Establishment—that Whittle's ideas were without merit. Therefore, the governing and influencing bodies in Britain's Establishment did not seriously evaluate his scheme, being convinced that piston engines were the only practicable source of airplane propulsion.

In 1930, since the Air Ministry was indifferent to his proposals, Whittle independently sought patent protection for the idea. One of the Flying School instructors—W.E.P. Johnson—had qualified as a patent agent and helped Whittle draft his British patent (347,206), for which the provisional specification was filed on 16 January 1930. This was the first patent in the world to describe a practical turbojet. Although not sponsored by the Air Ministry, it gave the inventor commercial rights, but the Crown retained "free user rights", making it unattractive to private companies to invest in the concept and, therefore, difficult to raise funding from them.

When the patent was granted in April 1931, the Air Ministry advised Whittle that it had no official interest in the patent because, in their view, "it had no strategic value". Therefore, secrecy was not applied, and it was not put on the Secret List. Instead, it was published, becoming available for perusal worldwide.

Whittle received a letter from the Patent Office in November 1933 advising that he should pay a certain sum if he wished to extend validity of his patent beyond January 1934. Sadly, Whittle could not afford the renewal fee and allowed the patent to lapse.

Whittle's proposal required a compressor having a pressure ratio of the order of 4:1 and an efficiency of more than 75%. At the time, the best aero-engine was possibly the Rolls-Royce Buzzard upgraded with a double-sided supercharger, to give the Rolls-Royce 'R'—a V-12 engine. This powered the Supermarine (Vickers-Armstrongs, Ltd) S.6B—a racing seaplane. When flown by John Boothman, it won the Schneider Trophy, in 1931. Two weeks later, George Stainforth flew the plane and was the first pilot to fly faster than 400mph.

For the Schneider Trophy, the 1931 piston engine configuration had a pressure ratio of 6:1 producing 2,350 hp (1,750 kW) at 3,200 rpm. For a short time between 1932 and 1933, the air, land, and water speed records were all held by machines propelled by R-powerplants. This was the only engine to hold all trifecta and was, of course, a source of great national pride. The legendry Malcolm Campbell subsequently held both land and water speed records, relying on R-powered engines. Whittle faced a major challenge to use gas turbine technology to beat the triumphs achieved with such piston engines.

His 1930 patent (347,206) featured a single-stage axial compressor feeding a single-sided centrifugal compressor. This is a design still used in small gas turbines and helicopter turboshaft engines.

However, when he embarked on building his prototype engine, the Whittle Unit "WU", he opted for a single, double-sided centrifugal compressor. He reasoned that a centrifugal compressor would be simpler and cheaper to produce. It would be more rugged, and more tolerant of varying air intake conditions (a potential cause of blade-stalling) and surges resulting from rapid throttle change (a potential cause of catastrophic blade failure). Moreover, the centrifugal compressor had the advantage of being known technology; sturdy; easy to construct; resistant to foreign object damage (FOD) and resistant to icing—a major concern to pilots.

Having completed four years of 'general duties', Whittle was expected to specialize, so between 1932 and 1933, he pursued focussed studies at the Officers' Engineering Course, at RAF Henlow.

His exceptional performance earned him a secondment to Cambridge University, between 1934 and 1936, where he took the Mechanical Sciences Tripos and gained a first-class honours degree in mechanical engineering after only two years. His academic degrees included M.A. (Hon) and M.I.Mech.E.

The RAF continued to support Whittle, firstly by granting him a post-graduate year to supervise work on his prototype "WU" jet engine and then, by assigning him to a Special Duty List, until his retirement from the Power Jets Company, in 1946.

4.7.4   The Power Jets Company. Overcoming Bureaucratic, Economic, and Technical Hurdles to Produce an Experimental Turbojet for Ground Testing

R. Dudley-Williams (Whittle's fellow cadet at Cranwell and fellow member of the Flying Boat Squadron at Felixstowe) together with retired RAF officer, J.C. Tinling, proposed a business partnership to progress Whittle's ideas. They secured financial backing from an investment bank, O.T. Falk and Partners and, with Air Ministry approval, the four parties agreed to form a private company—Power Jets Limited—on 27/1/36. Incorporation followed in March 1936.

Turbojet development in Britain was to be rescued from oblivion by private entrepreneurial endeavour. The commercial venture was based on assembling a fast aeroplane carrying 500 pounds of mail across the Atlantic in six hours.

Whittle was allotted shares in the Company, but these were held in trust for the Department. As a serving officer in the RAF, Whittle was permitted to act as honorary chief engineer, but only off-duty. Then, in 1937, the RAF appointed him to the Special Duty List. Officially, he was permitted to devote no more than six hours per week to his project, but this time-limitation was to be ignored since the RAF did not enforce his service-related duties.

Power Jets Ltd. opened its doors in 1936 and Whittle's quest to convert theories and design drawings into tangible hardware began. It was felt that the lapsed patent situation could be partially ameliorated by patenting a series of improvements to the original specification. In May 1936, three provisional patents were submitted.

Construction of a prototype 'Whittle Unit, WU' engine for bench-testing was assigned to the British Thomson-Houston (BT-H) Company—a subsidiary of America's General Electric Company—a steam turbine company, in Rugby. BT-H engineers were accustomed to large, stationary electric generators for power stations. They were unqualified for pioneering a new concept of engine requiring light weight, high precision, and revolutionary creep-resisting materials. For this reason, Power Jets themselves had to become experts at stitch- or seam-welding of very thin sheet metal.

On 12 April 1937, an experimental "bench" engine was tested, on the ground, at the Rugby works. This event marks the first run of a self-contained, liquid-fuelled, turbojet engine, thereby making aviation history.

In the design of the turbine blades, BT-H had not allowed for the vortex (whirlpool) flow of gases from a ring of turbine nozzle blades. Whittle brought this to their attention, and, in December 1937, Power Jets sought patent protection for Whittle's vortex discovery and solution (Patent 511278).

The Whittle Unit was powered, initially by diesel oil, and later by kerosene. By WU, Model 3, Whittle had abandoned a single combustion chamber in favour of ten small combustion chambers which conformed with the existing ten discharge ducts from the compressor. In this 'reverse flow' design of burners, heated air from the flame cans was piped towards the front of the engine, before entering the turbine area. This allowed the engine to be 'folded' with the ten flame cans lying around the turbine area, making for a shorter, lighter, and more compact engine. Ignition was provided to only two chambers and, via a series of inter-connecting tubes, one chamber lit up the next.

In 1938, due to its dangerous nature, research was shifted eight miles north to an abandoned foundry—the Ladywood Works—near Lutterworth, in Leicestershire. A reconstructed engine (Model 3) was delivered in September 1938 and serious testing began on 26/10/38. In May 1939, a new compressor impeller with 29, not 30, blades was fitted to avoid resonant coupling with the 10-blade diffuser system. In July, a shaft rotation of 16,650 rpm (94% of his original design specification of 17,750 rpm) was achieved.

Whittle aimed for an intensity of combustion never previously achieved. Early trials with fuel injection (atomisation) was found difficult at first, so fuel vaporisation was chosen as the best bet until something better came along.

It was the Shell combustion chamber, Type 75—with fuel injection—that liberated Power Jets from their difficulties. In July 1940, Isaac Lubbock (of the Asiatic Petroleum Company) working with Shell engineers, rescued Power Jets by developing high pressure atomizing burners, giving the desired outcome. Fuel was injected as a fine mist through a controllable atomising burner. WU Model 3, with this feature, was run on 9/10/40.

At the end of 1941, the Air Ministry authorised the building of a greatly enlarged research and development facility, together with factory, at a site nine miles north, at Whetstone, near Leicester.

In the development of turbojets, there was a recurrent problem with turbine blade breakage due to a hostile environment of high operating temperatures and the physical stresses associated with high rotation speeds. This was overcome by Leonard Pfiel's development of a high-nickel alloy, Nimonic 80, at the Wiggins Works, in Hereford (then part of the International Nickel Company, INCO). Nimonic turbine alloys contain nickel and chromium, with additives such as titanium and aluminium. They are super-alloys, tolerant to high-temperatures and resistant to 'stretching- or creep-fractures'. Nimonic 80 was first used in the W.2B engine in 1941.

4.7.5 German Developments with Jet Engines (Strahltriebwerk) and Aircraft

Since Whittle's 1930 patent was not on the Secret List, upon publication, it was noted by Trade Commissioners of most embassies in London, including the German embassy.

Copies of the 1930 patent were purchased by the German Trade Commission and despatched to German aeronautical research establishments, as well as aero-engine and airframe manufacturers. Whittle's patent was registered at the Berlin Patent Office on 15 August 1935. The German Air Ministry (Reichsluftfahrtministerium, RLM) was highly supportive of emerging technologies and willing to assign significant funds to promising developments, such as the jet engine. Before, and during WWII, the Germans were developing six alternatives to the piston-engine/propeller combination. The turbojet joined the queue of evaluation, along with the rocket, pulsejet, ramjet, ducted fan, and turboprop.

A German journal devoted to patents from Britain, giving notice of Whittle's patent, was published on 14/12/1931 and was made available in the library of the University of Göttingen, where Hans von Ohain was a physics student. Another recipient was the Aerodynamische Versuchsanstalt (AVA; the Aerodynamic Research Establishment) also in Göttingen. Head of AVA was Professor L Prandti who mentored von Ohain in applied machinery. Other tutors of von Ohain were Albert Betz and Walter Enke, who were Germany's top supercharger research scientists.

Research on jet engines in Germany began with von Ohain who completed his doctorate in 1935 and then moved to Heinkel in 1936. In parallel, Dr. H. A. Wagner, at the Junkers Airframe Division, turned his attention from a turbo-prop concept to that of an axial turbojet.

In 1934, whilst a doctoral student, von Ohain conceived the idea for jet propulsion. In 1935, he developed a theory of turbo-jet engines and then applied for two patents. The second—for a "Process and Apparatus for Producing Airstreams for Propelling Airplanes"—featured a single-sided, single-stage centrifugal compressor, back-to-back with a radial inflow turbine to drive it. The German Patent Office referenced Whittle's 1930 patent. It is a matter of dispute as to whether von Ohain was granted Reichspatent 317/38 on 10 November 1935. In any event, it is likely that he would have lost all rights to these inventions with Germany's defeat in 1945.

With self-funding, von Ohain attempted to build a working model with help from engineering craftsman, Max Hahn. Development costs of the jet engine proved to be prohibitive, even for the wealthy Dr. von Ohain. Therefore, in 1936, he and Hahn joined the Heinkel Company which had considerable financial resources and was designing aircraft for the re-emergent Luftwaffe. Ernst Heinkel's obsession with speed had been triggered when he witnessed the Schneider Cup race in 1927.

In his memoirs, Heinkel said that his company had built a hydrogen-fuelled, demonstration engine, the Heinkel Strahltriebwerk (jet engine) He-S.1, and it was bench-tested in September 1937. Others claimed March 1937. In fact, his first experimental device was not a self-contained turbojet since it relied on a separate electric motor to power the compressor, and not the turbine alone. Its main purpose was to assess flow characteristics.

Six months after the bench trials with He-S.1, its successor, the He-S.2, ran on its own power, fuelled by gasoline. By 1939, the engine had been further modified to become the He-S.3A.

On 27 August 1939, as Hitler's troops moved to the Polish border, a Heinkel He-178 V1 airplane, propelled by a turbojet engine, the He-S.3B, took to the air. It was the world's first turbojet-powered aeroplane. The maiden flight of Britain's first turbojet did not take place for another 20 months, on 15/5/1941.

The jet engine powering the He-178 was still in an embryonic stage of development and was only allowed to run, at flight-thrust, for about six minutes, possibly due to the risk of thermal degradation. The He-178 aircraft was never developed sufficiently to be dubbed "operational". It never reached mass production or active service because both the aircraft and its engine were deemed sub-optimum. Moreover, the Reich Air Ministry was already collaborating with other German engine manufacturers to develop axial-flow turbojets.

On 1st September 1939, less than one week after the maiden flight of the He-178, Warsaw was subject to aerial bombing and the Panzers crossed the Polish border. Two days later, Britain declared war on Germany, and the race between Germany and Britain, to build the first combat jet aircraft began.

In April 1936, Herbert Wagner at the Junkers Air Frame Company (Junkers Flugzeugwerken) began work on a turbojet comprising an axial compressor in combination with an axial turbine. The prototype failed its first run in 1938 due to problems with its axial compressor.

In July 1942, Germany conducted the first test flight of a jet fighter—the Messerschmitt Me-262, propelled by two axial-flow turbojets 004, built by Junkers Motoren (Jumo) and designed by Anselm Franz. The Me-262 beat the first Allied jet fighter prototype into the skies by about nine months. However, compared to Power Jets' W.1A, the Jumo 004A was much heavier in proportion to power, resulting in higher fuel consumption and high wing-loading.

Hitler, who was obsessed with 'wonder weapons' (Wunderwaffe), wanted an attack jet fighter-bomber, but flying ace, Adolf Galland, wanted a defensive interceptor jet fighter, believing it would be the saviour of his fighter force. About fourteen hundred ME-262s for different functions were built but only a quarter reached front line squadrons.

Since the German factories were being so heavily bombed, some production was assigned to the REIMAHG (Reichsmarshall Hermann Göring) aircraft factory, in a subterranean lair which had been a kaolin mine, in the Walpersberg Mountain. About 12,000 forced labourers were used to enlarge the tunnels and build aircraft, of whom about 1,000 died from malnutrition, disease and/or SS brutality.

The Me-262 was readied for operational service in July 1944 and became operational with the Luftwaffe in early October 1944. It was the only German jet fighter to fly in combat in WWII, against unarmed Allied reconnaissance aircraft and slow-moving bombers. It claimed 509 Allied kills against about 100 losses.

At the time, because of its advanced aerodynamic characteristics, the Me-262A-1a was the fastest aeroplane in the world, with a claimed top speed of 530 mph. It was known as 'das Schwalbe' (swallow). It had good firepower. Fortunately for the Allies, the Jumo 004B-1 turbojet engines—propelling the Me-262A—were insufficiently developed before full-scale production began.

In this regard, the Me-262A was sluggish in response to the throttle. Their pilots were unable to make high-energy manoeuvres for fear of stalling due to disturbances of the engine's intake airflow. Allied Mustangs took advantage of this vulnerability to shoot down the Me-262s. Neither could the German aircraft make sudden changes of air speed due to the risk of flameout. Overall, their poor handling characteristics drastically reduced their flexibility. They had a short service life and engine reliability issues, due in part to the lack of heat-tolerant alloys. The turbine blades were hollow, air-cooled blades. The turbines required overhaul after ten hours and were often scrapped after only twenty-five hours of flying.

At the end of WWII, both von Ohain and Franz were two of many exceptional, German scientists and engineers selected to bring their expertise to the USA. Dr von Ohain and Franz continued to work on engine issues at what became the Wright-Patterson Air Force Base, near Dayton, Ohio.

4.7.6   Impending War and Its Effect on Attitudes in Britain and America to Whittle's Pioneering Work on Turbojet Engines. Flight Engine, W.1 and the Experimental Aircraft, the Gloster Pioneer

Before the war, the comparative situation between Britain and Germany for mastery of jet engine development was stark. One contestant, namely Whittle, was operating on a shoestring, in a decrepit foundry, initially without much help from Government or industry. Most British aero-engine manufacturers were indifferent because they had vested interests in piston engines.

In the beginning, alongside budgetary constraints and bureaucratic inertia, Whittle could only work on the project in free time from his RAF commitments. However, in 1937, he was assigned to Special Duties which meant that he could fully devote his time to his engine.

On the other hand, Germany had at its disposal, the resources of great industrial concerns, with a host of full-time engineers and scientists, supported by the German government. By April 1938, the German Air Ministry had more than 2,000 engineers working on twelve jet projects.

Anticipation of World War II finally spurred the British Government into supporting Whittle's development work. In 1939, after protracted negotiations with the Air Ministry, the project was approved, and the Air Ministry placed an order for a "flight engine". The Gloster Aircraft Company was contracted to build an experimental aircraft. The Gloster Aircraft Company was formed in 1917 and became part of the giant Hawker Siddeley Group in 1935. It was finally rebranded, and its name disappeared in 1963.

A radially compact, axial compressor can be slung under a wing without much design modification. On the other hand, centrifugal compressors are bulky. For a single engine format, the engine is placed in the fuselage itself (with air-intake through the nose). For a twin-engine configuration, the engines are built into the wings.

By April 1941, the new, 'Whittle Supercharger', type W.1 engine was ready for testing. The Air Ministry insisted that it should undergo 25-h of rigorous assessment before it could be used for flight trials. Ultimately, it was fitted to a specially built Gloster E.28/39 airframe (indicating Experimental Project No. 28; ordered in 1939). When restricted to 16,500 rpm, it had a thrust of 3.8 kN (850 lbf). The 17-min, maiden flight of the propeller-less Gloster Pioneer took place on 15 May 1941, from RAF Cranwell, flown by test-pilot Gerry Sayer. Ironically, Cranwell is where Whittle had written his cadet thesis in 1928.

During the flight, the plane reached a speed of 345 mph (545 km/hour) in level flight, at an elevation of 25,000 ft (7,620 m). Within days, the piston-less engine was reaching 370 mph (592 km/h) surpassing the performance statistics of contemporary Spitfires, the propeller of which was driven by a Rolls-Royce Merlin engine, regarded as being one of the best piston engines in the world. The Merlin, however, weighed twice as much as the W.1, giving the same thrust as a rpm-restricted, W.1.

Over a limited number of days, ten hours of flying time were achieved over seventeen flawless flights—a feat unheard of in aviation history. Apart from a small crack in the outer casing of one combustor, no other engine defects materialized. The plane was a bliss to fly. Being propeller-less, visibility from the cockpit was perfect. The engine and plane exceeded all design specifications. The jet-powered plane used relatively less fuel at flight-ceiling altitudes.

One of our WWII allies, the USA, heard about the success of the project and asked for details and an engine that they could replicate, so they could jump-start their own jet program. In America, the technology was enthusiastically embraced. A period of co-operation between Britain and the USA on the development of the turbo-jet engine began.

Whittle shared his technology with Britain's Rolls-Royce and the American General Electric (GE) Company. In March 1941, it was agreed that the W.1X experimental engine should be shipped to the USA so that GE could build an equivalent, under licence, for war purposes only. At the end of WWII, the UK released America from that undertaking for the payment of $800,000, giving the USA a low-cost entry into gas turbines for civil aeroplanes.

After a period of development, a pair of General Electric engines, based on the W.1X, and W.2B/23 were fitted to a Bell airframe and the maiden flight of the Bell XP-59A Airacomet fighter aircraft took place on 2/10/1942, five to six months before the British Meteor jet fighter was airborne, in 1943.

4.7.7   Whittle's 2nd Generation Turbojet. Involvement by the Rover Car Company and Rolls-Royce. Nationalisation of the Power Jets Co.

By the end of 1940, before the E28/39 had flown with the W.1 engine, the Air Ministry decided to have a higher thrust engine (the W.2) developed.

The Government invited the Rover Car Company to prepare for a production line. Rover secretly started work on its own variant of Whittle's W.2 and W.2B engines. For instance, when designing the Rover W.2B/26, Adrian Lombard opted for a straight-through flow, with hot gases exiting the combustion chambers directly onto the turbine wheel, instead of being piped forward as per Whittle's reverse flow.

Since Rover failed to deliver parts of adequate quality, the W.2B/23 design and production facility at Barnoldswick were acquired by Rolls-Royce for their RB.23 Welland production unit. Fortuitously, Rolls-Royce already possessed a fully-fledged supercharger division. This fitted well with the turbojet project because they had better experience of meeting the problems of extreme heat on metal working parts.

In January 1943, a W2B was run for 400 h. Eventually, it went into production as the Rolls Royce Welland 1, being earmarked for the Meteor Mk 1 aircraft (Fig. 4.4).

In May 1943, Whittle was posted to RAF Staff College for a 3-month 'war course'. In the same month, the British W.2B Gloster F.9/40 (Meteor) twin-engine jet fighter, powered by Rolls Royce Welland engines (production W.2B's), was flown. It was state of the art in military aviation, offering all round visibility, smoothness, lack of both noise and vibration in the cockpit.

It commenced trials with the RAF to bring it to operational readiness. It became operational with 616 Squadron, at RAF Manston, on 27 July 1944. The Meteor was the only Allied jet to be operational in WWII, and 616 Squadron held the honour of being the first British unit to operate jet-powered aircraft. No jet-to-jet battles were experienced during WWII, but the Squadron achieved their first victory in August 1944, flying their Meteors to defend the nation against Hitler's pulsejet powered, V-1 flying bombs (the 'Doodlebugs').

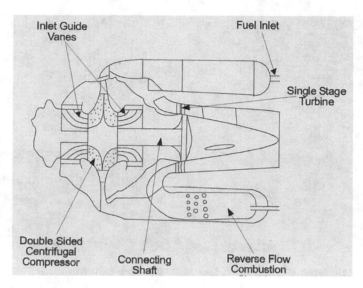

Fig. 4.4 Frank Whittle's W2B engine

Before his attachment to Power Jets ended in January 1946, Whittle was responsible for the development of the W.2/500 and the W.2/700 engines— the parents of subsequent Rolls-Royce engines. The best features of the W.2/500 and Rover's W.2B/26 were incorporated into a new design, the RB.37 or Rolls-Royce Derwent I (see Appendix C).

The nationalisation of Power Jets Ltd to create Power Jets (R&D) Ltd, in March 1944, hastened the disbanding of a dedicated team of engineers. Whittle had handed all his shares (worth £47,000 in 1944) and rights to the Ministry of Aircraft Production (MAP) sometime earlier.

In 1946, the Company was merged with the gas turbine section of the Royal Aircraft Establishment, being consolidated into the National Gas Turbine Establishment (NGTE) at Pyestock, near Farnborough. It was compelled to limit its activities to fundamental research and component development. It was stripped of its rights to design and develop new engines. Angry at this decision, Whittle and several co-workers resigned from the Company. He felt that "Power Jets was smothered to death by the Government".

For instance, promising projects arising from Whittle's 1936 patents, namely, the High-Bypass Turbofan and the Reheat/After-Burner Exhaust-Fan, were subsequently abandoned. He claimed that nationalisation seriously retarded Britain's post-war jet industry. Indeed, any modern turbojet, turbofan, turboprop, and after-burning turbojet (see Fig 4.5) can trace its

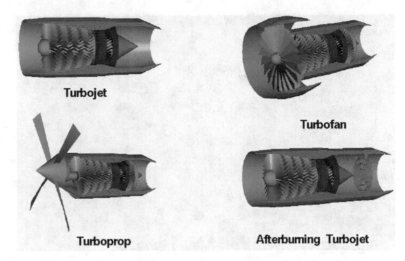

Fig. 4.5   Main types of jet engine. Courtesy NASA/JPL-Caltech

origins to Whittle's W1 engine (see Appendix C herein for Whittle's engine lineage).

At the front of a **Turbojet** is the compressor (cyan) followed by the combustion section (red) and a turbine (magenta). The **Turbofan** also has a fan (green) at the front, as well as a rear turbine (green). The fan and rear turbine are on a second shaft which passes through the main core shaft. The fan offers slightly more thrust. The **Turboprop** has a rear turbine (green) which is attached to a drive shaft which passes through the core shaft and is connected to a gear box. The gear box is connected to a propeller. This produces most of the thrust since most of the energy of the exhaust gases goes into turning the propeller drive shaft. The **Afterburning Turbojet** has a rear burner (light orange) that squirts fuel into the exhaust stream to generate additional thrust.

Responsibility for Britain's jet engine design passed to Rolls-Royce and Armstrong Siddeley. Licences were given to General Electric and Pratt & Whitney in America. Whittle saw the folly of the great 'jet giveaway' to the Soviets by Atlee's post-war Labour government who authorised Rolls-Royce to sell their most advanced turbojets to the U.S.S.R. The Soviets reverse-engineered Nene engines, building 39,000 derivatives without licence.

4.7.8 Whittle's Life After the RAF; Tributes and Civic Honours; Emigration to the USA; Whittle's Legacy

Whittle was awarded Commander of the Order of the British Empire (CBE) in the 1944 New Year Honours. Some days later, the public were made aware of the jet engine, making Whittle a national hero. In 1947, he was awarded Companion of the Order of the Bath (CB).

Whittle suffered mental torture and physical fatigue to achieve his goal. He suffered nervous breakdowns in December 1941, March 1944 and late 1946, all of which would have affected his temperament. On the grounds of ill health, Whittle retired from the RAF in April 1948, with the rank of Air Commodore. A Royal Commission recommended an ex-gratia payment of £100,000 and he was made Knight Commander of the Order of the British Empire (KBE). He was elected Fellow of the Royal Society and was also a Fellow of the Royal Aeronautical Society (FRAeS).

Gradually his strong religious beliefs eroded—they became inconsistent with his scientific teaching and approach. He became an atheist. By 1964, he had deserted his long-held socialist beliefs. In 1976, he married his second wife, the American, Hazel Hall. In the following year, Whittle emigrated to the USA, becoming NAVAIR Research Professor of Aerospace Engineering at the US Naval Academy, in Annapolis, Maryland. In 1978, the US Department of Transportation honoured him with the Extraordinary Service Award, the highest accolade the office can bestow. Whittle became friends with his German counterpart, von Ohain who was also living in the USA.

In 1986, Whittle was awarded the Order of Merit, a distinction the British monarch limits to twenty-four members. He was sitter for fifteen portraits now held at the National Portrait Gallery (WC2H 0HE).

Whittle had twenty-seven patents to his credit, ranging from jet engines to turbo-drills used for oil-field applications. An abridged list of Whittle's patents is given in Appendix 1 of the 1987 publication: "Jet—Frank Whittle and the Invention of the Jet Engine" by John Golley, in association with Sir Frank Whittle and W. Gunston, an expert on aviation history. Whittle's medals, professional and civic honours and membership of learned societies are also listed in "Jet".

He died in 1996 at his home in Columbia, Maryland. He was cremated in America and his ashes were flown to England, where they were interred in the chapel at RAF Cranwell. At a thanksgiving service, in Westminster Abbey, in November 1996, he was saluted by a fly-past of six jets.

Sir Frank Whittle was a shy, reserved, mathematical genius, and a technically proficient, "hands on" engineer, remembered for his modesty, his tenacity of purpose and his persistence in the face of rejection and bureaucratic roadblocks. He was an inspirational team leader. He was one of the greatest engineering minds of the twentieth century but initially, his futuristic ideas were dismissed as impracticable. To use a gas-turbine power plant in an aircraft was ground-breaking. He overcame numerous technical and funding hurdles which led him to assemble relatively simple mechanical components to achieve his goal.

The jet engine is a product of WWII but had no measurable effect on its outcome. Whittle was disappointed that his jet engine did not make a major contribution to Britain's war effort. Had the Air Ministry expedited his technological advances, the duration of WWII may have been shortened. Whittle's 1928 thesis and 1930 patent led to a revolution in the design of power plants for both military and civilian aircraft.

The jet engine came to prominence in the post-war years and Whittle's engine design had world-wide application, radically changing the speed at which we travel around the globe. Centrifugal compressors are still used in smaller gas turbine engines, but axial compressors are used for larger, more powerful engines.

His technology also played an important part in gas turbines used for non-aeronautical applications. For instance, when the main shaft of a turbojet is connected to a gear box, it can be used for the propulsion of ships and trucks. When the main shaft is connected to an electro-magnet, it will generate electricity.

In 2000, a memorial stone was unveiled in the Lady (RAF) Chapel in Westminster Abbey. The inscription on the stone reads: "Frank Whittle, Inventor and Pioneer of the Jet Engine". The Chapel also contains the graves of RAF leaders Lord Trenchard ("Father of the RAF") and Lord Dowding (Commander-in-Chief of RAF Fighter Command).

The Lutterworth Museum (LE17 4DY) now boasts a wealth of papers and memorabilia from the jet engine era. The Midland Air Museum at Coventry Airport in Baginton (CV3 4FR) plays host to the Frank Whittle Jet Heritage Centre. Next to Farnborough Airfield, in Hampshire, is the Farnborough Air Sciences Trust (FAST) Museum (GU14 6TF) which has, on display, two of Whittle's earliest engines.

At the Jet Age Museum (GL2 9QL) next to Gloucester Airport, is a replica of the Gloster E28/39—Britain's first jet aircraft, as well as two of the jet fighters used by the Allied forces during WWII, a Meteor T7 and a Meteor F8. The original Gloster E28/39 aircraft, together with the W.1 that powered it, can be seen at the Science Museum in London (SW7 2DD). The Whittle W.1X engine can be seen at the Smithsonian National Air and Space Museum/the Boeing Milestones of Flight Hall, Washington DC 20,597, USA.

5

Drawbacks with Industrialization. Sanitary Revolution Offering Technologies to Improve Public Health

The civilization of peoples can be measured by their domestic and sanitary appliances

(J G Jennings, 1850s)

5.1 Précis

As the Industrial Revolution roared into life, the urgency and clamour of factory work replaced the slower, seasonal rhythms of the countryside. The Industrial Revolution created enormous wealth, as well as generating inequalities in society, typified by dark satanic mills, child-labour, and workhouses. The working man was turned into an automaton, toiling to the tireless demand of the steam engine and its functional attachments. Some of his human dignity was lost.

Living conditions for many workers and their families were grim. Fortunately, alongside wealth creation and imperialism, ran an active radicalism of protest and humanitarianism. Social reformers established a link between poverty, inadequate living conditions, lack of clean water, poor sanitation, and disease. The way humans congregate and live their lives creates vectors for the transfer of microorganisms and viruses between them. A holistic approach to public health was required. Responding to various health crises, resulting from the mass movement, and crowding of citizens in towns and cities, our

© The Author(s), under exclusive license to Springer Nature
Switzerland AG 2022
J. Bailey, *Inventive Geniuses Who Changed the World,*
https://doi.org/10.1007/978-3-030-81381-9_5

sanitary engineers designed and built sanitary equipment and urban sewage systems copied by other countries as they industrialized.

It is notable that six extraordinary British men, namely John Harrington, Alexander Cummings, (Josiah) George Jennings, Thomas Crapper, James Newlands, and Joseph Bazalgette applied great ingenuity and simple engineering solutions to the task of human waste disposal.

For instance, Harrington invented the 1st flushing water closet; Cumming the S-bend. Jennings designed a 1-piece wash-out closet; Crapper a toilet cistern fitted with a floating ballcock.

In the latter half of the nineteenth century, Britain monopolized the international market for sanitaryware, pipes and fittings. Britain led a sanitary revolution, supported by legislation, offering a holistic approach to health. Consequently, between 1850 and 1900, life expectancy increased by ten years from 43 to about 53 years.

As urban population densities increased during the Industrial Revolution, the need to keep drinking water separate from human waste became more vital. It had been realized that water-borne bacteria spread disease. The requirement to dispose of human waste more efficiently and sanitarily became more crucial.

The ingenuity of visionary engineers made it possible to live more safely in cities. Liverpool would lead the way in urban and sanitary advancements to improve public health. The first integrated sewage system in the world was developed in Liverpool, in 1848, being overseen by Borough Engineer, James Newlands. The target of separating the provision of drinking water from the removal of human waste was achieved, so eliminating water-borne diseases, eradicating cholera. This transformed the health of the urban poor and saved countless lives.

In London, which had seven times more citizens than Liverpool, Joseph Bazalgette supervised the construction of an extensive, underground sewage system. At the time, it was the biggest civil engineering scheme in the world and is regarded as one of the greatest building achievements of Victorian Britain.

5.2   Background. The Liverpudlian Solution

Primitive latrines date back five millennia and relied on a constant stream of water to take away human waste. As regards terminology, the word "closet" is derived from the Latin word "clausa" meaning locked in or separated. "Toilet" originates from the French word "toilette", adapted to mean a place where one withdraws to perform necessary bodily functions, in private. It is also used to describe a fixed receptacle into which a person can urinate or defecate, the receptacle being connected to a system for flushing away waste into a sewer.

The need for disposing human waste more efficiently and sanitarily became more pressing as population densities increased during the Industrial Revolution. Over-crowded cities were ravaged by cholera and typhoid epidemics.

For instance, the population of Liverpool increased from about 77,000 in the year 1801, to 376,000 in 1851, swollen by mass emigration from Ireland during the 1845–52 potato famine. Life expectancy at birth was about twenty-three years. One in four babies died before their first birthday. Many Liverpudlians lived in squalid slums, sharing a privy connected to a cesspit that was emptied once or twice each year. The latter overflowed into the drinking water supply from boreholes, thereby polluting it with raw sewage. The whole urban area could become infected by disease and illness would then spread quickly.

Liverpool's Improvement of the Sewerage and Drainage Act 1846 was the first comprehensive sanitary act in Great Britain. A Tory-dominated borough council sought the provision of public services, including sewers, clean drinking water, comfortable housing, public baths and wash houses, recreational public parks, and easy-to-clean street paving.

James Newlands was appointed the first ever Borough Engineer. He recognised that public health could only be improved when drinking water was kept separate from the disposal of human waste. He supervised the construction of the world's first integrated sewer system.

Firstly, he studied the topography of Liverpool and its surrounds, making about 3,000 geodetical observations, from which he constructed a detailed contour map. From this, he was able to lay down sewers which discharged by gravity-feed. The sewers were brick-lined and egg-shaped with the tapered end pointing downwards to carry the waste slurry, in a self-scouring manner at a satisfactory speed, without hindrance. Outflows carried waste directly to the tidal River Mersey. Cesspits became redundant.

This public health revolution transformed the lives of the urban poor. By the time Newlands retired, life expectancy at birth, in Liverpool, had doubled. His approach became the model for sanitary improvements across the world and many engineers were inspired to follow his example.

5.3 John Harington (1561–1612)—1st Indoor Flushing Water Closet (WC)

John Harington was born at Kelston, near Bath. His mother served as maid of honour to Elizabeth before the latter assumed the throne. Elizabeth was named as John's godmother. Throughout his lifetime, he ambitiously pursued a favoured place at the Queen's court, becoming a courtier.

He was educated at Eton and obtained a bachelor's degree from Cambridge University in 1577/8. He entered Lincoln's Inn in London to study law but after the death of both of his parents, he abandoned his studies to claim his inheritance of the family estate in Somerset.

Basically, there are two things that an effective water closet (WC) is required to do. The first is to flush away human waste; the second is to seal off the smells and gases associated with that waste.

In 1592, Harrington invented the first flushing WC (called "Ajax—a "jakes" was a popular slang term for privy) which was installed at his home, at Kelston. Apparently, impressed by what she saw at his home, Queen Elizabeth I asked Harington to install a similar device at her palace, in Richmond, Surrey.

It was the forerunner of the modern flush loo, having a flush valve to let stored water flow out of a raised tank, into an oval bowl with an opening at the bottom, crudely sealed with a leather-faced valve. A wash-down design emptied the bowl.

However, the problem of objectionable, permeating, back-flow smells remained and, besides, few people could afford the cost of installation, as well as the need for about 7.5 gallons of water at each discharge. Therefore, the public continued to use the chamber pot which was often emptied from an upstairs window into the street below. Sometimes, upmarket commodes sent waste to the cellar.

5.4 Alexander Cumming (1733–1814)—Flush Toilet with an S-bend Outlet

Cumming was born in Edinburgh. He was apprenticed to a watchmaker. In 1775, he was granted the first patent (Patent Office specification 1105) for a flush toilet. His key, world-changing, invention was the S-bend. His solution was simplicity itself—simply bend the outlet pipe, under the toilet basin, into a S-shape. By so doing, water was continuously retained within the system and sewer gases were prevented from entering the room from a sewer or home septic tank.

Because the S-bend is replenished every time the bowl is flushed, it does not have time to become stagnant or smelly. The concept survives in today's plumbing, albeit often modified as a U- or J- shaped pipe trap.

In 1783, Cumming was a joint founder of the Royal Society of Edinburgh, himself becoming a Fellow.

'Sitting' toilets became a sign of European achievement and civilization. However, it has been noted that there are fewer bowel health issues in developing countries, where squatting in a primordial posture is the norm. Alexander Kira has argued that sitting toilets are ill-suited to human defecation because they result in straining which may be a cause of haemorrhoids. Proponents of squat toilets claim that, in such a position, the puborectalis muscle is better relaxed, unfurling the colon, facilitating bowel movement, resulting in health benefits.

5.5 Josiah George Jennings (1810–1882)—Sanitary Engineer; 1st Public Flushing Toilet

Josiah was born in Eling, at the edge of the New Forest, in Hampshire. He was one of seven children. He was educated at the local school. He worked for his uncle's plumbing business and then became a self-employed plumber, in London. In 1838, he set up a business in Lambeth, specializing in premium sanitaryware.

The Great Exhibition of 1851 was a shameless display of British greatness and inventiveness. It was housed in a massive glasshouse which became known as 'Crystal Palace'. About one third of the population attended. Profits from the Exhibition were used to fund the Victoria and Albert Museum, the Science Museum and the Natural History Museum.

It saw the first major installation of public toilets. Jennings fitted his 'monkey closet' flushing lavatories in the Retiring Rooms. Trying to keep the number of plumbing joints to a minimum, Jennings designed a 1-piece closet. These were "wash-out" closets, so-called because the solid waste was deposited in a porcelain bowl and then the waste was washed out, over a weir, into a trap.

Jennings' contraption was an enormous success. During the Great Exhibition, over 827,280 visitors paid one penny to use one. In 1852, he was granted a patent for a water closet in which the pan and trap were constructed in the same piece so that a certain quantity of water was retained in the pan (bowl) itself, in addition to that in the trap which formed the water joint.

Ceramic materials were perfect for water closets, resulting in tough, durable, easy-to-clean, hygienic bowls.

Jennings' plumbing skills, business acumen and inventiveness helped bring an end to the stinking closets that existed prior to his wash-out and wash-down closets. He observed that "the civilisation of peoples can be measured by their domestic and sanitary appliances". He became messianic about the civilizing effect of good sanitary practises.

His company won contracts with the British government, as well as businesses such as railways that operated internationally. He worked with the Admiralty and Royal Engineers.

In 1855, Jennings headed the sanitary commission that the British government—at the request of Florence Nightingale—sent out to Selimiye Barracks at Scutari, Turkey, to improve conditions at the military hospital. Liverpool's Borough Engineer, James Newlands, also spent time there as a government sanitary commissioner (see Sect. 9.5).

In Britain, Jennings advocated public sanitation projects such as underground public conveniences but the first, at the Royal Exchange, did not appear until three years after his death. Subsequently, his company would provide urinals and water closets for many public facilities throughout the world.

5.6   Thomas Crapper (1836–1910)—Toilet Cistern Fitted with a Ballcock

Thomas Crapper was born in Thorne, South Yorkshire. When he was fourteen years of age, Thomas travelled to Chelsea, London to become apprentice to his brother, George, who was a Master Plumber.

In 1861, he equipped his own premises with a brass foundry and workshops. He made several key inventions in sanitary engineering, including (i) the floating ballcock to improve the water tank (cistern) filling mechanism and storage, and (ii) the manhole cover. He promoted 'a certain flush with every pull'. The cistern was adopted to minimise wastewater. The ballcock is still used widely in toilets today, whilst manhole covers in Westminster still carry his name.

Thomas Crapper was an entrepreneur credited with manufacturing the first successful range of flush toilets. He received several Royal Warrants. In

1861, he was hired by Prince Edward to install lavatories in several royal palaces.

American servicemen, passing through England during WWI, coined the slang word "crapper" to mean toilet.

Thomas Crapper may have been the first plumber to set up a showroom for baths, toilets, and basins. Other sanitary-ware companies such as J G Jennings, T Twyford, E Johns, J Shanks, J Wedgewood and J and H Doulton began producing toilets, much as we know them today.

For instance, Thomas Twyford mass produced water closets at a price affordable to many. His most popular product was the Unitas which sold all over the world. From an over-head cistern, a 2-gallons, washed down flush removed solids from the basin, through a trap, leaving sufficient water in an after-flush chamber to receive the next soil deposit.

Flush toilets became both a cure and a curse. A small quantity of solid waste required ten to twenty times as much water to remove it. If flowing into cesspits, this increased the likelihood of unwanted overflows. The need for connection to an integrated sewerage system was paramount.

However, it was not until after WWI that all new housing in London and its suburbs was required to have an indoor flushing toilet. It had taken three and a half centuries, from John Harington's invention, for a flushing water closet to become widespread.

5.7 Joseph Bazalgette CB (1819–1891)—Civil Engineer; Urban Sewage System

5.7.1 Background

Joseph Bazalgette was born in Enfield, London. His French grandfather came to England in 1784. His father was a commander in the British navy. Joseph was of small stature, of delicate health and educated privately. His career began as a railway engineer from which he gained experience in land drainage.

In 1842, he formed a private consulting practice in Westminster, near Parliament Square. His proximity to the centre of politics was to prove propitious.

After his marriage, in 1845, he worked intensively for two years on the rapidly expanding railway network. This was at huge cost to his health, and he suffered a complete nervous breakdown from which it took him two years to recover.

5.7.2   Cholera in London and Its Eradication

In 1831, there was a global outbreak of cholera, killing over 6,500 people in London alone. The 'miasma theory' dominated public discourse, whereby it was believed that the disease was spread by 'fould air'—it was not the polluted water that was deadly, but the smell of it.

In London, in the early part of the nineteenth century, cesspits overflowed into underground rivers and the drinking water supply. Sewage found its way into open gullies which emptied into the River Thames. The river became an open sewer with dire consequences for the health of Londoners. When Queen Victoria came to the throne in 1837, only 50% of London's infants lived to their fifth birthday.

In his 1842 Report on the Sanitary Conditions of the Living Poor, social reformer Edwin Chadwick inveighed against cesspits and privies, saying that retaining human waste in them was injurious to health. Also, emptying them by hand labour, at night, and removing the contents by cartage was offensive and dangerous.

The first Industrial Revolution had led to a rapid increase in urban population concomitant with poor housing conditions, widespread disease, and poor health. Disease was treated as an issue of personal frailty rather than urban squalor. Chadwick believed that pauperism could be prevented by preventing disease. A second severe outbreak of cholera occurred in 1848, when 14,000 people died in London. This lent immediacy to sanitary improvements.

Prompted by Chadwick, the Public Health Act 1848 was passed. This made it unlawful (Clause 51) to erect or rebuild any home without a sufficient water closet (WC) or privy and ash-pit, furnished with proper doors. Clause 49 decreed that no new or rebuilt house was to be built without drains communicating with the sea or municipal sewer.

Between 1849 and 1854, the English physician, John Snow, used groundbreaking epidemiological studies to show that cholera might be a waterborne disease and well-pumps in Soho could be contaminated. Cholera epidemics—known now to be caused by the bacterium *Vibrio cholera*—were rife. These bacteria produce a potent enterotoxin that triggers the intestines to excrete vast amounts of highly contagious, watery diarrhoea. The victim suffers excruciating pain and death, possibly in hours. In 1853/4, the third cholera epidemic in London took another 10,500 lives.

5.7.3 London's Integrated Sewage System

In 1849, Bazalgette was appointed Assistant Surveyor to the Metropolitan Committee of Sewers, in London. In 1850, London was the biggest city on the planet with over 2.3 million inhabitants and over 300,000 inhabited houses. At the time, there was no unified system for handling sewage and drainage.

In 1855, the Metropolitan Board of Works was created. Championed by fellow, French-descended, Isambard Brunel (see Sect. 5.7.1), Joseph Bazalgette was appointed Chief Engineer. Over the next thirty years, he was to transform London. With seven times more citizens and dwellings than Liverpool to cater for, the construction of an integrated sewage system was a gargantuan task. His dedication to the project would eventually produce one of the greatest building achievements of Victorian Britain.

Tight-fisted politicians in government were urged to act against 'open sewers' but they prevaricated until the hot, sweltering summer of 1858, when the "great stink" occurred. The brand-new Houses of Parliament, being adjacent to the polluted R. Thames, were overcome with malodours. On 30 June, a sitting of parliament was suspended, and the building evacuated.

Desperate politicians voted a bill for a complete, publicly financed, sewage system. The bill passed through both Houses of Parliament in only 18 days, approving an expenditure of £2.2 m. It then commissioned Bazalgette to create an underground network of sewers. It was to be a 'combined system' carrying rainfall, as well as sewage. Whereas James Newlands discharged waste into the River Mersey in central Liverpool, waste in London was to be carried eastwards away from central London.

It was a colossal engineering project. Over a 16-year period (1859–75), he supervised the building of 1,100 miles of small, street sewers connected to homes, and 82 miles of main intercepting sewers—three to the north of the Thames, and two on the south side.

The construction was executed to the highest specification and built to the highest standards. About 2.5 million tons of earth had to be excavated using picks and shovels. About fitty-two acres of land were reclaimed from the river. Stone embankments were built to house the mega sewers, creating riverside vistas befitting a capital city.

With shallow gradients, and a series of pipes of ever-increasing diameter, gravity alone propels the waste, helped in four places with pumping stations. The pumping stations at Abbey Mills (nicknamed the 'cathedral of sewage') and Crossness were elaborately designed and lavishly decorated. They are both now listed buildings. The pump at Crossness was the largest in the world at the time. It was based on a rotative beam engine and built by James Watt to Bazalgette's own design. It was capable of pumping sixty-six tonnes of sewage every minute. The Crossness pumping station can be seen on open days (contact Crossness Engines; SE2 9AQ).

The sewer tunnels were up-side-down egg-shaped and brick-walled to increase velocity and carrying capacity. The construction made pioneering use of Portland cement which was water-resistant (see Sect. 14.2.4). About 318 m bricks and 0.9 m cubic yards of concrete/mortar were consumed.

On the south bank, the sewage flowed as far as Crossness (completed in 1865) and stored in giant reservoirs. On the north bank, sewage was carried eastwards as far as Beckton (completed in 1868).

Having reached the Thames Estuary, the whole waste was pumped into the tidal river, from where it was swept to sea on the ebbing tide. Hence, the sewage still passed into the Thames, but in less populated areas. After 1880, however, the solid waste was settled out at Beckton and Crossness, discharging only liquid waste into the Thames. After 1887, a fleet of vessels carried the solid sludge to the sea, where it was dumped.

Because of these measures, cholera in London was eradicated and the River Thames purified to a large degree. London's sewers were a triumph of Victorian engineering and, by their eventual completion, in 1885, served four million people. Daily, it dealt with an average of 420 million gallons (1.8 billion litres) of water and waste.

At the time, it was the biggest civil engineering scheme in the world. Bazalgette's system has survived into the twenty-first century. It was possibly the greatest single achievement in improving the health of Victorian Londoners. Through sanitation projects across the country, public health across Victorian Britain was transformed, with countless lives saved and the quality of everyday life improved, even in big conurbations.

The basic system is still operational 150 years later, but now, the destination of sewage is a treatment plant. At the time of writing, the population of London is about nine millions and the new, 20-mile long, £5 b Tideway or 'super sewer' is being built, together with a new sewage-treatment plant at Beckton which will be the biggest in Europe.

In 1875, Bazalgette was awarded the Companion of the Order of the Bath (CB) and, in 1883, he was elected President of the Institute of Civil Engineers. He retired in 1889 and died in Wimbledon in 1891. A memorial to Sir Joseph stands on Victoria Embankment.

6

17th and 18th Century Multi-disciplinary Scientists. Motion, Forces, Gravity and Light

Nullius in verba (Take nobody's word for it). (Motto of the Royal Society, 1660)
To explain all nature is too difficult for any one man or even for any one age. 'Tis
much better to do a little with certainty and leave the rest for others that come after
you, than to explain all things by conjecture without making sure of anything. (Isaac
Newton, unpublished notes from the preface to Opticks, 1704)

6.1 Précis

The task of scientists is to describe natural phenomena and to elucidate nature's laws. Three polymaths—Robert Hooke, Isaac Newton and Henry Cavendish—lived in the seventeenth century and were three of the most important scientists of the age, embarking on a host of ground-breaking multi-disciplinary studies. They are especially remembered for their theoretical postulations and experimental observations on gravity and light.

Robert Hooke formulated the law of elasticity and investigated capillary action. He used a primitive compound microscope to examine slices of cork tree bark. He coined the word "cell". He was described as England's Leonardo da Vinci. His *Micrographia* is famed for its engravings of the miniature world. He identified fossils as remnants of once-living creatures. He offered a wave theory of light; suggested matter expands when heated; air is composed of small particles and is involved with combustion; that gravity is applied to all celestial bodies.

© The Author(s), under exclusive license to Springer Nature
Switzerland AG 2022
J. Bailey, *Inventive Geniuses Who Changed the World*,
https://doi.org/10.1007/978-3-030-81381-9_6

Isaac Newton was the son of an illiterate farmer who achieved exceptional things. He is one of the most influential scientists ever to have lived, dominating the scientific view of the physical universe for about three centuries. Arguably, he had one of the greatest scientific minds in history—a creative genius in abstract thought and futuristic visions.

Just as the world faced quarantine during the COVID-19 pandemic in 2020, in 1665–6, Newton isolated himself during the Great Plague. During two extraordinary years, he developed laws of gravity and motion; a new theory of light and co-invented calculus. He used mathematics and scientific principles to describe a diverse range of natural phenomena not understood at the time. Commentators referred to this period as his "*annus mirabilis*" or his 'year of wonders'.

His scientific work revealed a Universe that obeys logical mathematical laws. His laws of dynamics and universal gravitation have a reach that extends to the extremities of the Universe. He succeeded in combining laws that govern the motion of objects on Earth with those laws that determine the motion of celestial bodies. He put forward a unified theory of the Universe, describing bodies moving with clockwork predictability, albeit on a stage of absolute space and time. When America's Mission Control directs a spaceship in the solar system, the trajectory will be determined, in part, using Newton's calculus, together with his theory of gravity and his laws of motion. We now know that gravity is both a great creator and destructor of planets and galaxies.

He related the observed ebb and flow of tides, as well as spring and neap tides, to the perturbing and varying gravitational forces exerted jointly by the Moon and Sun. He linked the precession of the equinoxes to the attraction of the Sun and the Moon on the Earth's equatorial bulge. He solved problems associated with fluids in movement and of motion though fluids. He calculated and determined experimentally, the speed of sound waves. When investigating the refraction of light by a glass prism, he demonstrated the divisibility of white light into several coloured rays which could not be further sub-divided, but which could be reconstituted.

Henry Cavendish approached every investigation with a strict quantitative examination. Like Robert Boyle, he was fascinated with gases, liberating them by heating solids or treating solids with acids. He investigated their properties, demonstrating that many were distinguishable from air which, at the time, was thought to be a unitary element.

He performed experiments to determine the density of the Earth. From his data, others were able to enumerate the universal gravitational constant—one of physics' fundamental constants. Cavendish co-founded the Royal Institution to introduce new technologies and diffuse scientific knowledge by public engagement. It is fitting that that Cambridge University's world-renowned physics laboratory is named the Cavendish Laboratory, where many Nobel laureates have studied and conducted pioneering research.

6.2 Robert Hooke FRS (1635–1703)—Polymath

Hooke was born in Freshwater on the Isle of Wight, where his Royalist father was curate for the local church. His father oversaw the local school and, because of Robert's frail health and persistent headaches, often taught him at home. Robert was a quick learner, skilled in drawing and loved making mechanical toys and models, as well as working on instruments, such as clocks. His father was convinced that his son was destined to be an artist or clockmaker.

When he was thirteen years old, his father died, and he was sent to London to apprentice with a Sir Peter Lely, dominant painter to the Court. This was a short-lived experience, and he went to study at the Westminster school. Here, he learnt the classical languages of Greek and Latin, and studied mathematics and Celestial mechanics.

He was an undergraduate at Oxford's Christ Church between 1653 and 1658 but never took a formal degree. However, after the restoration of the monarchy in 1660, Oxford's new chancellor would bestow an M.A. degree upon him, as an acknowledgement of established achievement.

From 1655, Hooke worked as an assistant to Robert Boyle for seven years to supplement his meagre funds. In 1659, Hooke designed the pump—the *machina Boyleana* or Pneumatic Engine—used by Boyle in his gas law experiments (see Sect. 7.2).

In 1660, he formulated the law of elasticity (Hooke's Law) which states that the force required to extend or compress a spring is proportional to the distance of that extension or compression. This law laid the basis for studies on stress and strain and for understanding elastic material. On a related subject, he invented a spring-regulated, pocket watch by attaching a balance spring to a balance wheel that produced a regular oscillation, allowing time to be kept accurately.

In 1661, he debated the rising of water in slender glass pipes, stating that the height to which the water rose was related to the bore of the pipe. This is due to what we now term 'capillary action'.

The Invisible College of natural philosophers and physicians was founded in 1660. When political changes made it difficult for College members to stay in Oxford, many moved to Gresham College, London, where they helped form the Royal Society.

Hooke had an unparalleled gift for creating and perfecting mechanical devices. For this reason, in 1662, supported by Boyle, he was appointed Curator of experiments for the newly formed Society. A year later, he became a Fellow. The purpose of the Royal Society is to advance scientific understanding of the world. Its motto is: "*Nullius in verba*—take nobody's word for it". He was Curator for 40 years.

Hooke used a primitive compound microscope to look at slices of the bark of a cork tree and saw a repeating pattern of honeycomb cavities. For the individual, empty compartments, he coined the term "cell" since they reminded him of monks' cells. As dead tissue, he was in fact looking at cell walls. In 1665, he was the first to view micro-organisms in the form of fungi.

He is author of one of the most significant scientific books ever written. The book—Micrographia—was published in 1665. It is famed for its spectacular copper-plate engravings of the miniature world, particularly its fold-out plates of insects, He outlined the crystal structure of snowflakes. In this publication, he demonstrated the power of the microscope, founding a new scientific discipline. He was able to bring together his highly developed skills in the design and construction of scientific instruments, with his abilities as an artist. He was once described as England's Leonardo da Vinci.

He observed the phenomenon of light diffraction and, to explain it, he offered a wave theory of light. He was the first to suggest that matter expands when heated; that air is composed of small particles separated by long distances and that air is involved with combustion.

He built some of the earliest telescopes and observed the rotation on their axes of Jupiter and Mars. During a lecture, in 1670, he said that gravity applied to all celestial bodies and that the gravitating power between bodies decreased with the distance between them. He suggested that, in the absence of such power, a celestial body would travel continuously in a straight line.

He was often overlooked in the scientific world, perhaps because of his caustic tongue. His scientific career was marred by arguments with other prominent scientists. For instance, he had several disputes with Isaac Newton (see Sect. 6.3.2) over recognition for his own work on planets and gravitation. He claimed that Newton had got the idea of gravity's strength being proportional to the inverse square of the distance between two objects from him. He became resentful that he was denied proper credit for his many achievements.

However, after a long period of relative obscurity, he is now recognised as one of the most important scientists of his age.

He correctly identified fossils as remnants of once-living creatures and concluded that some species that had once existed, must have become extinct. This was a very controversial proposal at the time. It was at odds with the literal Biblical teachings of the Puritans during The Protectorate (1646–1659), so he tried to get the history of the Earth examined in a non-Biblical way.

As a friend of Sir Christopher Wren, Hooke developed a side-line career as an architect, designing many of the buildings that replaced those destroyed in the Great Fire of London, in 1666. Thanks mainly to this work as an architect, he died, in 1703, a very wealthy man.

6.3   Isaac Newton PRS (1642–1727)—Polymath and One of the Greatest Physicists and Mathematicians

6.3.1   Background—Upbringing, Education, Decorations, Setbacks, and Honours

Observing the Julian calendar of the time, Isaac Newton was born prematurely on Christmas morning, in Woolsthorpe Manor, near Grantham, Lincolnshire. He was a tiny baby given little chance of survival. His illiterate but wealthy farming father had died before he was born. His mother remarried when Isaac was three, moving away to live with her new husband. He was raised by his maternal grandparents until his newly widowed mother returned with three stepsiblings.

As a pupil at the free grammar school—King's School—in Grantham (1655–59), he was described as idle and inattentive. Nonetheless, he was to go on and change the way the world was viewed and understood. He became one of the most influential men in history, dominating the scientific view of the physical universe for the next three centuries until the appearance of Albert Einstein and his General Theory of Relativity. During two extraordinary years (1665–7) Newton developed laws of gravity and motion; a new theory of light and co-invented calculus—a revolutionary new approach to mathematics. His pioneering work formed the basis of modern physics.

Disregarding his mother's request to manage the farm, Newton entered his maternal uncle's university college, Trinity College, Cambridge, in 1661, reading law. In the absence of financial support from his mother, he was classed as a "subsizer" meaning that he received free board and tuition in exchange for menial service. He soon became interested in the mechanics of Copernican astronomy of Galileo, as well as Kepler's optics and astronomical observations. His fascination in mathematics started in 1663. He graduated without honours or distinction in 1665.

During the Great Plaque of 1665–6, he decided to self-isolate, returning home to Woolsthorpe Manor (NG33 5PD; now owned by the National Trust) where some of his most profound observations were made. It was here that he advanced his generalized binomial theorem.

In 1667, he returned to Cambridge and was elected to a minor fellowship, at Trinity College. In 1669, cognizant of Newton's mathematical talents, Isaac Barrow resigned as the 1st Lucasian Professor of Mathematics in favour of Newton who was only 27 years old. This academic post is one of the most prestigious in the world and Newton's tenure lasted thirty-three years. Newton continued to lecture at Cambridge University until 1696. In 1672, for his mechanical ingenuity in constructing a reflecting telescope, he was elected Fellow of the Royal Society of London. He became its President in 1703 and was re-elected annually to the end of his life in 1727.

He was a relentlessly curious man with an obsessive desire to understand the world, working to the point of physical and mental exhaustion. Perhaps because of his unsettled childhood, however, Newton was an insecure man, fearful that others might steal his ideas. He suffered his first mental collapse in 1678. This was possibly caused by his feud with Robert Hooke who had allegedly accused him of plagiarism because Newton had not cited Hooke's 1674 publication dealing with "bodies in motion and moving in a straight line". In mid-1693, Newton suffered a second mental breakdown.

In 1689, whilst the Member of Parliament representing Cambridge University, he voted for the dethronement of King James II and the assumption of William of Orange and Queen Mary II.

In 1696, he left Cambridge and took up residence in London. He acquainted himself with the rich and powerful, thereby promoting his own image. In 1705, Queen Anne knighted Newton, not for his scientific achievements, but as a token reward for standing three times as Whig member of Parliament for Cambridge, as well as being a diligent Warden and then Master of the Mint.

At the time of his death in 1727, Newton was both wealthy and famous. He was buried in Westminster Abbey, with full honours, in a tomb that dwarfs those of many who held a higher place in society.

6.3.2   Light and Telescopes; Book I of Opticks

Our world is bathed in light from our nearest star, the Sun. Every scientist since Aristotle believed that white light is a basic, single entity. In 1665, Newton purchased a glass prism at the Stourbridge Fair, near Cambridge. He used the prism to investigate the refraction of light and reached a different conclusion about the indivisibility of sunlight. By "bending" sunlight, the process of refraction revealed its component colours. He demonstrated that white light is not 'colourless' but comprises a combination of rays with different wavelengths.

By about 1668, he had discovered that, when a tubular/circular beam of white light is targeted at a prism, at the point of minimum deviation, then the light exiting is neither white nor circular, but oblong and comprises several coloured rays (manifest in a rainbow). Blue light is bent (refracted) through a greater angle than is red. The colour sequence can be remembered by the mnemonic; *R*ichard *Of Y*ork *G*ave *B*attle *I*n *V*ain (=ROYGBIV). The acronym VIBGYOR gives the sequence in reverse order.

Furthermore, this multicoloured spectrum, produced by a single prism, can be reconstituted/re-synthesized into white light using a lens positioned between two prisms, the second prism being inverted with respect to the first. The lens focuses the spread of colours to a point and the colours are recombined to give a white outcome (Fig. 6.1).

He further showed that a beam of light of a specific colour did not change its properties (colour) and could not be further sub-divided, regardless of

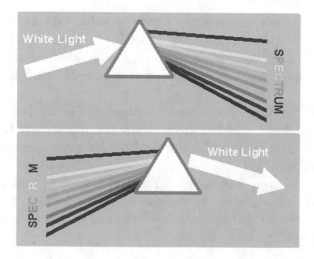

Fig. 6.1 Newton's experiments with prisms to reveal properties of light

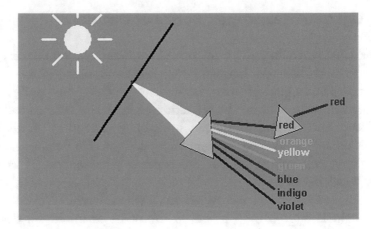

Fig. 6.2 Experiment to show that light of a specific colour cannot be sub-divided

whether it was reflected, scattered, or transmitted. This was a major scientific revelation, and his findings were communicated, by letter, to the Royal Society, in 1672 (Fig. 6.2).

'Dispersion' is the spreading of white light. The process of refraction in a glass prism is mimicked in drops of rain, giving a rainbow. Newton went on to argue that light consisted of streams of minute particles or extremely subtle corpuscles which were refracted by a change in the speed of light in a denser medium (e.g., glass). Both Hooke and Huygens challenged this view, arguing for a wave theory of light. Indeed, Newton himself verged on a wave-like behaviour to explain the repeated pattern of reflection and transmission by thin films.

Later, some physicists favoured a purely wave-like explanation to account for interference patterns and the general phenomenon of diffraction. However, in today's world of quantum mechanics, photons, and the idea of wave-particle duality, this is not far removed from Newton's understanding of light.

His book "Opticks" was largely written by 1692, but, because of widespread doubts / criticism from fellow scientists, its publication was delayed until 1704, a year after the death of his main scientific adversary, Hooke (see Sect. 6.2).

In 1668, by using one small flat mirror, together with one slightly larger curved mirror for focussing light, Newton created a more powerful (reflecting) telescope that was one tenth the size of a traditional refracting telescope using lenses for focussing.

6.3.3   Contemporaneous Views on Celestial Bodies—Copernicus, Kepler, Galileo and Newton. Conflicts with Religious Leaders and Beliefs

In 1543, Nicolaus Copernicus published his heliocentric model of the Solar System according to which, the Sun and not the Earth was the centre of the universe. However, his model did not explain certain phenomena, such as the apparent backtracking of planets, eclipses and planetary conjunctions.

In 1601, Johannes Kepler became the official mathematician to Holy Roman Emperor Rudolf II. By studying data from his predecessor, Tycho Brahe, Kepler was able to publish three empirical laws of planetary motion, becoming the founder of celestial mechanics. His first and second laws were published in 1609, the third, ten years later.

The laws were (1) planets move in elliptical orbits with the Sun at one focus; (2) a planet does not orbit the Sun at constant speed—what is invariable is that a line connecting a planet to the Sun sweeps out equal areas in equal time intervals, no matter where it is in its orbit and (3) the time taken for a particular planet to complete one orbit increases with its distance from the sun in an exact way.

Following from Kepler's Third Law, the ratio of the square of the orbital period (p) to the cube of half the widest axis (a) of the ellipse is the same for every planet. Observations do show that the p^2/a^3 ratio is indeed approximately the same for every planet, as it is for a moon or man-made satellite.

In 1627, Kepler published the Rudolphine Tables. This manuscript contained planetary tables and a star catalogue and facilitated the computation of the future positions of the known planets of the Solar System with great precision. It is possible that it could have predicted the recent Great Conjunction of Jupiter and Saturn, on the day of the winter solstice—21 December 2020. The two planets were indistinguishable at 13:30 GMT, when their angular separation was at its lowest, in the Earth's skies, since 1623. This has immense astrological significance since Jupiter will be in front of Saturn as the latter moves from the sign of Capricorn to the constellation of Aquarius.

Galileo Galilei was born in the same year as William Shakespeare (1564). He was always addressed by his forename. When he was only 25 years of age, he was appointed to the Chair of Mathematics at the University of Pisa. Here, and a year later, at Padua University, he studied the motion of terrestrial objects, free-falling balls, pendulums and projectiles. He was, perhaps, the first of the true experimental physicists—his research forming the basis of the science of motion, providing the foundation of the study of mechanics.

He also contributed to the development of gravitational theory and used his Galilean telescope to observe the stars and planets. Stephen Hawkins said that Galileo bears more of a responsibility for the birth of modern science than anybody else. Einstein called him the father of modern science.

Because of his acceptance of the Copernican model of the solar system, Galileo came into conflict with the authorities of the Catholic church in Rome and was placed under house arrest. Galileo's written works, together with those of Copernicus and Kepler, were placed on the Index of Forbidden Books.

Galileo argued that gravity was a kind of terrestrial force but seemed not to have extended this thinking to heavenly bodies. He established that all objects, regardless of weight or shape, fall at the same rate. (N. B. It is air resistance that differentially slows down different objects, so in a vacuum, objects fall at the same rate. An open or closed umbrella contains the same amount of matter but, when dropped from a tall building, the two fall at different rates due to air resistance).

Using inclined planes, Galileo demonstrated that a rolling spherical object accelerates as it rolls, gaining speed and momentum. The distance travelled is proportional to the square of the time of travel. Extrapolations from his experimental data suggested that an object, projected along an infinite, frictionless plane, will continue indefinitely at a constant velocity—his inertia principle.

Galileo died in 1642, the very year that Newton was born. Like Galileo, Newton developed an interest in astronomy. He combined theoretical hypotheses with experimental work and became an outstanding mathematician in his mid-twenties. He went on to use mathematics and science to describe a diverse range of natural phenomena not understood at the time.

When Newton was offered a fellowship at Trinity College in 1667, he was obliged to be ordained an Anglican priest within seven years of appointment. Unfortunately, he could not accept the Holy Trinity—that God existed in three persons, namely the Father, Son and Holy Spirit. He concluded that the dogma of a triune God was false doctrine.

He believed in a monotheistic God who was "Supreme Governour of the Universe". He refused, therefore, ordination in the Anglican church, becoming an Arian in about 1672. His unorthodox religious beliefs almost cost him his position at Cambridge University. Fortuitously, King Charles II granted exemption from the ordination requirement for Lucasian professors.

Although Newton was a devout Christian, he was vehemently anti-Catholic. He resisted attempts by James II of England to allow the admission of Catholics to the University of Cambridge.

Newton invoked God as an "agent" able to keep the planets in orbit, but his theory of gravity would sweep away the need for a divine hand in guiding the planets. In the mid-1660s, Newton would unify the work of Copernicus, Kepler, and Galileo into one scientific theory that stood the test of time, until Albert Einstein made us question the relationship between the force making objects fall, space and time.

Einstein jettisoned the accepted view that space and time are separate entities. He merged them together into a unified whole to give the fabric of the Universe called "spacetime". As the Earth moves through spacetime, its orbit traces out a spiral as it circles the Sun, and races into the future. It never returns to the same place because the Solar System is spinning around the centre of our galaxy, the Milky Way, itself constantly moving in the fabric of the expanding Universe.

Anecdotally, Newton's vision was triggered whilst he was sitting in his garden at Woolsthorpe Manor, in 1665, when an apple fell on his head, and this inspired his understanding of gravitational force. Indeed, an apple tree (a cooking variety known as the Flower of Kent) still grows in front of the Manor, in sight of Newton's former bedroom.

His theories of gravity were developed in 1666, when he was only 23 years old. His first question was: "Why should an apple always fall perpendicularly to the ground?". Why should it not go sideward or upwards, but always towards the earth's centre? Assuredly the reason is that the earth draws it, so there must be drawing power in matter. If matter draws matter, it must be in proportion to its quantity. Therefore, just as the earth draws the apple, the apple must draw the earth.

His next question was, "How far does this force roam?" Perhaps, this drawing power is not restricted to earth itself, but reaches as far as the Moon, influencing its motion. Yet the moon does not fall to the earth like an apple but travels around it. Why? Here was Newton extending his thoughts from terrestrial, to celestial phenomena.

6.3.4 The Principia (Mathematical Principles of Natural Philosophy) Books I to III

After twenty years of intellectual thought and two years of writing, Newton published, in Latin, the *Philosophiae Naturalis Principia Mathematica*, in

1687. It followed exhortation and financial sponsorship by E Halley who was Clerk to the Royal Society.

The Principia is a work in three books, preceded by an introduction containing definitions. Revised editions appeared in 1713 (with inputs from Roger Coates) and 1726. Based on the third edition, an English translation—The Mathematical Principles of Natural Philosophy—was published in 1729, two years after Newton's death. The Principia has been described as the single most influential publication in physics but is incomprehensible to most readers.

Book 1 deals mainly with objects moving without resistance—for example, in a vacuum; Book 2 with motion in resisting media—for example liquids; whilst Book 3 presents Newton's cosmology—his 'system of the world'.

6.3.5   Book I, De motu corporum (On the Motion of Bodies)

Book I outlines the three laws of motion; his own theory of calculus and gives the first account of his theory of universal gravitation—the innate attraction of all bodies to each other.

6.3.5.1   Laws of Motion; Quantitative Foundations of Classical Mechanics

Motion is the action of changing location or position. The general study of the relationships between motion, forces and energy is called "mechanics". Galileo realized that the success or failure of any scientific theory depends on observations and measurements. His approach to mechanics, however, did not pay any attention to either the forces or energies that may be involved. This sub-branch of mechanics is called 'kinematics'.

Newton advanced Kepler and Galileo's kinematic approach to the motion of objects. He adopted a 'dynamic' treatise, considering the effect of forces on motion. He built a system of mechanics based on the concepts of mass, momentum and force, expressing their inter-relationships through his three laws of motion. Everything in the universe is in motion. Newton's detailed exposition of the concepts of force and inertia is detailed in his laws of motion.

His first law of motion (aka Galileo's law of inertia) states that, if no forces are acting, a stationary object will remain at rest, whilst a body in motion will remain in motion in a straight line. Thus, if no forces are acting, the velocity (both magnitude and direction) will remain constant. (Think of a hockey puck moving frictionlessly across an air hockey table).

The first law may be restated in terms of the momentum (p) of an object, where $p = mv$ (m and v are mass and velocity respectively). Hence, the momentum of an object remains constant unless it experiences an external force.

Newton's second law defines the concept of force. The second law of motion states that, if a force is applied, there will be a change in velocity (manifest as the resultant acceleration [a]) that is proportional to the force (F), and in the direction in which the force is applied. If the mass (m) of the object is constant, then: $F = ma$.

The third law states that, every time a force is applied by one object on another, the other object pushes back with an equal and opposite force. This is often written as follows: "For every action (force) in nature, there is an equal and opposite reaction", but it does not mean the two forces cancel each other. Rather, if object 'A' exerts a force on object 'B', then object 'B' also exerts an equal force on object 'A' but in the opposite direction.

$$F_{12} - F_{21}$$

Action-reaction pairs always act on different bodies, never the same body. In the case of a space rocket, the action is on the exhaust gas and the reaction on the rocket. The rocket is thrust forward and the exhaust gases, backwards (see Sect. 4.7.2).

Classical mechanics, as described by Newton's laws of motion, become increasingly inaccurate when speeds reach substantial fractions of the speed of light, and when gravitational forces are extreme. Einstein's equations are then required to produce more reliable results.

6.3.5.2 A New Form of Mathematics—The Method of Fluxions (Published Posthumously in 1736), Better Known as Infinitesimal Calculus

Newton's laws of motion laid the foundations for classical dynamics. However, the mathematics to derive these laws—which include multiple variables and continuously changing quantities—did not exist. So, in the mid-1660s, he sketched out an entirely a new mathematical discipline which he called his "method of fluxions".

Between the mid-1870s and 1890s, Gottfried Leibniz developed his concepts of calculus. More of Leibniz lives in modern calculus than does

Newton. For instance, his mathematical notations were ultimately adopted (e.g., dx/dy for derivatives, and $\int y\,dx$ for integrals).

Both Newton and Leibniz claimed to have conceived calculus independently, using different notations. From their imaginative minds came the ideas that we now call differential calculus, integral calculus, and differential equations.

Differential calculus is used to discover the rate at which one value changes with respect to another value by examining the effect of infinitesimally small changes, and then allowing those small changes to approach zero. As these parts become so small, as to effectively vanish, the correct outcome is reached. It might be a rate of change of velocity (acceleration).

Integral calculus is the inverse of differential calculus. It might be used to summate a collection of increasingly small values or to determine the area under a curve. This is achieved by adding up the areas of small slices of the shape which are made thinner and thinner until their width tends to zero.

Calculus, this new type of mathematics, can be used to explain happenings in the universe, in mathematical terms. It is used in mathematical models to arrive at optimal solutions. Unlike the 'static' geometry of the Greeks, calculus allows physicists, mathematicians, and engineers to make sense of the motion and dynamic change in the world around us, including the orbits of planets and motion of fluids. Other disciplines using calculus include economists, statisticians, and those in the medical profession.

It led to another way (see Sect. 6.3.5.1) of expressing the second law of motion. That is, "The resultant force (F) exerted on a body is directly proportional to the rate of change of linear momentum (p) of that body". For instance, a motor car is propelled by the force exerted by the engine. When the accelerator pedal is depressed, the force increases, and the car's momentum rises.

This can be expressed in mathematical form, using calculus, whereupon the change expressed in the second law is most accurately defined in differential form (viz $F = dp/dt$ during time, t).

6.3.6   Book II, Part 2 of De motu corporum

Book II is largely concerned with motion through resisting media. Newton solved problems of fluids in movement, and of motion through fluids.

He used mathematics to model the movement of 'normal' liquids or fluids, observing that they have a constant viscosity (internal resistance to flow). Such (Newtonian) liquids adopt the shape of the container into which they are poured.

From the density of air, he calculated the speed of sound waves which he determined to be 298 m/s (actually 343 m/s at 20 °C in dry air, at sea-level). Newton had assumed that the propagation of sound waves in air occurs at constant temperature (isothermal). Laplace later rectified the deficiency in Newton's analysis by allowing for the fact that the temperature is not constant (i.e., adiabatic).

Newton attempted to determine the speed of sound experimentally. Standing at one end of the Cloister of Nevile Court, Trinity College, Cambridge, he clapped and listened for the return of the echo. To determine the time taken, he adjusted the length of a pendulum until its swing period was the same as the echo's return. Thus, knowing the time and distance travelled by the sound of the clap, the speed of sound under the prevailing conditions could be crudely determined.

Newton conducted a variety of experiments on pendulums. For instance, he took a pair of pendulums, each eleven feet long and weighted with a round wooden box, containing equal amounts of gold, silver, lead, glass, sand, salt, water and wheat. For the same length of pendulum, the periods of swing were identical. He concluded that this is only possible if the Earth attracts all the particles, in those diverse substances, in exact proportion to their quantities of matter.

6.3.7 Book III, De mundi systemate (On the System of the World)

In this book, Newton applied all his insights to the detailed motions of the planets and their satellites, including our own moon and even the comets. He explains the motions of heavenly bodies in terms of attractions propagated through empty space. He demonstrated the orbits of the six planets known at the time, as well as their satellites. He explained a wide range of unrelated phenomena, including a theoretical proposal for the eccentric orbits of comets; ocean tidal motions; the precession of the Earth's axis and irregularities in the motion of the Moon, as perturbed by the gravity of the Sun. He identified the oblateness of the Earth—that is, the flattening of the Earth at the poles and the bulge at the equator. He related the precession of the equinoxes to the attraction of the Sun and Moon on the Earth's equatorial bulge.

6.3.7.1 Centripetal Force; Law of Universal Gravitation

The issue of celestial movement vexed the intellectuals of Newton's day. Kepler believed that the Sun firstly pulled planets towards it, and then, when they got "close", repelled them. He maintained that this alternating pull and push was the force that created elliptical orbits. Newton became obsessed by the orbit of the Moon around the Earth. He reasoned that the influence of gravity must extend over vast distances and developed his thinking to account for elliptical orbits.

Whilst in a circular motion, the velocity (i.e., speed in a specific direction) of a circulating object is constantly changing, not because its speed is changing, but because its direction is changing. Because velocity involves both speed and direction, if something changes direction, whilst maintaining a constant speed, it is still undergoing acceleration. Centrifugal (translated to centre 'fleeing') "force" is an imaginary force, introduced to account for the effects of inertia in an accelerated reference frame, reflecting the tendency of an object to resist any change to its state of motion.

When an object is in a uniform circular motion, it wants to move away from the centre of the circle but is prevented from doing so by some actual force. To account for this, Newton introduced the concept of centripetal force. It followed from his first law of motion which says that, unless a force is applied, a moving body will travel in a straight line. For a planet orbiting a star, there must be a force towards the central star that is pulling the planet out of its straight-line path. This is centripetal force, counteracting the outward, centrifugal 'force' that is encouraging the body to continue in a straight line.

For a small globe being swung around, at the end of a chain, the centripetal force, channelled through the chain, holds it in a steady orbit. Cut the chain, however, and the circling stops, and the chain flies off in a straight line. Heavenly bodies are not tethered by chains, do not fly off but remain in fixed orbits. Why so? It is because an increase in velocity is manifest as centripetal (centre-seeking) acceleration.

As stated above, Newton introduced the notion that the force needed to produce centripetal acceleration is the centripetal force. Without this resultant force, a circulating celestial object would move off in a straight line. Gravity is the centripetal force that keeps the Earth and other planets orbiting around the Sun. Our moon may want to travel, at a steady rate, in a straight line but is pulled down by Earth's gravity so that it is constantly accelerating downwards, towards Earth, but is constantly missing it. The combined effects result in its orbit.

Newton also calculated the centripetal force needed to hold a stone on a sling, and the relation between the length of a pendulum and the time of its swing. Newton had succeeded in combining the laws that govern the motion of objects on Earth with those laws that regulate the motion of celestial bodies.

In 1684, after a visit by Edmond Halley, Newton reputedly sent Halley a paper setting out a mathematical proof that elliptical forms of planetary orbits would result from a centripetal force inversely proportional to the square of the radius vector. His treatise, entitled *De motu corporum in gyrum* (On the Motion of Bodies in Orbit) sketched a new system of celestial mechanics.

Using his three laws of motion, together with Kepler's third law ($p^2 = a^3$), Newton was able to find an expression to describe the force that holds planets in their orbits.

His Law of Universal Gravitation states that every mass (m_1) attracts every other mass (m_2) in the universe, and the gravitational attraction force (F) between the two bodies is proportional to the product of their masses, and inversely proportional to the square of the distance (r) separating the centres of the two objects:

$$F = Gm_1m_2/r^2$$

where G is the Gravitational Constant—a universal constant and the hallmark of gravity theory. Newton did not know the value of "G" which sets the size of the force. Henry Cavendish would later devise a pioneering experiment to facilitate the determination of "G" (see Sect. 6.4.4).

Gravitational force can only attract and never repels. The bigger the respective masses of objects 1 and 2, the bigger the attractive force. Since the Sun contains at least 99% of the mass of the solar system, it exerts the greatest gravitational force in the solar system and holds all other objects in orbit and governs their motion.

Although gravitation is the weakest of the fundamental forces, it is the only force that functions over vast distances. With a denominator of r^2, Newton's law is said to be an inverse square law. The bigger the distance between objects, the weaker the attractive force. Thus, if r doubles, the force decreases by a quarter. If we do have only three spatial dimensions, the inverse square law holds. However, if the universe were to have more than three spatial dimensions, the inverse square law would break.

Newton proved mathematically that any object moving in 3-dimensional space, affected by an inverse square law, will follow a path of one of the so-called conic sections—planes cut at different angles from a cone. Newton

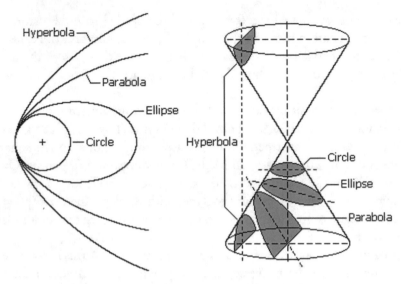

Fig. 6.3 Conic sections

was able to provide a mathematical basis for Kepler's Laws (1) and (3) [see Sect. 6.3.3]. Thus, planets follow elliptical orbits and not circular ones. Comets follow elliptical, parabolic, or hyperbolic orbits (Fig. 6.3).

In 1705, Halley predicted that a comet, seen previously in 1531, 1607 and 1682, would reappear about 76 years later, in 1758. It appeared as predicted and represents a triumph of scientific reasoning and Newtonian physics. It is given the official designation, 1P/Halley.

Nearly three hundred years later, when mission control directs a satellite or spaceship to a planet or moon, anywhere in the solar system, the trajectory will be determined, in part, by using Newton's calculus, his theory of gravitation together with his laws of motion. The film "Hidden Figures" depicts the contribution that NASA mathematician, Katherine Johnson, made, particularly to the challenging return orbit of astronaut John Glenn.

The simplest application of Newton's theory of gravitation is to the mutual attraction between two bodies. Mathematical solutions become more difficult when three or more celestial bodies are considered.

In wider space travel, the trajectory of a small body is determined by the simultaneous gravitational influences of other, much larger bodies. By exploiting Michael Minovitch's approach to gravity-propelled interplanetary space travel, it became possible to explore the entire Solar System without requiring rocket propulsion exceeding that necessary for the escape from Earth. By utilising his 'slingshot technique', the gravitational pull of a celestial object is utilised to accelerate and change the course of a spacecraft, in a calculated manner.

6.3.7.2 Acceleration Due to Gravity (g); Mass Versus Weight

The word "gravity" is derived from the ancient Latin word "gravitas" which means 'heaviness' or 'weight'.

As objects fall, they accelerate, gaining speed and momentum. At the Earth's surface, the mean free fall acceleration, produced by gravity, is 9.81 ms^{-2} which means that a falling object speeds up by 9.81 m/s for every second it drops. Variation in this value occurs due to altitude, latitude, and local geology. The Moon's surface gravity is only 1.6 ms^{-2}.

In his Principia, Newton defines "mass" as the 'quantity of matter that an object contains'. It is a fundamental characteristic of matter, independent of where the object is in the universe, and what gravitational forces it is subject to.

'Weight', however, reflects the downward force on an object. Thus, it will be lower when the gravitational pull is weaker—when farther away from the Earth's core, namely at high altitude and at the equator, where the surface bulges. A dumbbell has the same mass on the moon as it does the Earth, but its weight is less because the Moon's gravity is weaker.

6.3.7.3 Ocean Tides; Lunar Versus Solar Tides; Spring and Neap Tides; Tidal Bores

The Earth is orbiting the Sun at 67,000 mph or 30 kms^{-1}. At the equator, the Earth is spinning on its axis at about 1,000 mph or 460 ms^{-1}. Were there to be no gravitational attraction by the Sun and Moon, the water levels of the seas and oceans would be relatively constant since the action of the Earth's gravity pulling water inwards would be offset by the centrifugal force pushing it outwards.

In 1687, Newton postulated that the oceans' tidal ebb and flow, together with spring and neap tides, result from the perturbing and varying gravitational attraction exerted jointly by the Moon and Sun.

We now know that the Sun has 27 million times more mass than the Moon and is 390 times farther away from the Earth than the Moon. Taken together, the Sun's tide-generating force on the world's oceans is 44% of that of the Moon. Hence, the dominant force affecting the Earth's tides is the Moon which attracts water facing towards it, causing a tidal bulge beneath it. The ocean water—being a liquid—deforms more than the Earth's solid crust.

As the Earth rotates the Sun, and the Moon orbits the Earth, the first ocean bulge follows the course of the Moon, causing a high tide beneath its path.

Moreover, on the side of the Earth opposite the Moon, the latter's gravitational pull is at its weakest, so the Earth's centrifugal force dominates, tending to push water outwards, causing a second ocean bulge called an antipodal tide. Thus, two high tides occur each day, on opposite sides of the Earth.

It takes 24 h for a specific site on Earth to complete one revolution, from an exact point under the Sun, to that same point under the Sun (=a solar day). However, it takes 24 h and 50 min for that site to rotate from an exact point under the Moon to the same point under the Moon (=a lunar day). Thus, on consecutive days, high tides occur 50 min later. Conditional upon the shape of ocean basins, and ocean floor topography etc., most shorelines will experience two high tides and two low tides with a gap of 12 h and 25 min between high tides, and a gap of 6 h and 12.5 min between high and low tide.

The height of the tides will vary during the month because the distance between the Earth and Moon varies as the latter follows its elliptical path—the Moon taking 27.3 days to orbit Earth (=lunar cycle). The relative position of the Sun will also have a direct effect on the heights of daily tides, as well as the intensities of tidal currents.

High (*aka* spring = springing/lively) tides peak when the Sun, Moon and Earth are in alignment (syzygy) whereupon the solar tide reinforces the lunar tide. This occurs when the Moon is between the Sun and the Earth (i.e., New Moon) and when the Moon is on the opposite of the earth from the Sun (i.e., Full Moon). In these celestial situations, predictable tidal bores occur across the World. For instance, mega bores can be observed on the Qiantang River in China (the Silver Dragon); the River Severn and the Amazon River (the Pororoca).

One week later, when the Sun and Moon are at right angles (i.e., Half Moon) moderate (namely neap = without power) tides occur. By analysing English tidal observations with respect to the ratio between spring tides and neap tides, Newton determined the tidal force of the Moon to be 4.5 times that of the Sun. The true value, however, is 2.2.

6.3.7.4 Declination of the Sun; Relationship to Equinoxes and Solstices—Four Key Astronomical Events. Solar Noon

The declination of the Sun is the angle between the equator and a line drawn from the centre of the Earth to the centre of the Sun. As the Earth orbits around the Sun, the latter's angle of declination (δ) to the Earth's equator changes (Fig. 6.4).

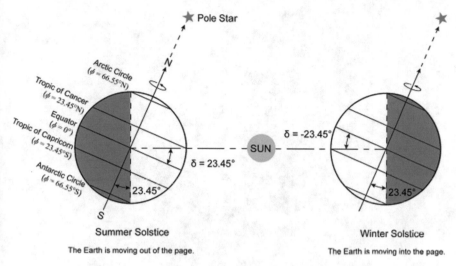

Fig. 6.4 Declination of the Sun. *Source* https://www.itacanet.org/the-sun-as-a-source-of-energy/part-1-solar-astronomy/ Fig. 1.3. Itacanet—all material is copy-left and freely reproducible

The Sun's changing declination occurs because the Earth is tilted on its polar axis, currently by about 23.45°. This possibly resulted from a collision—about 4.5 billion years ago—between Earth and another proto-planet, Theia. The debris from this collision may have coalesced to form our moon. The Earth's axial tilt can also be expressed as the angle between the Earth's axis and a line perpendicular to the plane of the Earth's elliptical orbit (the ecliptic).

In astronomical terms, the celestial sphere is an imaginary sphere which has an arbitrarily large diameter, being concentric to Earth. As a sphere, it has a northern and southern hemisphere, as well as north and south celestial poles. The north celestial pole is the point in the sky about which all the stars seen from the northern hemisphere rotate. The North Star (currently Polaris) is roughly located at this point (see Sect. 6.3.7.6). Polaris sits almost motionless during the course of the night, whilst other stars rotate from east to west around the North Star.

The ecliptic is the annual path of the Sun across the celestial sphere and is sinusoidal about the celestial equator. Twice during the year, the Sun crosses the celestial equator—once moving north along the ecliptic and, a second time, about six months later, moving south (Fig. 6.5).

As seen above, the celestial equator is an imaginary circle that is equidistant from the north and south celestial poles. When the Sun lies in its plane, day and night are of equal length. This occurs twice each year, when the Sun

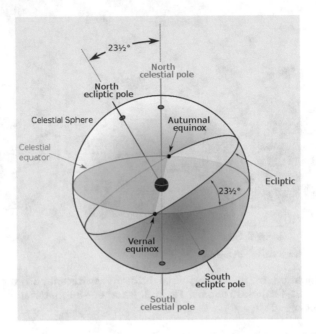

Fig. 6.5 Abstract celestial sphere showing when the Sun is at its minimum declination. *Source* Author N. Sanu—own work. https://commons.wikimedia.org/wiki/File: Celestial_sphere_with_ecliptic.svg. This file is licensed under the Creative Commons Attribution-Share Alike 4.0 International license. https://creativecommons.org/lic enses/by-sa/4.0/deed.en. https://commons.wikimedia.org/w/index.php?curid=92994350

crosses the celestial equator—on the vernal (or spring) equinox (19, 20 or 21 March in the Northern Hemisphere) and the autumnal equinox (22 or 23 September). The word 'equinox' is derived from Latin words *aequus' (meaning equal)* and *'nox'* (meaning night). In the Northern Hemisphere, the first day of astronomical spring is on the vernal equinox.

On these two occasions, the Sun is at its minimum declination of zero degree, whereupon it rises exactly east and sets exactly west. The centre of the Sun will set 12 h after rising, assuming a level horizon, no atmospheric refraction and the sun is a point. The Sun favours neither the Northern nor Southern Hemisphere.

By the same token, the Earth's rotational axis—this imaginary line through the north and south poles—is at 90° to a line between the centre of the Earth and centre of the Sun. Neither pole is tilting towards the Sun.

The English word 'solstice' comes from the Latin word *solstitum,* meaning 'sun standing still'. It suggests a brief pause as the Sun reaches its most extreme point (as experienced on earth) before its direction of travel is reversed.

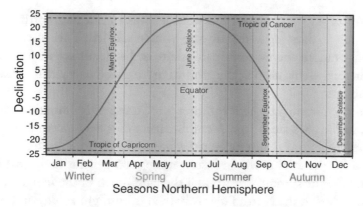

Fig. 6.6 Relationship between the angle of declination, the solstices, equinoxes and seasons in the Northern Hemisphere

On the summer solstice (the longest daytime/shortest night-time on 20–22 June) and winter solstice (shortest daytime on 20–23 December) the Sun is at its largest angle to the equator. On the December solstice, the Sun reaches its most southerly excursion—directly over the Tropic of Capricorn. Conversely, on the June solstice, it reaches its most northerly latitude, over the Tropic of Cancer—23.5° above the equator—when, at noon, the Sun is directly overhead.

If a graph is drawn of the angle of solar declination against time through the year, the result is approximately a sine wave with an amplitude of 23.45° (Fig. 6.6).

6.3.7.5 Effect of a Varying Axial Tilt on Seasons and Intensity of Sunshine. Reason Why Summer Season in the Southern Hemisphere is Cooler Than in the Northern Hemisphere

The Earth's tilt determines the seasons, and because of it, we have four benign seasonal shifts and two solstices.

Around the summer (June) solstice, the North pole is tilted towards the Sun and the Northern Hemisphere is facing the Sun to its fullest extent. It gets more of the Sun's direct rays, so June, July and August are the summer months in the Northern Hemisphere.

Because of the tilt of the Earth, on the noon of the summer solstice (20–22 June), the Sun appears 23.45° higher in the sky than it would if the Earth did not spin on a tilt. By the same token, the Sun appears 46.90°

higher in the sky, at noon, on the summer solstice, in the Northern Hemisphere, than it does, at noon, on the winter solstice. This means that the Sun deposits more energy per square metre of Earth surface in the Northern Hemisphere, thereby, heating the ground more effectively, yielding more heat in the summer.

In addition, in the northern summer, the Sun spends more time above the horizon, so the days are longer. A combination of longer days and more solar energy per m^2 means that temperatures in the summer are hotter than in the winter, so for the solar power industry, more solar generation is possible.

In June, on the summer solstice, all the land north of the imaginary Arctic Circle (23.4° from the North Pole) is bathed in sunlight all day and night, so the Sun never sets. North of the Artic Circle, the period of continuous light lasts for up to six months, equivalent to 4,380 h. Indeed, any point on the planet experiences about 4,380 h of daytime each year.

From the perspective of the North Pole, the Sun is always above the horizon in the summer, and below the horizon in the winter. This means that the region experiences up to 24 h of sunlight in the summer and 24 h of darkness in the winter. It has only one annual sunrise (at the March equinox) and one annual sunset (at the September equinox).

Around the winter (December) solstice, the North pole is tilted away from the Sun, so in the Northern Hemisphere, the sunlight is less direct. Thus, December, January and February are the winter months in the Northern Hemisphere. Theoretically speaking, 19th January is the coldest day in the Northern Hemisphere. Thereafter, heat from solar energy exceeds heat losses. At the equinoxes, there is no irradiance advantage between Northern and Southern Hemispheres (Fig. 6.7).

The Earth may follow an elliptical track around the Sun, but it is an eccentric path—the Sun not being at the centre of the ellipse. In 2021, for instance, the Earth was closest (perihelion at 147.1 million kilometres; 91.4 million miles) on 2nd January. It will be most distant (aphelion at 152.1 million kilometres; 94.5 million miles) on 3rd July in 2025. Being 3.1 m miles closer to the Sun on 2nd January, the whole Earth receives 7% more solar energy than it does on 3rd July. Since the Earth is moving faster at the perihelion than at the aphelion, the changes in solar declination happen faster in January than they do in July.

Proximity to the Sun does not guarantee more warmth to all parts of the World. In January, the Northern Hemisphere is pointing away from the Sun, reducing the amount of solar radiation it receives by about 50%. Being exposed to less direct sunlight, for a shorter time, winter conditions prevail.

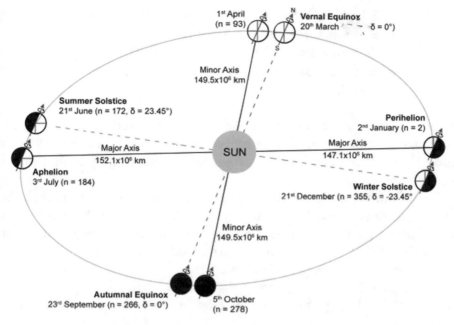

Fig. 6.7 How the Earth's axial tilt determines key astronomical events, the seasons and the amount of sunshine. (n = Day number, where 1st of January = 1; δ = angle of declination). *Source* https://www.itacanet.org/the-sun-as-a-source-of-energy/part-1-solar-astronomy/Fig. 1.2. Itacanet—all material is copy-left and freely reproducible

Photons of electromagnetic radiation emitted from the surface of the Sun, take an average of 8 min and 20 s to travel 93 million miles (150 million kilometres) to Earth. At noon, the Sun is at its highest point. In the early morning and late afternoon, particularly in the winter, the Sun's rays are more slanted, and travel farther through the atmosphere. Such visible light is more diffuse because some of it is absorbed, scattered and/or reflected by air molecules, water vapour and pollutants etc.

Strangely, in January, the summer season in the Southern Hemisphere is about 4 °C cooler than summer in the Northern Hemisphere. Therefore, there must be something countering the perihelion effect. It can be explained by the fact that the Southern Hemisphere is 80% covered by water. The vast oceans in the Southern Hemisphere moderate their higher intake of solar energy because water has a higher heat capacity and absorbs more energy than do the land masses in the Northern Hemisphere in its summer, so keeping countries in the Southern Hemisphere cooler.

The Earth's tilt varies between 22.0° and 24.5°, over cycles lasting 41,000 years. Changes of tilt will affect weather patterns and, hence, climate and eventually, local terrain.

6.3.7.6   The Earth's Oblateness. The Ecliptic Plane. How Precession Changes Our Alignment with the Constellations and the North Star. Significance of the Tropic of Cancer and Tropic of Capricorn

In astronomical terms, precession refers to any of several, gravity-induced, slow, and continuous changes in an astronomical body's rotational axis or orbital path. If the Earth were a perfect sphere, then no precession would occur.

The North Pole (*aka* geographic or terrestrial North Pole) is where all the man-made lines of longitude converge, and the latitude is 90° north. It is the northern point of the Earth's axis of rotation—the top of the Earth. It happens to fall in the middle of the Artic Ocean, mostly covered by an ice sheet which is not sitting on a land mass.

The current North Star (Polaris) sits above the pole, making it an excellent, virtually fixed point to use in celestial navigation in the Northern Hemisphere.

The Earth is not a perfect sphere but an oblate spheroid, bulging at the equator where its spin is greatest. It has an equatorial diameter about 43 km larger than the polar diameter. However, the equatorial bulge of only 0.3% is sufficient for the gravitational forces of the Sun, Moon, and to some extent, Jupiter, to apply torque to the equator, attempting to pull the equatorial bulge into the ecliptical plane of the Earth's orbit around the Sun, as seen from Earth. Acting like a spinning top, the rotating Earth resists this pull.

The plane of the ecliptic intersects the celestial equator (the projection of the Earth's equator on the celestial sphere) at an angle of about 23.5°—an angle which is approximately constant over millions of years. There are two points when the ecliptic intersects the celestial equator, and these are called nodes or equinoxes (see Sect. 6.3.7.4).

Earth's N/S axis is not fixed in space. In fact, the Earth's axis rotates (precesses) slowly westwards, tracing out a cone, like a spinning toy-top or gyroscope, moving 1° every 71.6 years, and 360°—the period of precession—in about 26,000 years (Fig. 6.8).

Because the ecliptical plane and the celestial plane revolve in opposite directions, the equinoxes do not occur at the same points on the ecliptic every year. The two planes make a complete revolution with respect to each other about once every 26,000 years. This movement of the equinoxes along the ecliptic is called the precession of the equinoxes. It refers to the rotation of the heavens by one full cycle. During this time, the constellations appear

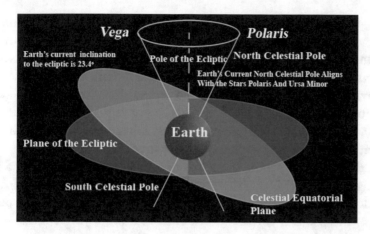

Fig. 6.8 How the Earth's 25,920-year precessional arc determines the North Star

to slowly rotate around the Earth, taking turns at being an unseen 'backdrop' to the Sun in the blue, dawn sky.

Currently, the north celestial pole points towards the star Polaris, in the constellation of Ursa Minor. As the present "North Star", on a clear night, Polaris is always visible in the northern hemisphere—it undergoes imperceptible movement whilst other stars move round it.

However, in about 12,000 years (i.e., almost halfway through a full 26,000-year cycle) the north celestial pole will be pointing towards the Vega star. After about 26,000 years, the north celestial pole will once again point towards Polaris. Presently, the south celestial pole points in the direction of the South Star, Sigma Octantis, which is barely visible, even from the Southern Pole.

For the vernal (March) equinox, it is currently Pisces that is behind the Sun during the daytime for that part of the Earth facing the Sun. Noting that the gap between constellations is not constant, the vernal equinox spends roughly 2,200 years in each of the twelve traditional constellations of the Zodiac, before moving on towards the west.

Over two thousand years ago, latitude 23.5° north was given the name "Tropic of Cancer", whilst latitude 23.5° south was given the name, "Tropic of Capricorn". This is because, 2,000 years ago, on the summer solstice, the Sun reached its most northerly latitude, and this occurred when the Sun was passing by the constellation Cancer. The winter solstice took place when the Sun was traversing the constellation Capricornus. The names have remained, even though, 2,000 years later, the Sun has shifted and now resides on the border between Taurus and Gemini on the June solstice, when the sun is directly overhead the Tropic of Cancer.

6.3.7.7   Zodiacal Constellations. Western Astrology Star Signs

Since ancient times, astronomers and astrologers have organised stars into constellations. During the second century CE, Claudius Ptolemaeus (Ptolemy) recorded 48 constellations, based on imaginary figures—animals or mythological figures—seen in the night sky (e.g., Leo, the lion; Taurus, the bull). The 'cardinal signs' (Aries, Cancer, Libra and Capricorn) are said to have 'initiatory force' and start each of the four solar seasons.

Over the years, forty more constellations have been added giving a total of eighty-eight designated constellations. The twelve (or thirteen if Ophiuchus is included) significant constellations that lie on the Earth's ecliptic plane are known as the zodiacal constellations, each one occupying a space of about 30° along the ecliptic.

These constellation stars are visible, in the night sky, with the naked eye, without the need for a telescope. They are within our own galaxy, the Milky Way. The stars are all at different distances from the Earth. For instance, of the stars in Cygnus, the swan, the faintest star is the closest and the brightest star is the farthest.

From Earth's perspective, the Sun, which also lies on the ecliptic, passes each of background Zodiac constellations at roughly the same time every month, making a complete circuit once per year. Thus, as the Earth orbits the Sun, after a period of about 30 days the Earth faces a different constellation during the morning sunrise, and thus passes through a different 'house' of the Zodiac. Therefore, we get a continually changing view of the stars in the Milky Way.

Because the Earth spins from west to east, the Sun, Moon and stars rise in an easterly direction and set in a westerly direction. At any one time, on a clear night, one should be able to see four zodiac constellations. During the night, these four will sink in the western sky, while others will rise from an easterly direction. Up to ten will be visible during an entire night, the few unseen ones will be below the horizon in the celestial sphere. Of course, the area of sky that one can see is determined by the latitude, as well as the direction that one is facing. The night sky looks slightly different each night because Earth is in a different spot in its orbit. The stars appear each night to move slightly west of where they were the night before.

In the highly approximated graphic below, the Sun's position, on the 21st of each calendar month, is shown on the orange, dotted ellipse. The zodiacal constellations are in a band beyond the orange ellipse. The passage of the Earth's annual orbit is represented by the inner blue ellipse.

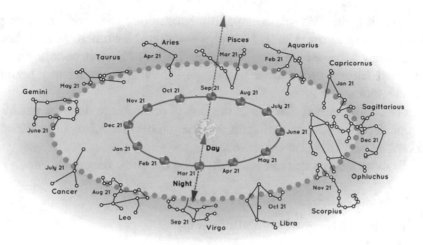

Fig. 6.9 Approximated graphic showing Earth's orbit in relation to the zodiacal constellations. *Courtesy* NASA/JPL-Caltech

According to the diagram, when looking at a clear night sky, from the Northern Hemisphere, on the 21st of September, the constellation of Pisces will be seen. Virgo will not be visible at night because it is on the other side of the sun. In theory, Virgo's stars will be visible during daytime, but in practise they are obscured by the brightness of the Sun (Fig. 6.9).

Part of Newton's Monument in Westminster Abbey is a celestial globe showing the constellations as seen by an imaginary person, in the centre of the globe, at the time of the Voyage of the Argonauts.

6.3.8 Summary

Newton, a solitary man with few friends and no known relationships, is arguably the greatest scientist who ever lived. He is certainly one of the most creative geniuses the world has ever seen, ranked amongst humanities greatest achievers in abstract thought and futuristic visions. His scientific work combined theoretical with practical skills and revealed a universe that obeys logical mathematical laws.

Not only was he a natural philosopher, physicist, mathematician, and astronomer, but an alchemist, historian and theologian as well. He had a particular interest in biblical dating and was always searching for symbolic relationships between the spiritual world and the everyday workings of nature. His theories on the universe, however, were contrary to the views of the Establishment in the seventeenth century.

Following the War of the Spanish Succession, peace on the European continent was restored in 1714 and, thereafter, Newtonian science became increasingly accepted on the Continent. Indeed, Newton became the most esteemed natural philosopher in Europe—a village farming boy becoming world famous.

It became accepted that it was the task of scientists to describe nature and to elucidate nature's laws, whilst it was the job of theologians or philosophers to explore why these laws were so.

His most famous epitaph was written by the poet, Alexander Pope

Nature and Nature's laws lay hid by night,

God said, Let Newton be! – and all was light.

6.4 Henry Cavendish FRS (1731–1810)—Natural Philosopher; Experimental and Theoretical Chemist and Physicist

6.4.1 Background

Henry was born in Nice where his parents were living at the time. He was grandson to two dukes. His father was Lord Charles Cavendish (third son of the 2nd Duke of Devonshire); his mother, Anne (née Grey) was the fourth daughter of the 1st Duke of Kent. When Henry was two, his mother died, soon after the birth of her second son. His father raised Henry and his newly born brother, Frederick.

Henry went to the fashionable Newcome's school in Hackney—a private school in London. When he was seventeen, he entered Peterhouse College, Cambridge but left in 1753 without taking a degree. Other eminent scientific and engineering Petreans include Charles Babbage (see Sect. 15.3.2) Christopher Cockerell, Lord Kelvin (see Sect. 7.4.1), John Kendrew (see Sect. 12.4) Max Perutz and Frank Whittle (see Sect. 4.7.3).

Cavendish returned to London to live with his affluent father, building himself a laboratory and workshop. He was eccentric, extremely shy and avoided social contact outside his circle of scientific acquaintances. His anthropophobia was particularly acute with women. He may have displayed features of what today is described as autism.

Like his father, he dedicated his life to science. When twenty-nine, he was elected to the Royal Society. At the age of forty, his father died, and Henry inherited an immense fortune which afforded him the luxury of pursuing his own scientific interests, allowing him to indulge his insatiable curiosity about the world. He was described by Biot as "the richest amongst the learned and the most learned of the rich".

He became distinguished for great accuracy and precision in his research. He approached every subject of investigation with a strict quantitative examination.

He saw parallels between a sting from an electric ray and a shock from an electric capacitor, postulating about the amount of electricity and the intensity from each. He studied the composition of atmospheric air; the properties of different gases; the constituents of water; the law governing electrical attraction and repulsion, and the calculation of the density (and, hence, weight) of the Earth.

6.4.2 Cavendish's Mimicry of Torpedo Fish; Artificially Produced Electricity. Electric Charge and Potential Difference

In 1773, Cavendish knew that the sting of a torpedo fish (electric ray) could numb a fully-grown man. He built an 'artificial torpedo fish' using a pair of Leiden jars (N. B. The first electrical capacitors were built independently by Ewald von Kleist {1745} and Pieter van Musschenbroek {in 1746}). A capacitor is a kind of temporary battery, keeping positive and negative charges separated by an insulator. Cavendish saw parallels between the fish's sting and an electric shock from a Leiden jar, albeit without the spark seen with the latter.

He pondered on the similarities and differences between "natural and artificial electricity". He reasoned that there was a difference between the amount of electricity (now called electric charge) and its intensity (now known as potential difference or voltage). The Leiden jar was high voltage, low charge whilst the Torpedo Fish sting was low voltage but high charge.

The separation of opposite charges requires energy. We now know that once apart, such energy gets stored in an electric field around each of the individual charges. The field spreads out from the charge in all directions, moving at the speed of light (in a vacuum), and literally goes on for ever. But fields get weaker with distance so a charge's attractiveness to unlike charges and repulsiveness to like charges declines over a short distance. The field has no effect on neutral objects.

The greater the number of negative and positive charges that are separated, the stronger the electric field that forms between and around them, and the more the resulting electric energy can be tapped. Everything from batteries to touch screens and body cells rely on these electric fields. That separation of charge is what creates voltage or potential difference. The positive and negative electrodes of a battery are separated by an insulating layer. Every time the battery is used, a little more of the separated charges get reunited via the circuit and consequently, the electric field between opposite charges gets weaker.

The electric ray has adapted muscle cells called electrocytes, arranged in columns, rather like a bank of batteries. When stimulated, a movement of charged ions across the cell membrane results in an electric discharge. The electrical potential is said to be between 20 and 50 V.

6.4.3   Cavendish's Experimentation on Gaseous Substances. Phlogiston Theory

Cavendish worked on gaseous substances liberated from solids by heat or acids. He called these artificial gases or 'factitious airs'. He demonstrated that many gases evolved in chemical reactions were distinguishable from atmospheric air.

At the time, many natural philosophers regarded atmospheric air as a unitary element. This arose from Aristotle's four elements of terrestrial matter which were air, fire, water, and earth. The often-cited demonstration of Aristotle's reasoning was that wood, if sufficiently heated, releases flames (fire) moisture (water) and vapour (air), leaving behind ash (earth).

Medieval alchemists introduced the concept of 'phlogiston'. They thought that the destructive power of fire was caused by an ethereal entity called 'phlogiston'—a colourless, odourless, tasteless substance. They believed that (i) all combustible material contained a fire-like substance that was released as the object burned or (ii) phlogiston was added from the air to the flame of a burning object. Phlogistonists used terms (see Table 6.1) no longer in use today because, ultimately, the identity of the substances in question were better elucidated, and phlogiston theory was finally debunked by Lavoisier (see Sect. 7.3).

Cavendish developed ingenious ways to capture and store gases released during chemical reaction, enabling him to precisely measure their volumes and weights and investigate their other properties. In 1766, by dissolving metals in acids, he produced 'inflammable air' (which Lavoisier would later call 'hydrogen'). Since the gas was colourless, odourless, and caught fire, Cavendish erroneously concluded that it was phlogiston.

Table 6.1   Phlogiston terms

Phlogiston terminology	Phlogisticated or noxious air	Dephlogisticated air	Fixed air	Inflammable air
Rationale	Saturated with phlogiston and unable to absorb anymore, so nothing could live or burn in it	Thought to be air deprived of phlogiston	Could be returned to / 'fixed' by, sorts of material from which it was produced	Caught fire in air
Modern description	Air exhausted of oxygen and carbon dioxide = nitrogen	Oxygen	Carbon dioxide	Hydrogen

When zinc, iron or tin were treated with either hydrochloric acid or sulphuric acid, the same gas (viz hydrogen) was produced:

$$\text{acid} + \text{metal} \rightarrow \text{salt} + \text{hydrogen}$$

Joseph Black had discovered that, if magnesia alba (magnesium carbonate) or chalk (calcium carbonate) were heated or treated with acid, a gas was produced that was denser than air and could neither sustain fire nor animal life. He showed that this gas was not atmospheric air in a different state of purity, but a stand-alone chemical substance.

Cavendish repeated the experiments conducted by Black, on metal carbonates exposed to heat or mineral acids. Cavendish collected the so-called 'fixed air' (=carbon dioxide) over mercury since carbon dioxide is soluble in water.

$$\text{metal carbonate} + \text{heat} \rightarrow \text{metal oxide} + \text{carbon dioxide}$$

$$\text{metal carbonate} + \text{acid} \rightarrow \text{metal salt} + \text{carbon dioxide}$$

He proved that the fixed air (CO_2) produced from mineral carbonates and bicarbonates was the same as the gas produced by the fermentation process (*aka* brewery gas). In 1766, the Royal Society awarded him the prestigious Copley Medal for his experiments on fixed air.

He also examined the 'airs' released by decaying animal and vegetable material. Whilst carbon dioxide was a constituent, he demonstrated that these 'airs' contained an inflammable gas identical to marsh gas (=methane).

Water was thought by the ancient Greek philosophers to be an element, together with fire, earth, and air. In 1781, Cavendish modified an experiment conducted by Priestley who had used an electrostatic machine to spark a mixture of atmospheric air and "inflammable air (=hydrogen) to produce water. Instead of atmospheric air, Cavendish passed an electric spark through a mixture of oxygen and hydrogen and dew formed on the inside of the vessel. He proved conclusively that this was water. He estimated that two volumes of hydrogen combined with one volume of oxygen. Although he may not have appreciated it at the time, Cavendish had proven that water is a compound, not an element. He delayed publication of his findings until 1784, by which time Lavoisier was aware of Cavendish's experimental results.

In 1785, Cavendish broadened the investigation by repeatedly sparking a mixture of atmospheric air with excess oxygen. The oxides of nitrogen produced were dissolved in an alkaline solution. Having removed the residual oxygen with potassium sulphide, the remaining gas represented 1/120th of the original air. This observation resulted from rigorous quantitative experimentation, using standardized instruments and methods, aimed at reproducible results.

In 1894, more than 100 years after Cavendish's observation, W. Ramsey and Lord Rayleigh showed that the residual gas was a new element, namely the inert gas, 'argon'—the first of the noble gases to be discovered. Since it was so unreactive, they named it after the Greek word for 'lazy'.

In 1787, Cavendish became one of the earliest outside France to convert to the new anti-phlogistic theory of Lavoisier.

6.4.4 The Cavendish Experiment—Measurement of the Force of Gravity Between Masses in a Laboratory; Density of the Earth

In 1687, Newton formulated the Universal Gravitation Equation: $F = Gm_1m_2/r^2$ (see Sect. 6.3.7.1). For non-astronomical objects, the gravitational force (F) is extremely small and requires an extremely sensitive apparatus to detect it. In 1798, Cavendish, aged 67, reported on an experiment with the objective of determining the density of the Earth. It was a simple experiment in principle, but there were numerous complexities that he overcame by meticulous attention to experimental detail.

He used a torsion balance that he had acquired and modified after the death of his friend and original owner, the Reverend John Mitchell. The balance consisted of a 1.8 m long wooden rod suspended from a thin wire at its centre and was free to rotate. Twisting the wire required torque that was a function of the wire's width and its composition. Torque changes are highly periodic and can be measured very accurately.

Two small ($m_1 = 0.73$ kg) lead spheres were suspended at each end of the rod. Alongside each small ball was placed a much larger ($m_2 = 158$ kg) stationary lead ball on a separate suspension system. The closest distance between their centres was $r = 8.85$ inch (22.5 cm).

Cavendish took exceptional precautions to isolate his experiment from external forces and disturbance. Thus, to minimise the risk of extraneous influences such as temperature, air movement and vibration, he enclosed the apparatus in a wooden box; the box was isolated in a shed, with him positioned outside it.

All four balls were pulled downwards due to the Earth's gravitational force but due to an even weaker gravitational pull, the smaller balls were attracted to the larger, stationary balls, thereby twisting the rod. The wire acted as a torsional pendulum. At some angle, the torsional torque was equal to the gravitational torque.

The inertia of the balls caused them to go slightly beyond the equilibrium point and this resulted in a harmonic oscillation around that point. The oscillation was measured by the light reflected from a mirror. Cavendish measured the deflection angle of the rod whilst it was oscillating and the time-period of oscillation. Using the equation of simple harmonic motion, Cavendish was able to determine the torque.

Having determined the gravitational force between the two lead balls, Cavendish then used Eratosthenes' value for the circumference (giving the radius) of the Earth to calculate the attraction between the balls and the centre of the earth. From that, by ratios, he was able to determine the density of the Earth relative to the density of water. His figure of 5.48 g/cm^3 is less than 1% smaller than today's standard value of 5.52 g/cm^3.

Since Cavendish's calculations were based on ratios, he did not need to enumerate the gravitational constant (G). In 1873, Cornu and Baillie published a value for the 'attraction constant'. Using data from the Cavendish experiment, the Universal Gravitational Constant was deemed to be 6.754×10^{-11} Nm^{2}kg^{-2}. The currently accepted figure is 6.67×10^{-11} Nm^{2}kg^{-2} or m^3/kg s^2 (6.7×10^{-8} cm^3/g s^2). It is equal to the force in newtons exerted on a pair of 1 kg masses, separated by 1 m. It is one of physics' fundamental constants, tiny in value and sensitive to the experimental apparatus and the environment in which measurement is made.

6.4.5 Cavendish Laboratory and the Royal Institution

Cavendish was a natural philosopher, considered to be one of the greatest theoretical and experimental chemists and physicists of his age. It is fitting that the University of Cambridge's great physics laboratory is named the Cavendish Laboratory, where many Nobel Laureates have studied and conducted pioneering research.

Since the awarding of the first Nobel prize in 1901, to 2020, thirty such people have won one of the illustrious prizes in physics (21), chemistry (6) or physiology or medicine (3). It was a generous gift from the 7th Duke of Devonshire—Chancellor of the University—that funded both the Laboratory and the Cavendish Professorship of Experimental Physics.

Cavendish was one of the leading British scientists who co-founded the Royal Institution of Great Britain (RI) in 1799. Its aim is to introduce new technologies and to diffuse knowledge by public engagement with science. To 2020, fifteen scientists attached to the RI have won Nobel prizes and ten chemical elements were discovered there.

7

Natural Sciences

In physical science, many of the greatest advances that have been made from the beginning of the world to the present time, have been made in the earnest desire to turn the knowledge of the properties of matter to some purpose useful to mankind. (William Thomson Kelvin, Lecture to the Institution of Civil Engineers, 3rd May 1883)

7.1 Précis

Scientific endeavour is a way of examining phenomena, submitting evidence for competing theories and putting rival interpretations forward for international debate. Doubt is at the heart of it. In 1966, during a speech entitled "What is Science?", given by Richard Feynman, at the fifteenth annual meeting of the National Science Teachers Association, he said, "Science is the belief in the ignorance of the experts". Scientific progress is made when scientists argue with one another—at their symposia and through their publications—and refinements to prevailing theories result.

Robert Boyle lived in an age called the 'period of scientific revolution'. Scientific understanding progressed radically, revealing some of Nature's secrets, helping to unlock the mysteries of matter. Boyle is most famous for his work on gases and particularly Boyle's Law. He demonstrated that sound is propagated through air and not a vacuum, whereas light and magnetic forces do travel through a vacuum. He showed that a portion of atmospheric air supports combustion. He put chemistry on a scientific footing,

© The Author(s), under exclusive license to Springer Nature Switzerland AG 2022
J. Bailey, *Inventive Geniuses Who Changed the World*,
https://doi.org/10.1007/978-3-030-81381-9_7

calling for controlled, methodical, experimental procedures, before reaching factual conclusions. Contrary to the practice of alchemists, he encouraged the reporting of results publicly so that other researchers could assess their reproducibility.

Joseph Priestley was a maverick theologian who possessed originality of thought, as well as the courage to promote unpopular views. He observed that plants have the capacity to restore to air, that which burning candles and breathing animals remove. Effectively, he had identified oxygen as an end-product of photosynthesis and established that oxygen was required for respiration.

In the eighteenth century, with his experimental skills and ability to design ingenious apparatus, he discovered or co-discovered as many as nine gaseous compounds, more than any other scientist. He helped repudiate the Greek theory of the four elements of creation, held to be air, earth, fire and water. He was the first to produce carbonated (soda) water.

At the age of 22, William Thomson, later Lord Kelvin, was appointed professor of natural philosophy at Glasgow University, where he established the first physical science laboratory for both teaching and research purposes. He brought together disparate areas of physics, synthesizing a view that many physical changes were energy-related phenomena. He was well versed in thermodynamics, electrostatics and magnetostatics. With Faraday's empirical evidence, he introduced the concept of an electromagnetic field and made substantive steps in mathematizing electric and magnetic phenomena, providing the groundwork for Maxwell's dynamical theory of electromagnetic field.

By combining scientific understanding with engineering skills, his contribution to the success of the 1866 trans-Atlantic telegraph cable was one of the outstanding applications of science to technology.

He helped to develop the second law of thermodynamics, and particularly, the explanation of irreversible processes. He formulated an absolute zero of temperature, when molecules would stop moving, and a hypothetical ideal gas has zero volume. The "kelvin" is one of the seven base SI units, being the unit of thermodynamic temperature.

7.2   Robert William Boyle FRS (1627–1691)—Father of Modern Chemistry

Robert was Anglo-Irish, born in Lismore Castle in County Waterford, Ireland, to an aristocratic family. He was the fourteenth child of the 1st Earl of Cork, one of the richest men in Ireland and England. Robert was shy, with fragile health, poor eyesight, and a stammer. His parents sent him to live with a poor Irish family to toughen him up and to learn Irish. When he returned home, he was tutored in Greek, Latin and French and sent to Eton College when he was eight years of age.

When he was twelve, his older brother took him on a lengthy tour of Europe. In Italy, at the age of fourteen, he learnt how Galileo Galilei had used mathematics to explain motion and this inspired him.

Whilst Robert was on his Grand Tour, his father died, leaving him a large country estate near Stalbridge in Dorset to which he moved in 1644. Living in Stalbridge House, he studied the scientific literature and, having equipped his own laboratory, started scientific experiments in 1646/7.

In 1654/5, aged 27/8, he moved to the university town of Oxford, renting rooms at University College. He never officially joined the university but found a scientifically convivial environment where he set up another personal laboratory. He employed Robert Hooke as his assistant (see Sect. 6.2).

Boyle lived in an age often called the period of "Scientific Revolution" when modern science came into existence. He wanted to put chemistry on a firm scientific footing.

In 1660, Boyle became an active member of the 'Invisible College' of natural philosophers and physicians who discussed scientific issues at Gresham College in London. After the granting of a Royal charter by Charles II, this became the Royal Society (of London for Improving Natural Knowledge).

He became convinced that chemistry is an important physical science, neither a practical art nor a mystical science. Whilst not the obvious candidate to launch a scientific revolution, his scientific friends persuaded him to release what he believed was a premature publication. In 1661, he reluctantly published "*The Sceptical Chymist*", marking the first big step from alchemy to chemistry.

The ancient Greeks—Empedocles and Aristotle—adopted the four-element model: earth, water, air, and fire, to explain the nature and complexity of all matter.

The ancients believed that cures for medical problems must come from plants or animals. To the alchemists, there was a sympathy between the microcosm of the human body and the macrocosm of nature. There was a concept that both the human body and the universe were composed of materials, or 'principles', or elements.

Paracelsus was an alchemist and physician, with a Faustian character, who pioneered the use of minerals and other chemicals (many poisonous) in medicine. He believed that there were three 'principles', arbitrarily named 'spirit of sulphur', 'philosophic mercury' and salt. Some alchemists believed that metals were compounds of these three principles and possession of pure forms of the principles meant that transmutation of metals, including lead to gold, would be possible.

Boyle rejected both the Aristotelian and Paracelsus models because they gave equally inadequate explanations of what happens when complex substances are attacked by fire and by powerful solvents.

He tried to construct a 'corpuscular' theory of matter. According to the Boyle corpuscular philosophy, God had originally formed matter in tiny particles of varying sizes, shapes texture and motion. These particles tended to combine in groups or clusters which, because of their compactness, had a reasonably continuous existence and were the basic units of chemical and physical processes. He argued that solidity and fluidity depend on the amount of relative motion of the constituent particles; the particles of solids being relatively quiescent.

In his mind, chemistry was the science of the composition of substances. Although not explicit, he described elements, compounds and mixtures and coined the term 'chemical analysis'.

Boyle suggested that a helpful classification of chemical compounds might be based on the differences between acids, alkalis, and neutral substances. He was the first to document the use of natural vegetable dyes as acid–base indicators, an important step towards the development of reliable pH indicators.

He called for a more in-depth method of scientific inquiry and the development of exhaustive proof. He applied an 'inductive approach' to science, where evidence came first, and hypothesis followed. He argued for controlled, methodical experimental procedures, espousing the view that observation, experimentation, and reliable measurement would further scientific-understanding and reveal Nature's secrets, to unlock the mysteries of matter. A rigorous scientific approach was adopted for his research and chemical processes were recorded in plain English.

It is his work on gases—particularly Boyle's Law—for which he remains most famous. In 1654, Otto von Guericke had invented the vacuum pump. Hooke improved upon von Guericke's design after which, he and Boyle performed experiments to study the properties of air and vacuum.

Boyle's Law, published in 1662, states that the volume (V) of a given mass of a gas is inversely proportional to the pressure (p) if the temperature remains fixed ($pV = $ a constant value). Essentially, when pressure on a gas is increased, its volume shrinks in a predictable manner. This was the first gas law. Later scientists demonstrated that it only applies to 'ideal' gases. It was not until 1787, that the second gas law was formulated by Jacques Charles.

Boyle discovered that air has mass and exerts a pressure which led him to conclude that air is composed of minute particles. He performed some exquisitely simple experiments to make profound experimental observations, thereby discovering several key physical properties of air.

For instance, he placed a bell inside a 28-L glass jar. The bell was rung with the help of a magnet outside the jar. Using the pump designed by Hooke, air was pumped out, and the sound of the bell grew ever fainter, demonstrating that sound does not travel through a vacuum, but is propagated through air (see Sect. 15.2.3).

Since the bell continued to oscillate (albeit silently to the listener) the magnetic forces continued to travel from the magnet through the vacuum. Everything inside the evacuated jar remained visible to the observer, showing that light travels through a vacuum (see Sect. 10.4.5). Only sound did not.

When a lighted candle was placed in the jar open to the atmosphere, the flame continued. However, it was extinguished in the absence of air. Thus, he showed that part of atmospheric air supports combustion.

In 1680, he coated a piece of coarse paper with phosphorus and then drew a sulphur-tipped, wooden splint through a fold in the paper. The frictional heat produced a flame—fire on demand. This combination of elements—phosphorus and sulphur—was one of the precursors to the chemical match.

Boyle was a hands-on scholar who produced a roadmap for conducting scientific research. His approach was to conceive an experiment to test a hypothesis; to assemble the apparatus to perform the experiment, and to precisely record the experimental details and results. He amassed extensive experimental data before reaching factual conclusions. He insisted that experimental procedures and results had to thoroughly documented, ideally reported publicly, so that other scientists could validate their reproducibility.

Over the years, he was offered numerous honours, including a peerage and bishopric, all of which he declined, preferring to remain a 'simple' gentleman. His Christian character shaped the way in which he conducted his scientific life. In 1680, he was invited to be President of the Royal Society but declined as he would have been required to take an oath, contrary to his strict religious beliefs.

7.3   Joseph Priestley LLD, FRS (1733–1804)—Maverick Theologian; Unitarian Minister; Natural Philosopher; Discoverer/Co-discoverer of Oxygen and Other Gases

Joseph was born in Birstall Fieldhead, Yorkshire, six miles south west of Leeds. He was the eldest son of a finisher and dresser of woollen cloth. His mother died after bearing six children in six years and his father remarried. In 1742, Joseph was adopted by his paternal aunt with whom he lived until the age of nineteen. She entertained Presbyterian clergy who were to influence Joseph's theological views.

In 1745, he attended Batley Grammar School for Boys, where he learnt Latin and Greek. A mercantile uncle planned to send him to Lisbon, so Joseph taught himself French, Italian, and German. He also sought instruction in advanced mathematics and philosophy, all of which demonstrated his intellectual talents.

Since his family were Calvinists—dissenters from the Church of England—he could not attend Oxbridge so, in 1751, he enrolled at Daventry Academy—a school for Dissenters. The Academy could not award degrees, but he did gain qualifications that allowed him to take appointments to minister at various non-conformist chapels around the country. He would go on to support himself by teaching, tutoring, and preaching in several small towns—all somewhat surprising pursuits since he had an unpleasant voice and a sort of stammer.

In 1761, he was offered a teaching post at Warrington Academy, established by dissenters, where he initially taught languages and history, and then later logic and natural philosophy (science). There, he began researching scientific topics, particularly those electrical and chemical. The Warrington trustees nominated him for an honorary law degree (LL.D.) from Edinburgh University. He was ordained a minister in the dissenting church in 1762.

Priestly travelled frequently to London where he met numerous men of science and independent thought, including Benjamin Franklin who became a life-long friend. Franklin was a passionate believer in American emancipation, becoming one of America's founding fathers.

In the 1740's, "electricians" exploited man-made electricity as a magical phenomenon, for parlour tricks. Franklin proposed a 'fluid theory' of electricity whereby electricity flowed from an area of electrical excess to a zone where it was deficient. Franklin also saw similarities between electricity produced artificially by machine and that produced naturally via lightening (see also Sect. 6.4.2). Franklin invented the lightening rod.

Franklin encouraged Priestley to further investigate electricity. Resulting from his extensive work on electricity, Priestley was made a Fellow of the Royal Society in 1766.

In 1767, whilst still at the Warrington Academy, he conjectured that the law describing electrical forces between stationary charges was, mathematically speaking, like gravity, that it was an inverse square law. This means that the force between two charges weakens by, say 25 (5^2) if the distance between them increases by 5. Whilst gravitational forces are always attractive, electrostatic forces are only attractive between opposing (unlike) charges, but repulsive between like charges. Priestley's inverse square law of electrostatics predates C-A de Coulomb's law by eighteen years. Coulomb devised sensitive apparatus that confirmed Priestley's law.

Since the Warrington Academy was not flourishing, Priestley returned to his native territory of Leeds, in 1767. Whilst living and working in Leeds, he began his most significant scientific research projects, namely those connected with the nature and properties of gases.

Until a minister's house became available, he lived in Meadow Lane, next door to a brewery where he discovered that 'fixed air' ('brewery gas', carbon dioxide, CO_2) was being formed during fermentation. He found that this gas was absorbed by water, giving soda water, so he became the 'father' of carbonation and spawned a highly profitable industry. The commercial potential of soda water was first exploited by J. J. Schweppe, in Switzerland, in 1783.

In 1771, by placing a burning candle in a bell jar, Priestley showed that the flame eventually went out. The air in the jar was depleted of 'goodness', and mice could not survive (respire) in such a depleted atmosphere. He observed that plants have the capacity to restore to air, that which burning candles and breathing animals remove. Effectively, he had identified oxygen as an end-product of photosynthesis and confirmed that oxygen was required for respiration. In 1774, he would devise a way of isolating this elusive gas (see later).

Other, notable, British researchers on gases included Joseph Black who, in 1754, had identified 'fixed air' (carbon dioxide) so called because it could be returned to, or 'fixed' by, the sort of materials from which it was produced. In 1766, Henry Cavendish produced 'inflammable air' (see Sect. 6.4.3) a substance that Lavoisier later called 'hydrogen', from the Greek word for 'water-maker'. In 1772, Black's young student, Daniel Rutherford, found that when he burned material in air, in a bell jar, and absorbed the 'fixed air'

(CO_2) in potash, a gas remained that asphyxiated mice placed in the jar. He called this 'noxious air' (now called nitrogen).

Priestley was awarded the 1772 Royal Society Copley Medal in recognition of the many useful experiments contained in his publication: '*Observations on Different Kinds of Air*' (also see *Philosophical Transactions*, volume 62, 1772, p 147–264). Five further volumes would follow, as well as numerous articles in the Royal Society's *Philosophical Transactions*.

In 1772, German-Swedish apothecary, C. W. Scheele, produced a gas from red-hot manganese dioxide. He called it 'fire air' because of the brilliant sparks that were produced when contacted with hot charcoal dust. In 1774, Priestley isolated the same gas (now known to be oxygen) by heating mercuric oxide. However, Scheele did not publish his findings until 1777, in a treatise called *Chemical Observations and Experiments on Air and Fire*.

The broad dissemination of printed literature helped fuel scientific curiosity, investigation, and experimentation. Many blossoming scientists from poor or modest backgrounds, however, had to find a benefactor to fund promising research. In 1773, the 2nd Earl of Shelburne, seeking an intellectual companion, employed Priestly to fulfil several roles, namely tutor for the Earl's children, scientific sage, and librarian for his estate at Bowood House, in Wiltshire. At the time, the two men held similar religious views. The Earl equipped a laboratory for Priestley and gave him an allowance of £40 per annum for experimental studies, so Priestly became the first professional chemist.

The ancient Greeks nominated air, earth, fire, and water as the four elemental components of creation. In the eighteenth century, the prevailing theory was that inflammable substances contained a hypothetical substance called "phlogiston" that was released from such items when they burned. It was believed that air could only absorb a finite quantity of phlogiston, whereupon combustion ceased (see Sect. 6.4.3).

In 1774, Priestly, using a 12-inch 'burning lens', focussed sunlight on a lump of mercuric oxide (calx), held in place by mercury in an inverted glass tube, itself positioned in a pool of mercury. The gas that was liberated caused a smouldering ember of wood to burst into flame. It also kept a mouse alive for about four times as long as a similar quantity of air. He concluded that it was the glowing ember that introduced phlogiston to the new gas which was devoid of phlogiston. He, therefore, called it 'dephlogisticated air', not realizing that he had isolated a new element.

Over a series of experiments, Priestly had found that air is not a single, elementary substance but a mixture of gases. Among them was this colourless and highly reactive gas, dephlogisticated air. After this discovery, Lord Shelburne took Priestley on a trip to the European continent. He met the French chemist, Antoine-Laurent de Lavoisier in October 1774, and they discussed Priestley's experiments. This meeting between the two scientists was highly significant for the future of chemistry.

In the following year, Lavoisier repeated some of Priestley's experiments, isolating a gas he called oxygen (from the Greek words for "acid forming") from mercuric oxide. He also conducted the experiment in 'reverse', heating liquid mercury in air. He meticulously weighed each set of reagents and products, finding that mercury took something in from air that mercuric oxide released when it was heated. Lavoisier eventually concluded that combustion, the conversion of metals into oxides, and respiration were all processes that involved combination with oxygen. He, thereby, discredited the theory of phlogiston and crushed any views on the Aristotelian view on the four elements.

Lavoisier proved that atmospheric air is largely a mixture of oxygen and 'azote' (nitrogen). He revolutionized chemistry by developing a quantitative approach to chemistry based on elements and compounds. He argued that an element could not be decomposed by any chemical means. He transformed the language of chemistry. Before he was guillotined in 1794 (because he was a tax collector and member of the bourgeoisie) he assigned the thirty-three known elements to one of four categories, namely gases, non-metals, metals, and earths.

Credit for the discovery of the ubiquitous element, oxygen, is now shared by three researchers, namely Scheele, Priestley, and Lavoisier. The intellectual works of these three men helped lay the foundation for that branch of science that we now call chemistry.

Priestley's best scientific work was conducted in the period 1772 to 1780, under Shelbourne's patronage. During these years, Priestley had turned his attention to the preparation and study of other gases, often collecting them over mercury, rather than water, in which they were soluble.

Priestley discovered/co-discovered and isolated a host of so-called 'airs' (gases) including 'marine acid air' (hydrogen chloride, HCl); 'nitrous air' (nitric oxide, NO); 'alkaline air' (ammonia, NH_3); 'diminished nitrous air' (nitrous oxide, N_2O); 'red nitrous vapour' (nitrogen peroxide, NO_2/tetroxide, N_2O_4); 'vitriolic acid air' (sulphur dioxide, SO_2); 'fluor acid air' (silicon tetra-fluoride, SiF_4); dephlogisticated air (oxygen, O_2). When living in America, he isolated 'heavy inflammable air' (carbon monoxide, CO).

No other scientist has discovered so many gases. Between 1774 and 1786, he published six volumes of "*Experiments and Observations on Different Kinds of Air*", repudiating the Greek theory of the four elements of creation. His overall success resulted from his ability to design ingenious apparatus, as well as his skill in its manipulation.

In the late 1770s, Priestley escalated his support for the American Revolution, as well as for highly unorthodox religious views. He hated all oppression and denounced both the slave trade and religious bigotry.

He became an embarrassment to Lord Shelburne and left his service in 1780. Ironically, three years later, Lord Shelburne—now Prime Minister—agreed the Paris Peace treaty which brought the American War of Independence to an end.

Priestley moved to Birmingham, where his brother-in-law offered him accommodation. He was appointed head of a liberal congregation called "New Meeting". In Birmingham, he came in to contact with such scientific and engineering luminaries as Erasmus Darwin (grandfather of Charles), James Watt (see Sect. 3.4) Matthew Bolton and Josiah Wedgewood. Many of the participants were self-made industrialists who met for intellectual and scientific discussion at the monthly Lunar Society. They had a sense of social responsibility with a simple faith that the good life must be based on material decency.

Liberally minded Priestley supposed that chemistry was part of a broader moral and political vision. He believed that giving people a direct understanding of natural phenomena would free them from a state of ignorance that allowed corrupt authority to consolidate power.

Priestley became increasingly demonised for his support for the French Revolution, as well as his controversial theology and radical politics. He came under attack from conservative commentators such as Edmund Burke (in his *Reflections on the Revolution in France, 1790)* and numerous satirists (see *DOCTOR PHLOGISTON, The PRIESTLEY politician or the Political Priest, 1791)* who linked Priestley with atheism and sedition.

In 1791, the "Church-and-King/Priestley Riots" took place. Some of Birmingham's working men, perhaps encouraged by Anglicans and royalists, attacked the properties of the wealthy, dissenting elite. It resulted in a mob burning down the New Meeting House, and the Priestly home. The family fled to Hackney, in London, where he joined another dissenting group. Following renewed vitriol against he and his family, his three sons decided to emigrate to Pennsylvania in 1793, one after naturalization in France.

After Britain and France went to war, in 1793, the British authorities began interning "internal enemies". In 1794, due to his outspoken anti-Establishment position on politics, Joseph and his wife, Mary, decided to follow their sons to America.

He was warmly welcomed and offered the chair of chemistry at the University of Pennsylvania which had been founded by his friend, Benjamin Franklin. Priestley declined but did build a house in the remote hamlet of Northumberland, Pennsylvania where he was to spend the next ten years. He was never naturalised as an American citizen.

Priestly continued his research, isolating carbon monoxide (which he called heavy inflammable air). He founded the Unitarian church in the USA. Unitarians support freedom of religious thought, basing their ideas on rational thought rather than external authority, founding their principles on conscience, thinking and life's experiences.

He died in 1804 and was buried, originally in the Quaker cemetery, before his remains were moved to Riverview cemetery in Northumberland PA. His house is now preserved as a museum and marked as a Historic Chemical Landmark.

Priestley was a down-to-earth Yorkshireman. He was both a natural philosopher, obsessed with an idea of unity between nature and theology, and a Christian philosopher, trying to prove religious beliefs through science.

In tribute to his work on gases, the undergraduate chemistry laboratory at Leeds University was renamed the 'Priestley Laboratory'.

7.4   William Thomson, Baron Kelvin of Largs, OM, GCVO, PC, FRS, FRSE (1824–1907)—Mathematical Physicist, Engineer, and Inventor. The 'Cable Empire'

7.4.1   Background, Education, and University Professorship

William was born in Belfast, Ireland. He was the fourth of seven children. His mother died, in 1830, when he was only six years of age and he was raised by his dominant father, James, who brought up his family in a strict Presbyterian fashion. He was also taught by his father who was a professor of mathematics and engineering in the Collegiate Department of the Royal Belfast Academical Institution, and then professor of mathematics at Glasgow University, whereupon James and his six surviving children relocated to Scotland, in 1832.

In 1834, at the unusually young age of ten, William matriculated to study at Glasgow University but did not graduate. (N.B. At that time, universities in Scotland competed with schools for the most academically able adolescents). In the 1838–9 session, he studied astronomy and chemistry. In the following year, he took natural philosophy courses, which included a study of heat, electricity, and magnetism. He read "The Analytical Theory of Heat" by J. B. J. Fourier and suggested that Fourier's mathematics could be used to study other forms of energy besides heat, such as electrical currents. Thomson flourished at the university and throughout his stay showed a precocious ability in mathematics and physics.

In 1841, he enrolled at Cambridge University, graduating with a BA in 1845. He embarked on a comparative study of the distribution of electrostatic force and the distribution of heat through a solid. This led him to conclude that the two are mathematically equivalent. His publication "On the uniform motion of heat in homogenous solid bodies, and its connection with the mathematical theory of electricity" was the foundation of his later work involving electric and magnetic fields. Moreover, J Clerk Maxwell's acknowledged that it provided the groundwork for his theory of electromagnetism (see Sect. 10.3.6).

His interest in the French approach to mathematics and scientific methods inspired him to travel to France. He went to the Collège de France, in Paris, where he gained practical experience in the laboratory of physicist and chemist, Henri-Victor Régnault.

In 1846, at the age of twenty-two, he became professor of natural philosophy at Glasgow University, a post he held until 1899. He soon established the first physical science laboratory at a British university, introducing laboratory experiments into degree courses. The laboratory gradually evolved into a research laboratory with strong industrial applications. Thomson said that, "*If you cannot measure it, then it is not science*".

During his fifty years, Thomson taught over 7,000 students from all over the world. He published 661 scientific papers and applied for a total of seventy patents. He was an applied scientist, ideally suited to technological application. It is said, however, that he was a poor expositor, unable to coherently organise in written form, his brilliant solutions to technical problems.

He applied a few basic ideas to several areas of study. He brought together disparate areas of physics, namely magnetism, electricity, thermodynamics, mechanics, hydrodynamics, geophysics and telegraphy. This played a pivotal role in the synthesis of mid-nineteenth century science which viewed many physical changes as energy-related phenomena.

7.4.2 Electromagnetic Field. Electric Telegraph. The Cable Empire

Thomson had a deep understanding of electromagnetism and thermodynamics. Thomson met with Faraday every year between 1845 and 1849 to discuss their mutual interests. Together, they introduced the concept of an electromagnetic field. Thomson believed that magnetism was essentially rotational in nature. In August 1845, Thomson mathematically analysed Faraday's magnetic lines of force and then wrote to him suggesting that his calculations predicted that magnetic fields should affect the plane of polarised light. Some weeks later Faraday confirmed that magnetism and light are related, discovering a phenomenon known as the Faraday effect (see Sect. 10.3.12).

Cyrus West Field, together with other entrepreneurs, created the American Telegraph Company. He was the dogged promoter of the trans-Atlantic telegraph cable project between Newfoundland and Ireland. In 1856, Thomson was employed as scientific adviser to help solve the electrical and engineering challenges facing the project. A series of trans-Atlantic sub-marine cables were laid between 1857–8 and between 1865–6.

Thomson designed and patented (1858) his telegraph receiver—a mirror galvanometer—that was an extremely sensitive device, capable of detecting very faint telegraph signals. It was used in the first successful, sustained telegraph transmission along the Atlantic cable, capable of receiving a character every 3.5 s.

It can be seen (Object 1925–179) at the London Science Museum (SW7 2DD). Its disadvantage was that it required two operatives—one to read and call the signal; a second to write down the message. It was superseded, in 1867, by his 'siphon recorder' which required only one operator. His galvanometer and recorder were in high demand as a network of submarine cables was built to span the globe. Sales of these two patented devices, together with his involvement with two engineering firms guaranteed him a comfortable income.

Successful, land-based telegraph masts were widespread. By 1852, 4,000 miles of telegraph cable had been installed over the British railway network (see Sect. 4.4.6). However, trans-Atlantic electrical signals were blurred and indistinct. Initially, it was thought that this could be resolved by increasing voltage, by using more powerful batteries. However, the 1857 and 1858 expeditions failed because of weaknesses in the cable, or it short-circuited to the ocean as the voltage was increased beyond its design limit.

Based on his theoretical approach, Thomson recommended changes to the subsequent cable specification. For instance, the quality of the copper wire was improved to reduce its resistivity. The thickness of the copper conductor was increased from 107 to 300 lbs per nautical mile in the new cable.

The manufacturer of the cable—the *Tel*egraph *Con*struction and Maintenance Company (Telcon)—continued to use gutta-percha thermoplastic gum (see Sect. 4.6.3) as insulator but the cable was strengthened to bear 4.64 times its own weight in the deepest water. Brunel's Great Eastern was chartered by Telcon since it was the only British navy ship big enough to carry the heavy cable required (see Sect. 4.5.9).

A clear and crisp message was delivered in July 1866. The cable supported the transmission of eight words per minute. The eventual success of the 1866 project was one of the outstanding applications of science to technology. It was based on Thomson's accurate mathematical modelling of a durable submarine telegraph cable. It would confer world-wide social and material benefits on mankind. In 1866, as a reward for his contribution, Thomson received a knighthood. The work also brought him considerable wealth, arising from the royalties from the abovementioned, and related inventions.

Thomson's scientific and practical inputs helped Great Britain capture a pre-eminent place in world communication during the nineteenth century. By 1902, Britain had hundreds of submarine cables, including crossings of the Atlantic and Pacific Oceans, facilitating communication to North and South America, Africa, India, East Asia, and Australia. Porthcurno, in Cornwall, became a key sub-marine communications station for cables travelling eastwards, notably to India (Visit PK Porthcurno Telegraph Museum, TR19 6JX).

The intricate and far-reaching network of telegraph cables tied together Britain's vast empire, connecting England with its numerous colonies and trading partners. It gave London central control of its Empire and facilitated expeditious two-way messaging. It became a key component of economic inter-connectedness, promoting globalisation.

Today, undersea cables are the vital unseen plumbing of the Internet carrying about 98% of the world's data and web traffic. The messages travel at close to the speed of light across the sea floor, through optical fibres about the thickness of a horse's hair. A number of these are packed loosely in a tube of Petroleum jelly. Surrounding this are tubes of copper, aluminium, strands of steel wire and final layers of water-impermeable plastic, giving a deep-sea cable about the diameter of a garden hose. This "undersea web" follows the centuries-old trading routes with signal boosters every 100 km.

Currently, there are about 750,000 miles of cable running between continents. In 2020, Google announced plans to build a new undersea cable connecting New York with both Bude in Cornwall and Bilbao in northern Spain.

7.4.3 Second Law of Thermodynamics; Irreversibility of Natural Processes, and the Principle of Entropy Growth

The word 'thermodynamics' has Greek origins and is derived from 'therme' = heat and 'dynamis' = power or motion. In 1848, Thomson referred to thermo-dynamic engines, by which he meant steam engines, that convert heat (thermo) to motion (dynamics). It became a branch of physics dealing with the transformation of heat to, and from, other forms of energy.

It is a study of the effects of temperature on physical systems at the macroscopic level, as well as energy conversion, most typically through terms of heat and work. It describes what is possible and not possible during energy conversion processes, as well as the direction of a process.

Over a 100-year period, at least ten scientists played key roles in developing thermodynamics. There are just three leading concepts in this branch of physics, namely energy, entropy and absolute temperature.

Having assimilated the theories on thermal phenomena by Joseph Fourier, Nichols Carnot and James Joule, Thomson helped to develop the second law of thermodynamics, and particularly the explanation of irreversible processes.

It is a matter of experience that, "all natural or spontaneous processes (taking place without external interference) are irreversible". That is the basis of the second law of thermodynamics. For instance, a gas expands spontaneously into a region of lower pressure until the pressure distribution is uniform. Complying with the same law, heat is conducted spontaneously along a metal bar, hot at one end and cold at the other, until the temperature is the same throughout. The reverse of these two processes never occurs spontaneously.

By external application, however, it is sometimes possible to re-establish the original condition. For instance, a gas that has expanded into a bigger volume can be restored to its original volume by compression. To achieve this, work must be performed on the gas, resulting in the production of heat and a rise in temperature.

However, any attempt to fully convert the heat generated into work will not be totally successful; only a fraction will be converted. This failure, in practice, to convert heat completely into work leads to another way of expressing the second law: "It is impossible to construct a machine functioning in cycles which can convert heat into the equivalent amount of work without producing changes elsewhere".

This demonstrates that heat is 'different' from other forms of energy. Mechanical work, electrical work, the work of magnetic forces etc. can be fully converted into heat. However, from the point of view of transformation into other forms of energy, heat is exceptional since a fraction of heat must inevitably be transferred to a lower-temperature heat sink.

The equilibrium state of a system is the one in which the constituent particles are in the most random, chaotic arrangement that is possible for that situation. For instance, left to the elements, sandcastles crumble; panes of glass shatter and buildings collapse. Scientists now employ "entropy" to measure a system's progress towards equilibrium. It was Rudolf Clausius who conceived the word, it being derived from the Greek word "trope", meaning transformation. The entropy of a system is related to the number of ways in which molecules can be arranged within the system and is a measure of the bias towards dissipation/disaggregation.

In the 1870s, Boltzmann realized that in a gas, more states appear random than seem orderly. He enunciated a relationship between entropy and the number of possible states. Refinement by Planck, in 1913, led to the 'simplified' Boltzmann-Planck equation: -

$$S = k_B \ln(W)$$

where S = entropy; W = the number of possible arrangements of a system and k_B is the Boltzmann constant.

Nature abhors inbalance, whether it be mechanical, pressure, thermal, chemical or electrical. Wherever it can, it will drive a change to a neutral situation, resulting in an increase in entropy, S. For an isolated system, S can only stay the same or increase with time. For example, if a system has a hot part and a cold part, heat transfer will occur from hot to cold part, until thermal equilibrium is reached, and entropy has attained its maximum value.

If the system is not isolated, its entropy can decrease, but only at the expense of an increase in entropy in the environment to which it is linked. This is another expression of the second law of thermodynamics.

Thus, the second law of thermodynamics can be expressed in several different, but equivalent ways. Simply put, these include:

- All natural or spontaneous processes are irreversible.
- The total entropy in the universe can never decrease.
- Heat can only flow from a hot body to a cold body.
- Heat engines cannot be 100% efficient.

Thomson's thermodynamic theories centred on the 'dominance of the energy concept', upon which, he believed, all physics should be based. He said that the two laws of thermodynamics expressed both the indestructability and dissipation of energy.

The Emden statement can be reformatted as follows: "In the huge manufactory of natural processes, the principle of energy conservation does the book-keeping, balancing credits and debits, but the principle of entropy occupies the position of manager, for it dictates the manner and method of the whole business".

If everything tends from order to disorder, then there is a difference between the past and the future. This suggests that there is a direction to the passage of time with the 2nd law of thermodynamics introducing an arrow of time.

7.4.4 Absolute Zero of Temperature; Nernst's 3rd Law of Thermodynamics

One of the most important results of Thomson's toils were his ideas for an absolute zero of temperature, first formulated in 1848 but revised in 1854. This is the theoretical temperature at which all hypothetical, ideal gases (i.e., those for which there are no forces between molecules) have zero volume. The theoretical temperature was determined by extrapolating the ideal gas law. Using four methods, Thomson's French mentor, Régnault, calculated absolute zero to be -272.75 °C. By international agreement, it is now taken as -273.15 °C.

Thomson was unhappy with this "ideal gas" approach to a thermometry problem. He proposed a temperature scale in which a unit of heat transferred from an object at temperature 'T' to an object at temperature 'T – 1' would produce the same amount of work, regardless of the value of 'T'. Such a scale is independent of any material properties of the thermometric substance and, in that sense, it is 'absolute'. The kelvin scale, based on this concept, is named after him.

The kelvin is one of the seven base units in the Système International d'Unités (SI) and is the unit of thermodynamic temperature. It is a fundamental physical quantity and must equal zero at the point where entropy equals zero. In 1967, the triple point of water (i.e., the temperature at which ice, water and water vapour are in equilibrium) was adopted as a reference temperature. The triple point of water is defined as 273.16 K or 0.01 °C.

Nernst's third law of thermodynamics (1906) implies that objects can never be cooled down to absolute zero. Despite numerous attempts, absolute zero has never been reached. Nevertheless, scientists have devised increasingly clever ways of extracting energy, and temperatures of billionths of a kelvin have been reached.

In classical physics, at absolute zero (0 K or −273.15 °C) atoms and molecules have no kinetic energy, so molecular motion ceases and a state of 'perfect order' has been reached. The particle constituents can become no colder. In quantum mechanical terms, however, the system is effectively sitting in its ground state and has its lowest internal energy. At these low temperatures, superconductivity of solid crystals and superfluidity of liquids (e.g., helium-3) manifest themselves.

7.4.5 Adiabatic Gas Expansion via Either Decompression or 'Throttling'. Joule–Thomson Effect

Collaboration with other scientists was important to Thomson. He and James Joule worked together on several fruitful projects.

When a gas expands via a decompression process, and does work against an external pressure, its temperature will fall, providing the process is performed in a heat-insulated system (i.e., adiabatically).

If a gas, at high pressure, is passed through an insulated constriction, such as a throttling valve or porous plug, to a lower pressure region, then the temperature of the gas will change. This is known as the Joule-Thomson effect, based on studies they jointly made in 1852.

Depending on the start pressure and temperature, and the final pressure, the final temperature can increase, decrease, or remain constant for a specific gas. The limiting line, where a temperature-increase changes to a temperature-decrease, is called the inversion line. Maximum inversion temperatures for nitrogen, hydrogen and helium are 623, 200 and 24 K, respectively.

For most gases (excepting hydrogen and noble gases such as helium and neon) cooling occurs when 'throttling' is performed at room temperature. For most gases, the external work done in the expansion process through the constriction is negligible, but the internal work in overcoming inter-molecular forces in a non-ideal gas is sometimes considerable, resulting in Joule–Thomson cooling. (Witness a tyre valve getting cold when air is released from a bicycle tyre). Hydrogen and helium are different because forces between molecules (H_2) or atoms (He) are unusually weak.

This cooling phenomenon has proven useful in the advancement of mechanical refrigeration systems, as well as liquefiers, air conditioners and heat pumps.

In 1916, in honour of Thomson (Lord Kelvin), the American Kelvinator Company was named after him. In 1918, it introduced the first refrigerator with any type of automatic control.

7.4.6 Thermo-electricity, Inter-convertibility of Thermal Energy and Electrical Energy: Seebeck-Effect and Peltier-Effect

In 1787, Volta noted the conversion of heat into electricity at the junction of two different types of wire. This observation was repeated in 1821 by Seebeck. He recorded that a compass needle was deflected by a closed loop, formed by two different metal wires, with a temperature difference between joints. This so-called Seebeck-effect is now applied in thermocouples, to measure temperature differences, and in thermo-electric power generators.

In 1834, Peltier reported that, when a current is made to flow through a junction between two different conductors then, depending on the direction of the current, heat is generated or absorbed at the junction. The so-called Peltier cooling effect by electricity is now applied to refrigerators.

Joule suspected that these two thermo-electric effects were due to the inter-convertibility of thermal energy and electrical energy. In 1854, Thomson argued that the Seebeck and Peltier effects are related, indicating that any thermo-electric material can be used to either generate power in a temperature gradient or pump heat with an applied current.

7.4.7 Other Honours and Distinctions; His Resting Place

William Thomson was elected fellow of the Royal Society of Edinburgh in 1847 and the Royal Society of London, in 1851, receiving its Gold Medal in 1851 and its Copley Medal in 1883. In 1871, he was elected president of the British Association for the Advancement of Science.

In 1892, he was the first scientist to be raised to the peerage with the formal and legal title of Baron Kelvin of Largs (N.B. The River Kelvin is a tributary of the R. Clyde flowing through Kelvingrove Park and past Glasgow University; Largs is a town on the Firth of Clyde where Thomson had built his country house, Netherhall, in 1875). Thereafter, he was normally addressed with the appellation, 'Lord', to indicate membership of one of the five ranks of peer.

In 1896, he was made Knight Grand Cross of the Royal Victorian Order (GCVO). In 1902, he was awarded the Order of Merit and appointed Privy Counsellor (PC).

He died at his home, Netherhall, near Largs, in 1907. He was buried with national honours, alongside other scientific luminaries, in the nave of Westminster Abbey.

Thomson was a humble man with an insatiable desire to further the advancement of science and technology. Despite being a great theorist, he was very practically minded. He paved the way for global communication pathways. The Hunterian Museum, at the University of Glasgow (G12 8QQ) has a permanent exhibition of his work.

8

History of the Atom, 1803–1932

Whenever a theory appears to you as the only possible one, take this as a sign that you have neither understood the theory nor the problem which it was intended to solve. (Karl Raimund Popper, Objective Knowledge: An Evolving Approach, Oxford University Press, 1972)

8.1 Précis

For over two millennia, philosophers and scientists had theorized about the composition of matter. Between 1803 and 1932, four British scientists (viz Dalton, J. J. Thomson, Rutherford and Chadwick) put forward an increasingly sophisticated architectural model of matter.

John Dalton found that the total pressure created by a mixture of gases is the sum of their partial pressures. In 1803, he asserted that all matter, whether gas, liquid or solid, is composed of small, indivisible particles. All atoms of one element are identical to each other but different from other elements. A chemical reaction is simply a rearrangement of atoms. According to his law of multiple proportions, in a chemical reaction, atoms combine in small whole number ratios.

In 1884, Joseph J. Thomson succeeded his tutor, Lord Rayleigh, as Cavendish Professor of Experimental Physics at the University of Cambridge. Thomson was awarded the Nobel Prize in Physics, in 1906, for his theoretical and experimental investigations into the conduction of electricity by gases.

J. Bailey, *Inventive Geniuses Who Changed the World*, https://doi.org/10.1007/978-3-030-81381-9_8

In an evacuated glass vessel, he produced cathode rays—negatively charged particles in transverse motion. Thomson identified a universal constituent of matter and effectively overturned Dalton's indivisibility hypothesis. This fundamental particle was later given the name "electron".

Electrons are the most useful of the sub-atomic particles because of their detachability. Background scientists such as Davy and Faraday exploited their loose boundedness to produce electricity. Applied technologists such as W. Thomson, Bell and others utilised them to bring about the electric telegraph, telephones, electric generators and motors, and the electronics industry— encapsulating smart phones and computers. In microchips, electrons are shuffled around in a way prescribed by computer code. The chemical properties of elements are defined by the properties of orbital electrons, whilst radioactivity is determined by the nature of the nucleus.

F. Aston, working with J. J. Thomson, identified two isotopic forms of neon. Aston would go on to use their electro-magnetic focusing technique to identify 212 of the 254 naturally occurring, stable isotopes. He was awarded the Nobel Prize in Chemistry, in 1922.

Ernest Rutherford was a student of J. J. Thomson, and possibly the greatest experimentalist since Faraday. He was the central figure in the study of radioactivity. He studied two of the three types of radiation emitted by uranium which he named α-rays and β-rays. Using similar techniques to Thomson, he showed that α-rays are helium ions (He^{2+}) whilst β-rays are, like cathode rays, electrons.

Rutherford and F. Soddy collaborated on the theory of radioactive disintegration and the transmutation (rearrangement) of nuclei and atoms. For his contribution, together with his investigation into the chemistry of radioactive substances, Rutherford received the Nobel Prize in Chemistry in 1908.

Contrary to Dalton's view of the atom, he demonstrated that an atom can be 'destroyed'. Loss of an alpha particle means a lowering of the mass number by 4 and atomic number by 2. He observed that any quantity of radioactive isotope takes the same amount of time for half of it to decay. This constant rate of decay can be used as a 'clock' to gauge the age of such things as rocks, fossils etc. via radiometric dating.

In 1911, Rutherford proposed a 'nuclear model' of the atom in which its mass is concentrated in a dense, positively charged nucleus, with electrons orbiting at some distance from the nucleus. This crude template was refined both by N. Bohr, who proposed that electrons orbited at set distances from the nucleus and, later, by E. Schrödinger, who suggested regions of space where an electron will probably be.

In 1919, Rutherford investigated the nuclear transformation of a non-radioactive element to another element. The result of this prompted him to postulate that the hydrogen nucleus is a primordial, fundamental particle which he dubbed the 'proton'. It is no longer thought to be indivisible.

He further observed that about one half of the mass of the nuclei he had investigated could be ascribed to protons. He and Bohr theorized about the existence of a particle which would nullify the repelling effect of positively charged protons confined in an atomic nucleus. In his final years of research, he used particle accelerators to fuse together small nuclei, noting the high level of energy associated with the nuclear conversions.

Frederick Soddy, Niels Bohr, as well as eleven other collaborators and students researching with Rutherford would eventually receive Nobel prizes in their own sphere of scientific expertise.

James Chadwick and Rutherford studied the transmutation of non-radioactive elements such as beryllium ($_4Be^9$) by bombarding them with α-particles. Chadwick proved that radiation emitted from beryllium was more energetic than could be accounted for by γ-rays. Particles ejected from various targets were uncharged; had a mass slightly heavier than a proton and were more penetrating than protons. He demonstrated experimentally the existence of 'neutrons', what he thought was a new fundamental particle. So important was his finding that, three years later, he was awarded the Nobel prize in physics, in 1935.

His work marked the start of nuclear physics. Accelerated neutrons do not have to overcome any repulsive barriers of charged particles. This feature provides a tool for inducing atomic disintegration since neutrons are capable of penetrating and splitting the nuclei of the heaviest elements. When there is a critical amount of fission material, and other conditions are met, a self-sustaining chain reaction can occur with the release of a vast amount of atomic energy. Neutrons can mediate a nuclear chemical reaction and this property led to the atomic bomb and, later, to nuclear power production.

Between 1897 and 1932, the experimental research of Thomson, Rutherford and Chadwick identified three sub-atomic particles, namely the electron, proton, and neutron.

8.2 John Dalton FRS (1766–1844)—Chemist, Meteorologist and Physicist. First Modern Atomic Theory

Dalton was born in Eaglesfield, Cumbria. His father earned a modest income as a handloom weaver, owned his own house and a small amount of land. He and his wife belonged to the Society of Friends (Quakers). John received his education from his father and John Fletcher, a teacher at Pardshaw Hall, a stronghold of Quakerism.

In 1778, when Fletcher retired, John became an assistant teacher at the extraordinarily young age of twelve years. To improve his income, he became a farmhand. In 1781, he left his native village and returned to teaching, assisting his cousin run a Quaker boarding school in Kendal. He achieved promotion to School Principal, resigning and leaving the school in 1793, when he was twenty-seven.

As a non-conformist Quaker, he was barred from joining most English universities. However, he was able to take up a teaching post at the New College (Manchester Academy) in Moseley Street, Manchester—a 'dissenting' academy that was a successor to the Warrington Academy (see Sect. 7.3) where Priestley had worked in the 1760s. For seven years, he was tutor in mathematics, natural philosophy, and chemistry, both theoretical and experimental. It offered him a stable salary and a well-equipped library.

He was an avid reader and deep thinker. In 1794, for topical discourse, he joined the Manchester Literary and Philosophical Society ('Lit and Phil')—the second oldest learned society in Britain, founded in 1781. He was quick to present his first research paper—on colour vision. He and his brother were both colour-blind. Dalton made a careful study of the defect and identified the hereditary nature of red-green colour blindness. For many years, the condition was known as "Daltonism".

For about thirty years, he lived in George Street, opposite the 'Lit and Phil'. He led a modest and unassuming existence, in a room, in the home of the Unitarian minister, William Johns and his wife. Membership of the 'Lit and Phil' granted him access to laboratory facilities, where he conducted much of his original research. He became its secretary in 1800, and was its president from 1817, to his death in 1844.

In 1800, when the finances of New College deteriorated, he resigned and began a new career as a private tutor. He was always a teacher and part-time scientist.

He was a pioneering meteorologist, partially responsible for the transition of meteorology from folk law to a serious scientific pursuit. He kept daily records of the weather for fifty-seven years, amounting to over 200,000 observations, including wind velocity and barometric pressure.

Having been raised on the edge of the Lake District, regularly holidaying there, and now living in Manchester, he was extremely familiar with rain. His fascination and interest in the weather, the atmosphere, water, the nature of water vapour, and clouds led to his research into the nature of gases. In turn, this became the foundation upon which he built his atomic theory.

In 1801, Dalton found that the total pressure (P_T) exerted by a mixture of gases which do not interact, is the sum of the pressures which each gas would exert, if it were present alone in the entire volume occupied by the mixture. The pressures exerted by each gas separately are known as their partial pressures (p_1, p_2, p_3 etc.). Thus,

$$P_T = p_1 + p_2 + p_3 + \cdots .$$

He derived this law from his work on the amount of water vapour that can be absorbed by air at different temperatures. Now that tables exist giving the pressure of water vapour at different temperatures, it is straightforward to determine the pressure of a dry gas from the measured pressure of 'wet' gas.

Dalton performed a variety of experiments on gases, and it drew him to the conclusion that a gas was composed of infinitesimally small indivisible particles. He then asserted that all matter, whether gas, liquid or solid, was also composed of small indivisible particles.

Over two millennia before Dalton's research, the Greek philosopher, Democritus of Abdera (460–370 BCE) had reasoned that all matter could be divided and sub-divided into smaller and smaller units until one particle could not be further divided. He called this particle an atom or atomos (meaning uncuttable/indivisible).

Dalton was to revive this idea and he is credited with the first modern atomic theory, the rudiments of which he presented at a 'Lit and Phil' meeting, in 1803. Like many contributions to the advancement of science, it was a hypothesis that explained many historical and contemporary observations and facts in a simple and reasonable way.

For instance, in 1785, A.-L. Lavoisier had shown clearly that there is no change in mass during a chemical reaction. That is, the mass of products is equal to the mass of reacting substances.

J. L. Proust's law of definite proportion/constant composition (1799) states that different samples of a given substance contain its component elements in a fixed proportion by mass. Thus, any sample of water, from any source, contains hydrogen and oxygen in the same proportion, namely 1:7.94 by mass.

Based on the laws of conservation of mass and definite proportion, Dalton proposed that all matter is composed of atoms which are indivisible and indestructible building blocks. All atoms of one element are identical, but they are different from those of any other element. He believed that these elementary particles could not be metamorphosed, one into another.

Atoms of one element can combine with atoms of another element in small, whole number ratios, to give a 'compound atom' or 'compound'. In chemical reactions, atoms are combined, separated, and/or rearranged but atoms are neither created nor destroyed.

According to Dalton's law of multiple proportions, if atoms of element A, combine with element B, then the combination will occur in small, whole number ratios. Some of the possible compounds that may be formed can be represented as a binary compound (e.g., AB), a ternary compound (e.g., A_2B or AB_2) or quaternary (AB_3) etc. Dalton's approach worked well for oxides of nitrogen (namely, NO, N_2O and NO_2) but not for water for which he gave the formula: HO and not H_2O.

In 1808, Dalton consolidated his revolutionary atomic theory and law of multiple proportions in a publication entitled "A New System of Chemical Philosophy". The quantitative significance of these fundamental laws assisted the establishment of the composition and formulae of numerous inorganic and organic compounds, laying the foundation for the advancement of chemistry in the nineteenth century.

In 1808, scientists were aware of thirty-six elements. Dalton recognised that each element has different atoms, and each has a signature atomic weight. He was the first to publish a chart of relative atomic weights for twenty elements, including hydrogen, oxygen, nitrogen, carbon, sulphur, phosphorus and a range of metals. Dalton used circles containing geometric shapes to represent atoms of specific elements. This symbolization may relate to his only pastime, that of lawn bowling. Berzelius replaced these shapes with letters in 1818.

Like all theories, Dalton's atomic theory has grown and changed with the times and has had to be modified to accommodate facts unknown to Dalton. For instance, the indivisibility of an atom was proved wrong since it has been further subdivided into sub-atomic particle such as electrons, protons, and neutrons. However, an atom is the smallest particle that takes part in a chemical reaction, so this part of his theory is still valid today.

J. J. Thomson and Aston demonstrated that atoms of the same element can have different atomic masses because the numbers of neutrons in their nuclei are different (see Sect. 8.3.3). Atoms may not be destroyed by a chemical reaction but can be destroyed by nuclear fission to give smaller atoms. Moreover, atoms can be created by nuclear fusion to give larger atoms.

In 1822, without his knowledge, Dalton was proposed for fellowship of the Royal Society, and was awarded its Royal medal in 1826.

From his humble beginnings, in a village near Cockermouth, Dalton rose to become an eminent scientist, becoming the most internationally famous Mancunian living at the time. He had asked an almost childlike question about the weights involved with the construction of simple molecules, and this resulted in an atomic theory. This is the essence of science: ask an impertinent question, and you are on the way to a pertinent answer.

Although a quiet, retiring man, when he died, he was accorded a civic funeral with full honours. His body lay in state in Manchester Town Hall, where more than 40,000 people filed past his coffin, in homage to the man who had put the city of Manchester firmly on the scientific map. He was buried in Ardwick cemetery—the final resting place for some of the town's most influential and notable residents. In the 1960's, the headstones were removed, and the site became the Manchester College/Nicholls Community Football Centre.

8.3 Joseph John Thomson OM FRS (1856–1940)—Atomic Physicist

8.3.1 Background

Thomson was born in Cheetham Hill, a suburb of Manchester. He was the eldest son of a bookshop owner. He showed great academic promise and, in 1870, was admitted to Manchester's Owens College which had been founded to provide instruction, in the branches of education taught at English universities, to youths of 14 or older. C. T. R. Wilson, also a Nobel Laureate, was another alumnus of Owens College.

Thomson's parents wanted him to become an apprentice to a locomotive manufacturer, but the plan was not realized due to the death of his father in 1873.

In 1876, he entered Trinity College, Cambridge, where he received his B.A. in mathematics in 1880. In this same year, he began working at the Cavendish Laboratory, Cambridge University, working under the tutelage of Lord Rayleigh. He became a Fellow in 1881 and received his M.A. with honours two years later. In 1884, he was elected Fellow of the Royal Society and, in that same year, at the age of twenty-eight, became Cavendish Professor of Experimental Physics, succeeding Lord Rayleigh. Although clumsy with his hands, he had a genius for designing apparatus and diagnosing any problems with it.

8.3.2   Cathode Rays. Subsequent Exploitation of Moving Electrons. "Plum Pudding" Atomic Model

Under normal conditions, a gas is a poor conductor of electricity. However, if a sufficiently high voltage is applied, a current can be passed between a cathode (negative electrode) and an anode (positive electrode) held in a gas, ionised at reduced pressure (less than 10^{-6} atmosphere).

If this process is conducted in a sealed glass vessel (e.g., a Crookes' tube) and the anode is perforated, then a stream of 'rays' from the cathode travel through the anode and the emerging rays can be directed through a slit. When the invisible cathode rays strike the centre of the end glass wall, the glass fluoresces, the colour of which depends on the nature of the glass. If the glass end-wall is coated with zinc sulphide, then a very pronounced fluorescence results.

In 1897, Thomson performed a series of experiments designed to study the nature of the electric discharge in an appropriately evacuated cathode ray tube (CRT). If an electric field is applied to the rays generated from the cathode, then they will be deflected towards the positive plate. The amount of deflection at the end of the CRT can be measured using an attached scale (Fig. 8.1).

If, instead of an electric field, a magnetic field is applied, deflection will also occur. The beam will move in an arc whilst passing through the magnetic field. If both electric and magnetic fields are applied simultaneously, their

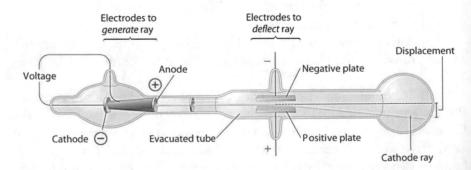

Fig. 8.1 Joseph J. Thomson's experiments with a cathode ray tube

relative magnitudes can be adjusted until their respective deflections are equal and opposite, and nil deflection occurs.

Knowing the deflection solely in an electric field, the potential difference between the cathode and anode, as well as the electrical and magnetic parameters under which nil deflection occurs, the ratio of the charge (e) to the mass (m) of the particle can be determined. Thomson found that the value of e/m to be 1000 times larger than the e/m value for the lightest particle known at the time, namely the hydrogen ion.

On the assumption that the charge (e) for both particles were the same, then the mass of the new particle was of the order 1,000 times lighter than the hydrogen atom. More refined methods of measurement by others showed it to be 1,837 times lighter.

The path of the cathode ray was the same regardless of the metallic nature of the cathode, or the residual gas in the CRT, suggesting that the cathode rays were of the same form, irrespective of their origin. They were sub-atomic particles in motion.

John Dalton (see Sect. 8.2) had previously reasoned that atoms could not be broken down into anything simpler. Thomson, however, had demonstrated that atoms were divisible, and his experiment had released, from atoms, 'corpuscles', that were a universal constituent of matter, namely a fundamental particle. In later years, other British scientists—Rutherford and Chadwick—were to provide suggestive evidence for other possible fundamental particles—protons (see Sect. 8.4.3) and neutrons (see Sect. 8.5).

In 1874, G. J. Stoney had proposed a standard unit quantity of electricity to be an 'electrine' (changed to 'electron' in 1891). 'Electrons' became the accepted descriptor for the cathode rays studied by Thomson.

Thomson had shown that tiny components of atoms conduct electricity. In 1906, he was awarded the Nobel Prize in Physics, in recognition of the great merits of his theoretical and experimental investigations on the conduction of electricity by gases.

In conventional electric circuits, electrons dumbly follow a path predefined by a conducting wire. Over a 100-year period, Thomson and others were to demonstrate that the path of electrons could be controlled within vacuum tubes, or, in a vacuum-less, solid state environment, such as transistors (composed of semiconductor metalloids germanium or silicon) and microchips (where a combination of light and chemical etching can give an integrated circuit).

On a silicon microchip, it is possible to engineer a collection of between a few hundred, to tens of billions, tiny circuits. Computers run on a binary system of 1s and 0s representing "on" and "off" states. Essentially, the chip's job is to shuttle electrons around in a way prescribed by computer code. In April 2021 about 100 billion microchips were sold globally. These technical progressions represent the development of the electronics industry, leading to the digital world of mobile phones, media players and computers. All the data on computers – numbers, text, pictures, music and images – are stored and processed in the way outlined.

In 1904, Thomson postulated a 'plum pudding' model of the atom. He imagined the atom to be a sphere of positive charge, with overall neutrality achieved via negatively charged electrons dotted around inside, rather like fruit in a Christmas plum pudding. Ernest Rutherford was to challenge this model in 1911 (see Sect. 8.4.3).

8.3.3   Collaboration with F Aston. Positive Ray Spectrograph. Isotopes

In 1906, having received his Nobel prize for cathode rays, Thomson moved his research studies onto positively charged ions, or positive rays. If a glass discharge tube is fitted with a perforated cathode, then the radiation passing through the holes in the cathode is found, by the effect of electric and magnetic fields, to consist of positively charged particles much heavier than the electrons of cathode rays.

In 1909, Francis Aston accepted an invitation from Thomson to work as his assistant. They noted that when the gas in the discharge tube is neon, two different deflections were observed. When expressed in whole number atomic mass units (amu), the positive particles were found to have masses of 20 and 22. The two forms are called isotopes of neon. The whole number principle was of vital importance in the eventual derivation of the structure of the atomic nucleus.

The two isotopes have the same atomic number; have the same number of protons; occupy the same place in the periodic table, but one has two more neutrons than the other. They explained the difference between the intensities of the two peaks in the spectrum by suggesting that 90% of the neon has a mass of 20 atomic mass units and 10% has a mass of 22 amu, giving an overall atomic weight of 20.2 amu.

This application of the "positive-ray spectrograph" proved the existence of isotopes for a non-radioactive element. Aston refined the arrangement of the electric and magnetic fields to create a mass spectrograph (later called mass spectrometer), recording a spectrum of mass values on a photographic plate. His electro-magnetic focussing technique was used to identify a total of 212 of the 254 naturally occurring, stable isotopes, an achievement for which he was awarded the Nobel Prize for Chemistry, in 1922. Later mass spectrometers relied on detection via an oscilloscope to become a fast, versatile analytical tool.

8.3.4 National Honours and Legacy

In 1908, Thomson was knighted for his contribution to science and, in 1912, he received the Order of Merit. He is buried in Westminster Abbey near two other influential scientists, namely Newton and Darwin.

Three years before J. J. Thomson's death in 1940, his son, Sir George Paget Thomson and C. J. Davisson were jointly awarded the Nobel Prize for Physics for their independent discovery of the diffraction of electrons by crystals. Like W. H. and W. L. Bragg, Joseph and George Thomson were a Nobel prize winning, father and son duo.

J. J. Thomson was respected and well-liked. He was a gifted teacher and nurturer of scientific talent. Undergraduates and graduates came from around the world to study with him. As a testimony to his influence, some of those assisting with his research would eventually receive Nobel prizes themselves. For instance, laureates in physics were: C. G. Barkla (British. X-ray spectroscopy, 1917); N. Bohr (Danish. Atomic structure and quantum theory, 1922); E. V. Appleton (British. Physics of the upper atmosphere, 1947) and M. Born (German. Quantum mechanics, 1954). The following colleagues were awarded the Nobel Prize in Chemistry: E. Rutherford (New Zealand. Radioactive substances, 1908) and F. W. Aston (British. Isotopes, 1922).

8.4 Ernest Rutherford, 1st Baron Rutherford of Nelson, OM FRS Hon FRSE (1871–1937)—Nuclear Chemist and Father of Nuclear Physics

8.4.1 Background and Education

Rutherford was born in Spring Grove, fourteen miles southwest of Nelson, South Island, New Zealand. His Scottish father was originally a wheelwright and engineer and later a flax farmer, and miller. His English mother was a schoolteacher at the Spring Grove school. At the time, there was no independent New Zealand citizenship—colonials such as the Rutherford's were regarded as British subjects and British citizens.

Ernest was the fourth of twelve children, having both high intelligence and sporting prowess. His inventiveness was honed on the challenges of assisting on the family farm. Without the money, they had to think. He read his first scientific book at the age of ten and, having been so enthralled, he carried out experiments as the book had instructed.

He won a school scholarship to board at Nelson College. In 1889, he next won a scholarship to Canterbury College, in Christchurch. In 1892, he obtained a B.A. and, in 1893, he graduated with a M.A. with a double first in mathematics and physical sciences. Two years later, he obtained a bachelor's degree in chemistry and geology.

In 1895, he observed that a magnetised needle, when placed in a magnetic field produced by an alternating current, lost some of its magnetism. This made the needle a detector of electromagnetic waves, specifically radio waves. Rutherford's receiver apparatus was simpler than Hertz's approach (see Sect. 10.4.6). He was awarded a research fellowship which allowed him to carry out post-graduate studies under the tutelage of J. J. Thomson, at the Cavendish Laboratory, University of Cambridge. He spent his first year in Cambridge increasing both the range and sensitivity of his radio wave detector. It had commercial potential, but Rutherford lacked the entrepreneurial skills of the Italian inventor Marconi who invented the wireless telegraph, in 1896.

Rutherford then accepted Thomson's assignment to study the electrical effects of X-rays on gases. The discoverer of X-rays—Röntgen—had noted that they had the ability to make gases conduct tiny electrical currents when the applied voltage was considerably below the undesirable spark-producing level. This, then, could be a way of creating and studying ions under controlled conditions. This research project was an ideal grounding for Rutherford. It led him to pursue other radiations that might produce ions, firstly ultraviolet, then emissions from uranium. The latter would be his initiation into the study of α- and β-rays.

8.4.2   Period at McGill University. Disintegration of Elements. Natural Decay of Radioactive Isotopes. α-Rays and β-Rays. Chemistry of Radioactive Substances. Half-Life. Nobel Prize

J. J. Thomson said that, "He had never met a student with more enthusiasm or ability for research than Mr Rutherford". In 1898, Thomson recommended Rutherford, aged just twenty-seven, be appointed professor at Montreal's McGill University. It had a world-leading physical laboratory, and it was here that Rutherford carried out the work that eventually would lead to his 1908 Nobel Prize in Chemistry.

Several great discoveries were made internationally in a period of a few years, beginning in 1895. This is when Röntgen, in Germany, discovered X-rays. In 1896, Becquerel, in France, inherited his father's supply of uranium salts which fluoresced when exposed to light, implying the need for a stimulating excitation energy input. Next, he found that, even without exposure to light, they emitted a radiation that passed through black paper and other opaque materials and exposed a photographic plate.

Also, in France, Marie Curie began a systematic investigation of 'Becquerel radiation'. With her husband, Pierre, the Curies separated natural pitchblende (a uranium-rich mineral) into factions. In one of the factions, she found a component which she named *polonium* after her Polish homeland. This was the first element discovered through its property of 'radioactivity'. In the same year, 1896, the Curies isolated another new radioactive element which they named *radium*. Radium's most stable isotope, radium-226, decays into radon-222 and emits α-radiation. A "curie" is defined as the quantity of radioactive material in which 3.7×10^{10} atoms disintegrate per second.

Starting in 1898, Rutherford studied two of the three types of radiation emitted by uranium. By placing successive layers of aluminium foil over a sample of uranium oxide, Rutherford discovered that one component was less penetrating than the other. For convenience, he named the former, α-rays, and the penetrating one, β-rays.

Thomson had shown his student—Rutherford—how to produce cathode rays (see Sect. 8.3.2). By using the same techniques, Ernest demonstrated experimentally that the mysterious α-rays were positively charged particles. In a magnetic field, α-rays deviate but to a lesser degree than β-rays, and in the opposite direction. In 1907, he and Hans Geiger would prove conclusively that alpha particles are helium atoms stripped of their two electrons (namely, doubly charged He^{2+} ions).

β-rays are like cathode rays in their general properties and this, coupled with measurement of their charge/mass ratio, showed that they consist of a stream of electrons.

In 1899, when working with R.B. Owens, Rutherford found that salts of radioactive thorium emit both α- and β-rays and generate a radioactive gas (loosely called 'emanation') that is more radioactive than its metallic parent, thorium ($_{90}Th$). They had discovered a new element with many of the properties of the 'noble', inert gases. In fact, it was an isotope ($_{86}Rn^{220}$) of gaseous radon which was called, for some time, *thoron*. In 1900, Rutherford identified a second isotope of radon ($_{86}Rn^{222}$).

In that same year, 1900, Rutherford was joined by F. Soddy, a chemistry graduate from Oxford and the two collaborated on the theory of 'radioactive disintegration and the transmutation of atoms' which regards radioactive phenomenon as an atomic, not a molecular, process.

Together, they discovered several new radioactive elements. In 1904, when he delivered the Bakerian Lecture to the Royal Society, in London, Rutherford summarized four key families of radioelements with parents: radium, thorium, uranium and actinium. Each headed a series of stepwise radioactive disintegrations. Later it was realized that the radium series was a continuation of the uranium series. All the disintegration pathways occur with a sequence of α- and β- emissions, ending with inactive lead. For instance, the decay cascade of uranium-238 culminates in lead-206; actinium-227 in lead-207 and thorium-232 in lead-208.

When the nuclear binding energy of an isotope is insufficient, its nucleus disintegrates. When large radioactive atoms emit α-particles, they become smaller atoms which means radioactive elements change into other elements when they decay. Contrary to Dalton's view of the atom (see Sect. 8.2), they had demonstrated that an atom can be 'destroyed'. Loss of an α-particle means a lowering of mass number by 4, and of atomic number, by 2.

Curiously, after chemical treatment, some radioelements lost their radioactivity but eventually regained it, whilst others, initially strong, gradually lost activity. This led to the concept of 'half-life'.

Rutherford observed that any quantity of radioactive isotope takes the same amount of time for half of it to decay. This time interval is called the half-life. The half-life is important for three main reasons. Firstly, it is unique for each radioelement and serves as an identifying tag. Secondly, it indicates the stability of a specific isotope. For instance, radium-226 has a half-life of 1,602 years, whilst radon-220 (Rutherford's emanation) has a half-life of only 55.3 s.

Third, the half-life has significant practical applications. The constant rate of decay can be used as a 'clock' to gauge the age of such things as rocks, fossils, bones, ancient pieces of art etc. For example, the stepwise breakdown of uranium-238 occurs by a radioactive series or decay series of fourteen steps to stable isotope lead-206. Using the ratio of $_{92}U^{238}$ to $_{82}Pb^{206}$, the age of rocks can be closely approximated. Another example would be the incorporation of carbon-14 in plant tissue during photosynthesis. Its radioactive decay to stable isotope carbon-12, following the death of the plant, can be used to determine the age of the dead tissue.

In their last collaborative paper, Rutherford and Soddy deduced that the amount of energy released during the transmutation of one gram of radium, through its entire disintegration series to a non-radioactive final product, is about 2,000 times that produced when one gram of water is formed from the energetic reaction between hydrogen and oxygen which is sometimes explosive. They concluded that, the energy, latent in the atom of a radioactive or nonradioactive element, must be enormous compared to that rendered free in ordinary chemical change.

In 1908, for his pioneering research at McGill University, Rutherford was awarded the Nobel Prize in Chemistry, in recognition for his investigations into the disintegration of the elements, and the chemistry of radioactive substances. However, it is his subsequent research at the Victoria University of Manchester for which he is most famous.

8.4.3 Period at the Victoria University of Manchester. The Gold-Foil Experiment. Rutherford's Nuclear Model Versus Bohr's Planetary or Solar Model

Rutherford yearned to be closer to British and European scientific centres. A year before the Nobel prize award, Rutherford returned to Great Britain to lead the physics department at the Victoria University of Manchester.

In 1909, two of his researchers —Hans Geiger and Ernest Marsden— carried out one of the landmark experiments in science to test Thomson's 'plum pudding' model of the atom (see Sect. 8.3.2). Geiger and Marsden used a sample of radium to provide a stream of α-particles which they projected at a strip of highly rolled gold foil, radially surrounded by a luminescent screen to detect scattered α-particles. Every α-particle that arrived at the zinc sulphide screen caused a scintillation, thought to be the next best thing to "seeing" it.

The bombarding α-particles had a mass about four times that of a hydrogen atom and carried a double positive charge. Based on Thomson's 'plum pudding model' of the atom, they anticipated a slight deflection of the α-particles. In practice, most of the α-particles passed straight through the gold foil without any deflection, but a few were deflected, some (0.01%) at over 90°. For a small number, their motion was reversed, the deflection being 180°, suggesting a 'backward bounce'. They must have encountered an extremely powerful force within the foil. Since they were positively charged, they must have been repelled by another positive charge, concentrated in a tiny volume. It led Rutherford to exclaim that "It was almost as incredible

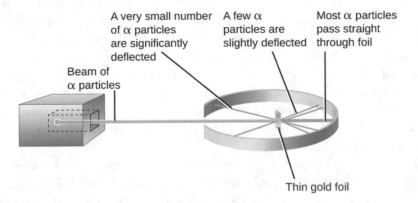

A very small number of α particles are significantly deflected

A few α particles are slightly deflected

Most α particles pass straight through foil

Beam of α particles

Thin gold foil

Fig. 8.2 Geiger/Marsden/Rutherford gold-foil experiment. *Source* https://courses.lum enlearning.com/astronomy/chapter/the-structure-of-the-atom/. cc licensed content, shared previously. https://openstax.org/books/astronomy/pages/5-4-the-structure-of-the-atom#term206. Chapter 5 (Radiation and spectra); Sect. 5.4 (Structure of atoms); Fig. 5.14. https://creativecommons.org/licenses/by/4.0/

as if you had fired a 15-inch (artillery) shell at a piece of tissue paper and it came back and hit you" (Fig. 8.2).

Thomson's theoretical sphere of positive charge was too diffuse to cause the observed result. A new theory of atomic structure was required. In 1911, Rutherford proposed a new model for the atom in which (i) most of the mass of the atom was concentrated in a tiny, dense, positively charged nucleus, whilst (ii) most of the atom's volume was empty space, with electrons orbiting randomly at some distance from the nucleus. His 'nuclear' model introduced the concept of a nucleus at the core of the atom with electrons circling around it. (N.B. The diameter of a nucleus ranges from about 1.6 fm (10^{-15} m) for a proton in 'light' hydrogen to about 15 fm for uranium—the heaviest, naturally occurring element. A typical atom has a diameter of 10^{-10} m).

In 1911, having completed his Ph.D. in Denmark, Niels Bohr was invited to conduct post-doctoral research with Rutherford. According to Bohr, Rutherford's nuclear model of the atom would be unstable since randomly orbiting electrons would, by their motion, accelerate, thereby emitting electromagnetic radiation continuously—sapping their energy—and spiral inwards, attracted by the positively charged nucleus.

Moreover, since chemicals burn with a characteristic-coloured flame, the pattern of energy released by electrons in the chemical reaction must be the same for every atom of that element. Therefore, electrons cannot be spatially arranged in a random manner. They must have fixed levels of energy within each atom. The solar or planetary model took form, albeit not in defined plane, as are the planets of the Solar System.

Based on new ideas from nascent quantum theory, championed by the revolutionary work of Max Plank, and wave-particle duality (Sect. 10.4.11), Bohr proposed that the exact path of electrons around the nucleus was restricted to well defined orbits. The orbits exist at set distances or 'shells', defined by their intrinsic energy. Each orbit has a certain energy associated with it. Movement of an electron between levels requires it to either absorb or emit a quanta of energy (as light or heat) equivalent to the difference in energy between the two levels. In 1922, Bohr was awarded the Nobel Prize in Physics for his services in the investigation of the structure of atoms and of the radiation emanating from them.

Each orbit holds a maximum number of electrons. In the first shell, no more than two electrons can be accommodated; the 2nd shell, up to eight; the 3rd, eighteen and the 4th, thirty-two. The nth shell can hold a maximum of $2n^2$ electrons. Atoms with less than the maximum number of electrons in their outer shell are less stable than those with full outer shells.

Elements that have the same number of electrons in the outermost shell appear in the same column of the periodic table of elements (see Mendeleev below) and tend to have similar chemical properties. To achieve 'full' shells, atoms will lose or gain electrons to form ionic compounds or share electrons with other atoms to form covalently bonded substances. Thus, chemical properties are defined by the orbital electrons, whilst radioactivity is determined by the nature of the nucleus.

8.4.4 Mendeleev, Moseley's Atomic Number, Chemical Periodicity, Schrödinger Model

In the 1860's, De Chancourtois (1862) Newlands (1864) Odling (1864) Meyer (1864) and Mendeleev (1869) independently proposed a *periodic* classification of the elements. The elements were arranged in order of ascending atomic weights. In the 1871 version of Mendeleev's table, each of the sixty-three elements, known at the time, was assigned to one of eight groups in which the members of each family group had common physical and chemical properties. However, there were some gaps in the groups and some anomalies, including 'pair reversal' (e.g., argon and potassium; cobalt and nickel; tellurium and iodine).

The infrequent misfits and misplaced order would be corrected following the research conducted by H. G. J. Moseley who joined Rutherford's team in 1910. Henry spent time with W.H. and W.L. Bragg (Sect. 12.2) learning X-ray spectrometric techniques, before ultimately returning to his *alma mater*, Oxford University.

A monochromatic X-ray beam can be generated by electron bombardment of a metal target in an X-ray tube. When these X-rays are focussed on a sample material, inner shells of electrons are excited into higher energy states. Subsequently, X-rays of other energies—characteristic of the elements of the sample—are emitted. A photographic plate can be used to record the resulting spectrum. Moseley measured the wavelength, and hence the frequency, of the sharp K and L lines. For the thirty-eight elements he studied, he found that each one had its own signature—the frequency of X-rays produced by each element being dependent on its position in the table of elements.

He found that, by plotting the square root of the frequency of the lines against a suitably chosen integer, Z, a set of (almost) straight lines was produced. Experiments by Rutherford and his students enabled Moseley to conclude that this integer, Z, represents the number of positive charges (protons) contained in the nucleus. He called this the atomic number, and it identifies the element's correct pigeonhole in the periodic table.

Flaws in previous periodic tables disappeared. It resulted in the modern form of the *periodic law* which states that "When elements are arranged in order of increasing atomic number, there occurs a periodic repetition of physical and chemical properties". It is the positive charge on the nucleus and the number of electrons in the neutral atom that determine both the sequence in which the elements occur and their properties.

Displays of chemical periodicity include atomic radius; ionization energy; electronegativity and electron affinity. The format of the periodic table of the elements had been evolving for over 100 years but, in 1914, it was starting to reach a layout with which we are all familiar. Moseley saw gaps in his new periodic table and predicted the existence of four new elements between aluminium (atomic number 13) and gold (atomic number 79). He predicted that these four elements—when discovered—would have 43, 61, 72 and 75 protons.

Qualitative analysis stems from the appearance of the X-ray spectrum. X-ray spectroscopy makes it possible to identify the elemental make-up of just about any substance in a non-destructive manner.

Sadly, during WW2, Moseley, aged 27, was killed, in battle, in Gallipoli, depriving the world of one of its most creative experimental physicists, and a potential Nobel laureate.

In 1926, calculations by Erwin Schrödinger suggested that it was unlikely that electrons would be located a fixed distance from the nucleus, as proposed

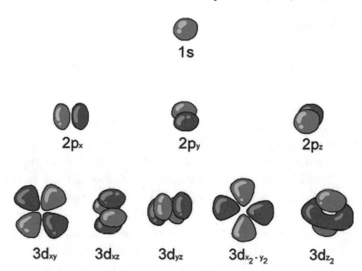

Fig. 8.3 Quantum mechanical model of electron clouds

by Bohr (see Sect. 8.4.3). Schrödinger's "quantum mechanical model" is a radical departure from the Bohr model. Solutions to the Schrödinger wave equation introduce an 'uncertainty' factor, leading to a probability assessment of finding an electron at a given point around the nucleus. In this model, a region of space (cloud) represents the location of the electron—high density where the electron is most likely to be, and low density where it is least likely to be. Electron clouds are called orbitals, named s, p, d, and f (Fig. 8.3).

8.4.5 Rutherford's Period at the Cavendish Laboratory, Cambridge. Nuclear Transformation of Non-radioactive Elements—Artificial Transmutation. The Proton and Neutron

When his mentor, J. J. Thomson, retired as the Cavendish Professor of Experimental Physics at the University of Cambridge, in 1919, Rutherford was appointed as his successor.

In 1919, during his last year at Manchester, Rutherford decided to move away from studying the natural decay of radioactive isotopes, to the study of the artificial, nuclear transformation of a known, non-radioactive element to another known element. He found that, by bombarding stable isotope nitrogen-14, with energetic alpha particles, he artificially produced a nuclear reaction. The nucleus of the highly radioactive intermediate, fluorine-18, ruptured and this resulted in another element, oxygen-17, and fast-moving nuclei of hydrogen:

$$_7N^{14} +_2He^4 \rightarrow \left[_9F^{18}\right] \rightarrow _8O^{17} +_1H^1$$

Co-worker Blackett used a cloud chamber to prove that the α-particle did not casually chip away a proton from the surface of a nitrogen nucleus. Rather, it penetrated the nitrogen nucleus, forming a composite (fluorine) nucleus that disintegrated into an isotope of oxygen, liberating a proton. Many years later, Blackett was awarded the 1948 Nobel Prize in Physics for his work in the fields of nuclear physics and cosmic radiation.

Effectively, the nucleus was split, and the atom disintegrated. It was not a ricocheting effect—one element was transmuted to another. The result also showed that the liberated hydrogen nuclei, originally must have been part of the nitrogen nuclei (and by inference, probably other nuclei as well). Rutherford decided that a hydrogen nucleus was possibly a fundamental building block of the nuclei of all elements. In 1920, he postulated the hydrogen nucleus to be a new primordial, fundamental particle which he dubbed the *proton*.

In the same year, Rutherford observed that about one half of the nuclear mass of most of the elements he had investigated could be accounted for by protons. He, therefore, suggested that particles of zero charge and of a mass like a proton might be present in the nucleus. He called such a particle a "neutron".

Rutherford performed calculations on the stability of atomic nuclei. He realized that, unless some binding particles were present, repulsion between positively charged protons would occur, causing the nucleus to disintegrate.

In 1921, he and N. Bohr theorized about the existence of *neutrons* which could somehow neutralize the repelling effect of the positively charge protons by creating an attractive nuclear force. Eleven years later, in 1932, Rutherford's Assistant Director of Research, James Chadwick, proved experimentally, the existence of neutrons (see Sect. 8.5).

In 1933, Irène Joliot-Curie (daughter of Pierre and Marie Curie) and her husband, Frédéric showed that other 'light' elements could be transformed with alpha particles, as Rutherford had done with nitrogen-14. They reported that boron and aluminium were made radioactive by bombarding them with alpha particles from polonium to produce two new radioactive isotopes, N^{13} and P^{30}, respectively. For this discovery they were awarded the Nobel Prize in Chemistry, in 1935.

One of the main limitations of this 'shot gun' approach is that a positively rich nucleus is being bombarded with positively charged particles. 'Heavy' elements with large nuclei will repel such particles. Therefore, other ways had to be found to transform heavy elements (see Sect. 8.5).

8.4.6 Rutherford's Public Decorations and Legacy

In 1903, Rutherford was elected Fellow of the Royal Society and served as its President from 1925 to 1930. In 1914, for his contribution to science, Rutherford was knighted and, in 1925, inducted as a member of the Order of Merit. In 1931, he was created to Baron Rutherford of Nelson.

At his passing, in 1937, he was honoured by being interred with some of the greatest British scientists. His ashes are buried in the nave of Westminster Abbey, just west of Isaac Newton's tomb, and by that of Lord Kelvin. Element-104, rutherfordium, synthesized in 1964, is named after him.

He was an unprepossessing, unpretentious man. His pathway from rural child is extraordinary. He is to the atom what Newton is to dynamics; Faraday is to electricity; Darwin to evolution and Einstein is to relativity. He was possibly the greatest experimental physicist since Faraday. His successful experiments were characterized by their primitive simplicity, elementary apparatus, executed with the minimum of fuss and the minimum of error.

During his career in Montreal, Manchester and Cambridge, Rutherford surrounded himself with outstanding scientists with enormous potential. He helped to steer at least thirteen of these students or collaborators towards their greatest achievement as Nobel laureates in physics or chemistry.

These include Soddy (chemistry of radioactive substances and nature of isotopes; Chemistry Prize, 1921) Aston (isotopes/mass spectrograph; Chemistry Prize, 1922) Bohr (structure of the atom; Physics prize, 1922) Chadwick (neutron; Physics Prize, 1935) G. P. Thomson (diffraction of electrons by crystals; Physics Prize, 1937) de Hevesy (isotopic tracers; Chemistry Prize, 1943) Hahn (fission of heavy nuclei; Chemistry Prize, 1944) Appleton (demonstrated the existence of the ionosphere; Physics Prize, 1947) Blackett (cloud chamber/cosmic radiation; Physics Prize, 1948) Powell (photographic method of studying nuclear processes; Physics Prize, 1950) Cockcroft/Walton (transmutation of atomic nuclei; Physics Prize 1951). Kapitsa (superfluidity of liquid helium; Physics Prize, 1978).

8.5   James Chadwick CH, FRS (1891–1974)—Physicist and Discoverer of the Neutron

Chadwick was born in Bollington, near Macclesfield in Cheshire, a corner of the cotton industry. He was a shy child from a working-class family. His father was a cotton spinner, his mother a domestic servant. In 1895, his parents moved to Manchester, where his father was going to run a family laundry business. James was left behind with his maternal grandparents. His academic talents caught the attention of his teachers at Bollington Cross school. He won entry to the prestigious Manchester Grammar School. However, since his parents could not afford the modest fees, he went instead to Manchester's Central Grammar School for Boys.

In 1908, he won a scholarship to the Victoria University of Manchester where he intended to read mathematics but, by accident, ended up studying physics. Rutherford had just returned to Great Britain from Canada, where he had been awarded the Nobel prize in chemistry. In Chadwick's final undergraduate year, he was assigned a research project by Rutherford. This was to devise a means of comparing the amount of radioactive energy of two different sources.

Having graduated with a first-class honours' degree, Chadwick continued to work in Rutherford's laboratory until 1912, when he was awarded a master's degree in physics for his work on the absorption of γ-rays by various gases and liquids.

In 1914, he won a scholarship enabling him to join Professor Hans Geiger at the Physikalisch-Technische Bundesanstalt (PTB) in Berlin. Between 1907 and 1912, Geiger had worked with Rutherford at the Victoria University of Manchester (see Sect. 8.4.3). Geiger was instrumental in developing a special kind of ionisation chamber. It was an electrical counter for ionised particles which, when perfected, became the universal tool for measuring radioactivity. In Germany, Chadwick was able to conduct research on β-radiation using the recently developed Geiger counter. He established the existence of continuous β-ray spectra.

Unfortunately, at the commencement of WWI, Chadwick was arrested and held for four years in an internment camp for British internees at Ruhleben. It is said that, whilst detained, he formed a science club and performed experiments on improvised materials, including a toothpaste being sold in Germany that happened to be weakly radioactive due the presence of thorium chloride.

At the end of the war, he returned to England and re-joined Rutherford in Manchester, taking up a part-time teaching post. Several months later, the two moved to Cambridge, where Rutherford had been offered the chair at the Cavendish Laboratory. Chadwick completed his doctorate in 1921, submitting a thesis concerned with atomic numbers and nuclear forces.

In 1923, Chadwick was appointed Assistant Director of Cavendish Laboratory. He and Rutherford resumed studies on the transmutation of non-radioactive elements by bombarding them with α-particles (see Sect. 8.4.5). They extended their investigation from the demolition of nitrogen to the nuclei of other light elements such as boron, fluorine, sodium, aluminium and phosphorus.

In 1927, he was elected Fellow of the Royal Society. In his research, he continued to probe the nucleus. Neutrons, by their nature, do not leave observable trails of ions when they pass through matter, nor do they make tracks in a cloud chamber. Geiger's latest counter was potentially a major improvement over the scintillation techniques used at the Cavendish Laboratory, but it also detected α-, β- and γ- radiation emitted by the radium that was normally used at the Laboratory.

At the beginning of 1932, Chadwick became aware of the research conducted by Irène Curie (daughter of Pierre and Marie Curie) and Irène's husband, Frédéric Joliot.

The Joliot-Curie duo used polonium which only emits α-particles. They had directed this α-particle radiation, from polonium, to a plate of beryllium. This resulted in so-called "beryllium radiation" which, when passed through paraffin wax or other hydrogen-containing matter (e.g., gelatine), liberated fast-moving protons that travelled about 26 cm in air. The Joliot-Curie duo postulated that bombardment of paraffin wax, by polonium γ-rays, with quantum energies of about 50 meV, were responsible for dislodging the protons.

Chadwick demonstrated that beryllium radiation could dislodge particles, not only from paraffin wax, but also other light elements such as hydrogen, helium, lithium, beryllium itself, boron and nitrogen. However, the ejected particles from light elements had a range of travel much less (ca. 3 mm) than the protons from paraffin and were recoil atoms of the element through which the beryllium radiation had passed.

Both Rutherford and Chadwick were doubtful that the almost mass-less photons of γ-radiation were sufficiently energetic to bring about these effects. They concluded that the γ-ray hypothesis was untenable. They were convinced that this property of triggering the motion of small atoms of matter must involve particles.

Chadwick measured the energy (52 meV) of the knock-on protons. He also determined the velocities of the protons ejected from paraffin, as well as recoil atoms of nitrogen observed in an expansion chamber.

Applying the laws of conservation of momentum and energy, he worked backwards to demonstrate that the projectiles were particles with a mass slightly heavier than a proton. They were uncharged, being able to pass through materials without any electrostatic scattering. Combining experimental observation with theoretical calculation, Chadwick demonstrated the existence of neutrons.

Chadwick's explanation was that α-particles reacted with beryllium forming carbon nuclei and neutrons—particles having an atomic weight of one, and zero electric charge. The high speed α-particles were able to overcome the repulsive forces of the beryllium nucleus. The beryllium nucleus (containing 4 protons) absorbs an α-particle (comprising 2 protons), to give a carbon nucleus with 6 protons, releasing energetic neutrons by a process symbolized as Be(α, n)C:

$$_4\text{Be}^9 + _2\text{He}^4 \rightarrow {} _6\text{C}^{12} + _0\text{n}^1$$

Later experimentation showed that these particles are more penetrating than protons. They could pass through 10–20 cm of lead, whereas protons of the same velocity are stopped by a thickness of 0.25 mm of lead.

Some researchers, at the time (including Rutherford) speculated that neutrons were paired proton-electron combinations, but in a more 'intimate' way than they are in hydrogen atoms, to give 'neutral doublets'. However, Werner Heisenberg argued that a proton-electron doublet was not possible and, therefore, this was a new, possibly fundamental, particle.

A coherent new picture of nuclear and atomic physics was developing. With this revised view of the nucleus, a carbon atom is shown schematically in Fig. 8.4).

When radium-226 decays into radon-222, the α-particle released has an energy of over 4 meV. This makes it sufficiently energetic to penetrate target nuclei and disrupt them. Rutherford surmised that other particles, such as protons, could be used as "munition bullets" if they were sufficiently energetic. Unfortunately, radioelements are not so generous with energetic protons as they are with α-particles.

Particle acceleration would require the application of heavy electrical engineering. Encouraged by Rutherford, two Cavendish physicists-engineers—J. Cockcroft and E. Walton—built a linear accelerator that produced a beam of protons with an energy rating of 1 meV. By 1932, they had developed a voltage-multiplying circuit which accelerated a 40 kV

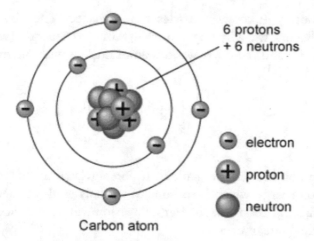

6 protons
+ 6 neutrons

⊖ electron

⊕ proton

⬤ neutron

Carbon atom

Fig. 8.4 Schematic of a carbon atom

protons beam exiting a hydrogen discharge tube. A section of the particle accelerator can be seen at the National Museum of Scotland (Edinburgh EH1 1JF; museum reference T.1975.25).

They were persuaded by quantum mechanical reasoning that "tunnelling" would lower the energy required for an incident, positively charged particle to overcome the Coulombic barrier of a target nucleus. Theory suggested that an energy of 300 keV might be sufficient for protons to penetrate a nucleus of boron, whilst an energy of 150 keV might split a lithium nucleus.

Hence, the beam was directed at a lithium target. Lithium nuclei absorbed a proton to transiently give beryllium-8 which then spontaneously disintegrated in to two helium nuclei, with a range in air of 8 cm. This was the first transmutation using artificially accelerated particles and for this work, they were jointly awarded the Nobel Prize in Physics, in 1951. Whilst Rutherford had brought about a transmutation of a nucleus, Cockcroft and Walton had gone further by splitting the atom.

$$_1H^1 + _3Li^7 \rightarrow \left[_4Be^8 \right] \rightarrow _2He^4 + _2He^4$$

M. Oliphant, working with Rutherford and P. Harteck, modified Cockcroft and Walton's accelerator to produce a greater flux of protons. Having been gifted with a few drops of heavy water (containing "heavy" hydrogen comprising a nucleus with one proton together with one neutron), they set about using heavy hydrogen nuclei (also known as deutons, diplons, and now, deuterons) as both missiles and targets.

Two previously unknown particles were detected. The Cavendish team postulated that one was a hydrogen nucleus with two neutrons (viz hydrogen-3; tritium; $_1\text{H}^3$) and the other was a helium nucleus with only one neutron (viz helium-3; $_2\text{He}^3$).

$$_1\text{D}^2 +_1\text{D}^2 \rightarrow \left[_2\text{He}^4\right] \rightarrow {_1\text{H}^1} +_1\text{H}^3$$
$$\rightarrow {_2\text{He}^3} + _0\text{n}^1$$

Their work on heavy hydrogen and fusion is related to the production of nuclear fusion energy, which is still one of the holy grails of energy research. Any mass "missing" from the nuclear conversion will have been converted to formidable amounts of energy.

In the late 1920s, E. Lawrence, at the University of California at Berkeley, was developing a circular, proton-accelerating machine that he called a "cyclotron. In 1935, Chadwick was appointed to the Chair in Physics at the University of Liverpool. In the same year, he was awarded the Nobel Prize in Physics for the discovery of the neutron. He used part of the monetary prize to purchase a cyclotron which he knew would revolutionize experimental nuclear physics. The accelerators at the Cavendish and Berkeley were the progenitors of both linear and circular particle accelerators, including the one at CERN which is several kilometres long.

Chadwick's research paved the way towards the fission of numerous elements since accelerated neutron particles do not need to overcome any repulsive barriers of charged particles. This feature provided a new tool for inducing atomic disintegration since neutrons are capable of penetrating and splitting the nuclei of even the heaviest elements. In 1938, German chemist, Otto Hahn, and Fritz Strassmann were the first to split fragile U^{235} with neutrons.

Enrico Fermi found that "slow", low energy neutrons were efficient projectiles for bombardment experiments—they spend longer in the target nucleus than "fast" neutrons. The possibility of capture will depend on whether the nucleus has an appropriate, vacant 'pigeonhole' for neutrons. For instance, bombarding uranium-235 with slow neutrons results in the transient formation of uranium-236 nuclei. These promptly shatter or undergo "fission", giving two fragments of roughly equal mass, a spontaneous decomposition process that is accompanied by the release of vast amounts of energy (viz 2.5 million times that evolved by burning the same weight of coal, and about 12 million times that evolved by exploding the same weight of nitro-glycerine).

$$_{92}\text{U}^{235} + _0\text{n}^1 \rightarrow \left[_{92}\text{U}^{236}\right] \rightarrow {_{56}\text{Ba}^{139}} +_{36}\text{Kr}^{94} + 3_0\text{n}^1$$

Neutrons born in one fission event can trigger more fissions, thus sustaining a nuclear chain reaction. When there is a critical mass of fission material, and other conditions are met, a self-sustaining chain reaction can occur, generating a formidable amount of nuclear energy. At the start of WW2, scientists were the first to see the potential military applications. Later, when it was possible to carefully control neutron losses and gains, it resulted in the production of usable heat and the development of the nuclear power industry.

In 1940, Jewish refugees Otto Frisch and Rudolf Peierls were co-workers at the University of Birmingham. They penned a memorandum to the British government explaining the idea of a uranium fission "super bomb" and said it was conceivable that Nazi Germany was itself developing such a bomb.

Chadwick was a member of MAUD, a British scientific working group tasked with determining the feasibility of an atomic bomb. The Committee concluded that nuclear weapons were possible and even inevitable. Their findings also energized the American effort, providing an impetus to President Roosevelt to authorise the American atomic bomb programme which eventually became the Manhattan Project.

Chadwick's research group deduced that the critical mass of uranium-235, for a nuclear detonation, was about 8 kg. The process of nuclear enrichment began. It soon became clear that the UK lacked the necessary resource to build a bomb, so British expertise was subsumed in the American nuclear programme. Between 1943 and 1946, Chadwick was Head of the British Delegation to the Manhattan Project in Los Alamos, New Mexico. This project cost about $2 b and involved an Anglo-American network of government / industry / academia and armed forces.

In 1945, Chadwick was knighted for his achievements in physics and his war-time contribution. He returned to Britain, from America, in 1946. The Vice Chancellor of Liverpool University wrote that "he had never seen a man so physically, mentally and spiritually tired as Chadwick …, having suffered almost insupportable agonies of responsibility, arising from his scientific work".

He became British scientific advisor to the UN Atomic Energy Commission. He was an outspoken advocate for the UK to acquire its own nuclear stockpile. In 1970, he was made a Companion of Honour.

After WW2, John Cockcroft was appointed the inaugural director of the Atomic Energy Research Establishment (AERE) at Harwell. He played a key role in the specification for the nuclear power station, Calder Hall 1, at the Windscale site. This was a gas-cooled, graphite moderated, natural uranium reactor. In 1956, it was the World's first reactor to provide electricity for the national grid.

8.6   Summary of British Advances in Our Understanding of the Structure of an Atom, 1803 to 1932

For over two millennia, philosophers and scientists had theorized about the composition of matter. Between 1803 and 1932, three British scientists, together with one offspring of British parents (born in New Zealand and conducting experimental research in Manchester) put forward an increasingly sophisticated architectural model of matter. They helped unlock the secrets of atoms.

By 1932, their model had developed from 'billiard-ball' atoms (Dalton, 1803), to a plum pudding structure (J. J. Thomson, 1904), to 'diffuse' atoms with a tiny, positively charged nucleus (Rutherford, 1911) comprising protons (Rutherford/Bohr, 1913) and neutrons (Chadwick, 1932) with electrical neutrality achieved by electrons (J. J. Thomson, 1897) orbiting the nucleus.

In the 1930s, atoms were viewed, not as solid balls composed of one material, but as tiny solar systems with different components. If the nucleus were the size of a fly, then the atom would be the size of a cathedral.

Collectively, their experimental research suggested the existence of three fundamental sub-atomic particles which had no detectable internal structure and could not be broken down into anything smaller. These were the electron, proton and neutron. As is the history of the atom, our thinking has advanced since 1932, so we now believe that, of the three particles, only the electron is a fundamental particle because protons and neutrons can be sub-divided.

Key developments in atomic theory are diagrammatically presented in Fig. 8.5.

All particles that make up an atom are, by definition, sub-atomic. The mathematical laws that describe these particles and how they behave is now the subject of quantum mechanics. Most sub-atomic particles can only be observed naturally in cosmic rays or, artificially, in particle accelerators.

As with all scientific topics, one should never close the book on any field of enquiry worthy of investigation. This is still work in progress. The next chapter on atomic structure is waiting to be written.

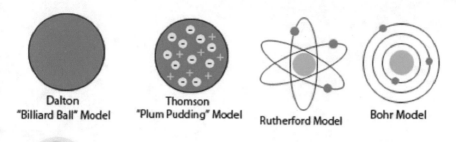

Dalton
"Billiard Ball" Model

Thomson
"Plum Pudding" Model

Rutherford Model

Bohr Model

Quantum Mechanical
Model

Fig. 8.5 Advances in our understanding of atomic structure

9

Life Sciences Leading to Health Care, Dental Hygiene, Disease Control, Hospital Sanitation and IVF. Great British Physicians and Nurses

The prevention of disease is one of the most important factors in line of human endeavour. (Charles Mayo. Collected papers of the Mayo Clinic and Mayo Foundation, 1913)

9.1 Précis

For centuries, people considered the heart to be the source of vitality and innate heat, as well as the seat of intelligence. In 1628, William Harvey published his "Anatomical Studies on the Motion of the Heart and Blood in Animals". He established that blood circulates round the body and passes through the lungs, where it is revitalised. From his experiments and the evidence of what he could see and feel, he developed his theories. In this regard, Harvey was one of the first scientists in the medical field.

Until the Elizabethan era, people had relatively healthy teeth. By 1750, however, all levels of society became consumers of refined sugar. Bacteria that colonize the dental surface convert fermentable sugar into an acidic film that attacks tooth enamel. This led to a serious decline in the nation's dental health. In 1770, William Addis, whilst in Newgate prison, originated a design for a 'mouth broom' (toothbrush) that he would mass-produce once he had served his sentence and was released. When effectively applied, brushing can control dental hygiene. Addis is credited with being the 'father' of dental

J. Bailey, *Inventive Geniuses Who Changed the World*, https://doi.org/10.1007/978-3-030-81381-9_9

hygiene for the British public, and the toothbrush has become one of life's essential, personal care items.

Protection and recovery from disease, particularly world-wide epidemics such as smallpox, were helped by revolutionary new medical advances such as vaccination (Jenner), sterile surgery and antiseptics (Lister). Since their adoption, millions of lives across the world have been saved.

Thousands of years ago, the smallpox virus emerged and began causing illness and the deaths of millions. Edward Jenner was a country doctor who was the pioneer of the world's first vaccine—a safe smallpox vaccine. After an extensive international vaccination programme, this ancient human scourge has been eradicated. Jenner was hailed as the 'father' of immunology and it is said that his vision has saved more lives than the work of any other human being. Vaccination has become a highly effective method of preventing a host of infectious diseases. We cannot develop the specific immunity necessary to protect us from a specific pathogen unless we are either infected with it or vaccinated against it. A desperate search for a SARS-CoV-2 vaccine has now been realized. Coincidentally, research workers at the Jenner Institute, in Oxford, have been involved in one of its developments.

Vaccines are probably the most successful medical intervention in history. It is concerning, therefore, that 'anti-vaxxers' promote, on social-media, misinformation about vaccinations, resulting in vaccination hesitancy and growing numbers of unprotected children. For instance, in England, in 2020, take up for the measles, mumps and rubella (MMR) vaccine is below the level needed to provide widespread (herd) immunity in the community. Charlatans also use social media to peddle well-meaning but useless remedies and therapies for a variety of ailments.

Britain survived two horrific world wars unconquered, with its democratic institutions intact. However, such victories were achieved at the expense of enormous human and economic sacrifices. Our compassion for those in distress and suffering, after military action in a previous century, was reflected in pioneering new nursing methods.

Like most women of her time, Florence Nightingale was denied tertiary education and deterred from entering a profession. During the Crimean War, in the 1850s, she ministered to those soldiers suffering from battle injuries, as well as typhus, cholera and dysentery. Compassionate by nature, she was immortalised as the "Lady with the Lamp".

She kept meticulous records about the deaths of soldiers and their causes. Being a gifted statistician and a champion of data-visualisation, she was able to present these novel methods of communication, to the Royal Commission on the Health of the Army, to show that most of the mortalities were from preventable diseases, poor nutrition and/or unsatisfactory hospital conditions.

Upon her return home from Crimea, she turned her attention to nursing and sanitary concepts in British hospitals, setting up a training school for nurses. Having the attention of Queen Victoria and eminent members of the Cabinet, she helped bring about a seismic shift in the UK's sanitation and public health programmes. Florence was the first woman to receive the Order of Merit.

During Victoria's reign, medical care was reshaped, and changes implemented to bring the health of the nation into the modern age. For instance, Joseph Lister observed that some patients, particularly those with open wounds, underwent an operation successfully, only to die from a post-operative infection, known as 'ward fever'. He believed that infection was invading externally. Many surgeons wore dirty aprons; did not wash their hands before operations and used unclean surgical instruments. Recovering patients might be placed on bed linen stained with blood and other bodily fluids from previous patients.

Lister reasoned that the way to stop post-operative infections was to prevent germs entering the wound. Applying a dilute solution of carbolic acid (phenol) he experimented with the soaking of dressings, the washing of hands, and the dipping of surgical instruments. Over a 4-year period, his antiseptic procedures brought about a significant reduction in mortality rates. As many of his techniques were adopted by other surgeons, he became known as the 'father of antiseptic surgery'. Refinements to his techniques led to sterile surgical procedures with which we are more familiar, and the saving of an incalculable number of lives.

During WWI, Alexander Fleming served as a practising bacteriologist studying wound infections. He demonstrated that the direct use of strong antibiotics on deep wounds often did more harm than good. He was convinced that antibiotic agents should only be used if they acted in a way that is complementary with the body's natural defence agents.

By a chance observation, he discovered that a certain mould culture prevented the growth of staphylococci. He named the active substance 'penicillin' but was unable to isolate it in adequate quantities. A multi-skilled team at Oxford University, led by Florey and co-worker, Chain, improved the extraction and purification process making clinical trials possible. Because of its military importance, further upscaling took place in America. By D-Day, enough penicillin was available to treat troops suffering from bacterial infections.

It became known as the 'miracle or wonder drug'. It proved to be the most efficacious life-saving drug in the world, making possible the treatment of a wide range of previously untreatable bacterial infections, saving millions of lives. In 1945, Fleming, Florey and Chain jointly were awarded the Nobel Prize in Physiology or Medicine.

For infertile couples, we want to give the gift of parenthood. After a decade of collaborative studies into in vitro fertilization (IVF), co-workers Patrick Steptoe, Robert Edwards and Jean Purdy facilitated the child of the century, the first 'test tube baby', Louise Brown, who celebrated her 40th birthday in 2018.

Her artificial conception, outside the womb, was a momentous achievement, sometimes equated with other major firsts in medicine, such as the application of vaccinations (Jenner) and the discovery of penicillin (Fleming). It was associated with a moment of national pride, demonstrating the country's excellence in medical research and innovation.

Initially, Edwards and Steptoe faced a backlash from religious groups, the media and even parts of the scientific community. The two co-workers argued that assisted reproductive technology merely 'gives nature a nudge in the right direction'. Ultimately, they succeeded in bringing about a change in moral attitudes to IVF, after which millions of infertile couples across the world were able to harness IVF as their last chance at parenthood. Because Nobel prizes are not awarded posthumously, and Steptoe died in 1988, only Edwards received the Nobel award in Physiology or Medicine, in 2010.

9.2   William Harvey (1578–1657)—Physician, Anatomist, and Physiologist

During the reign of Queen Elizabeth, Folkestone's most famous son was born. William was the eldest sibling of seven brothers and two sisters. His father was a farmer who also had a prosperous carrying business between the Kent coast and London. His father became a jurat and Mayor of Folkestone.

At the age of ten, William attended the elite King's School in Canterbury, which was adjacent to an apothecary. In 1593, aged only fifteen, he won a scholarship to Gonville and Caius College, Cambridge where he read philosophy and had an interest in medicine. The College Master, John Caius, persuaded English doctors to make use of a law which allowed them to examine, each year, the bodies of four executed criminals.

In 1602, Harvey qualified as a doctor at Padua University, then the greatest medical centre in Europe. His anatomy tutor was the renown Hieronymus

Fabricius. In that same year, he returned to England and took a second MD from Cambridge University. In 1605, he was admitted as a Fellow of the London College of Physicians and, in 1607, he was appointed Head Physician to St. Bartholomew's Hospital—one of two institutes that treated the sick poor. He became recognised as the best physician in London and for this reason, in 1608, he became Court physician to James I and, later, his son, Charles I (1632).

In a witchcraft court case, Harvey prevented four women from being burnt as witches. He worked adroitly within the system to banish barber-surgeons, quack physicians, tooth-pullers and apothecaries selling dangerous medicines. However, it was not until 1745 that King George II officially separated the professions of 'barber' and 'surgeon'. The King established the London College of Surgeons for those who were university trained but a century would pass before the Medical Act 1858 regulated the qualifications of practitioners in medicine and surgery.

Harvey worked tirelessly in his home laboratory. Sometimes he carried out vivisection. This is how he was able to demonstrate that, when the heart contracts, blood is pushed round the body.

For centuries people considered the heart to be the source of vitality and innate heat; the seat of intelligence, the emotions and sensation; the organ most related to the soul. In the second century, the Greek, Galen of Pergamon—physician to emperors in Rome—put forward a treatise on the heart and blood circulation. Galen was the originator of the experimental method in medical investigation. He believed that blood was continuously generated and consumed at the periphery.

After fifteen years and hundreds of experiments, Harvey challenged Galenic physiology. The human body contains about five litres of blood. The body simply could not produce or consume that amount of blood. Therefore, the blood had to circulate, an announcement he made, in 1616, during his Lumleian lecture, at the Royal College of Physicians. In 1628, Harvey published his famous "*Anatomical Studies on the Motion of the Heart and Blood in Animals*" (*De motu cordis et sanguinis*). He established that blood circulates round the body and through the lungs where it is revitalised.

He cut parallel veins and arteries, showing how the blood from the cut ends flowed in different directions. He tied off veins and demonstrated how they became swollen away from the heart, and that when the ligature was removed, blood drained towards the heart.

He showed that flaps inside veins (venous flaps) are cardio-centrically oriented, allowing the free passage of blood in one direction but strongly inhibiting backflow of blood in the opposite direction. He determined that the heart pumped 0.5–1.0 L of blood per minute. (N.B. Modern values, for this fist-sized powerhouse, are about 5 l/min {=7200 l/day; 2.6 million l/year} when at rest, and 25 l/min during exercise, as both the heart rate and stroke volume increase).

The heart is a 4-chambered, hollow organ. The left side is separated from the right by a muscular wall, the septum. The right and left sides are further sub-divided into two top chambers (the atria) which receive blood from veins, and the two bottom chambers (ventricles) which pump blood into arteries.

Harvey showed that the heart is a pump, propelling a defined amount of blood, in a closed system, through two main circulation loops in the body. He concluded that, during the active phase of the heartbeat, when the muscles contract (systolic phase), the heart decreases its internal volume, and blood is expelled with considerable force from the heart. The systolic pressure (typical 120 mm Hg) is a measure of this force of expulsion. The diastolic pressure (typical 80 mm Hg) is the blood pressure between beats.

Looking at the heart in situ, (mirror image of the diagram below), all deoxygenated blood in the right ventricle goes, via the pulmonary artery, to the lungs where it is oxygenated. Oxygen-rich blood passes through the pulmonary vein to the left atrium. When the atrium contracts, blood flows into the left ventricle.

All the oxygenated blood in the left ventricle is sent, via the largest artery—the aorta—into other arteries, eventually reaching smaller veins and returning via two large veins to the right atrium and then right ventricle. The superior vena cava (or precava) conveys deoxygenated blood from the upper body; the inferior vena cava (or post cava) from body parts below the diaphragm. In this way, the circulation is complete, and the blood has returned to its starting point.

The left ventricle is the thickest, most muscular chamber since it pumps oxygenated blood to tissues all over the body. By contrast, the right ventricle pumps blood solely to the lungs (Fig. 9.1).

Harvey's theory of circulation was opposed by conservative physicians. In 1649, in response to criticism of his theory, by French anatomist, Jean Riolan, Harvey published his "*Two Anatomical Exercises on the Circulation of the Blood.*

During the English Civil War, Harvey remained quietly loyal to King Charles. On the King's defeat, Harvey retired to Oxford for three years. He became Warden of Merton College and, during this period, he worked on his treatise of embryology. As an example of oviparous reproduction (when embryonic development occurs within eggs hatched outside the mother's body) he studied the chicken reproductive cycle. For his studies on viviparous

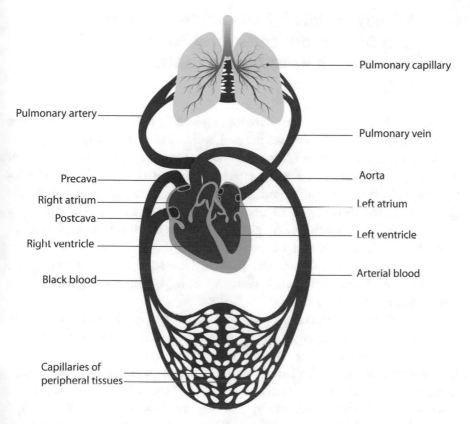

Fig. 9.1 Harvey's study on the motion of the heart and blood in animals

reproduction, in which embryonic development occurs within the mother's body, he chose deer.

Throughout his life, from his wide-ranging experiments and the evidence of what he could see and feel, he evolved his theories. In this regard, he is one of the first scientists in the medical field.

By Harvey's death in 1657, his theory of circulation had established a consensus amongst scholars. In 1674, by Harvey's legacy, the Free School for poor boys was established in Folkestone. It later became the Harvey Grammar School.

9.3   William Addis (1734–1808)—Inventor and Mass Producer of the 'Modern' Toothbrush Resulting in Better Dental Hygiene

About five thousand years ago, Babylonians and Egyptians used frayed twigs as natural, tooth cleaning devices for oral health. Later, herbal, chewing sticks made from the Arak tree (*Salvadora persica*) served several purposes. For instance, the frayed ends exerted mechanical control over plaque by friction between the plant fibres and the surface of gums and teeth; whilst the chemical compounds in the natural juices freshened the breath and had a therapeutic effect, particularly on gingivitis. These dry tooth sticks, known as miswak, are still used in many Islamic cultures.

The first bristle toothbrush was probably invented in China during the Tang Dynasty (seventh to tenth century) using bamboo or bone and the coarse hair of cold-climate hogs.

Until the Elizabethan era, people had surprisingly healthy teeth. Then, late in the fifteenth century, sugar started to appear in the western diet. Initially, it was rare and exotic, being a luxury item afforded by the rich.

Sugar became cheaper in Britain with the unacceptable use of slaves, taken as human cargo, from west Africa to the sugar cane fields in the Caribbean. Slavery is so abhorrent because it involves the brutal treatment of fellow human beings. Other commodities such as coffee, tea, cotton, and tobacco became affordable because of the cheap labour of Africans, brought against their will to the New World, or, by their descendants.

Sadly, slavery was found in nearly every ancient civilization, including the Babylonian, Egyptian, Persian, Greek, Chinese, Roman, Mughal, and Islamic empires. In the ninth century, when the invading Vikings raided coastal towns of the British Isles, the Norse warriors took captive, thousands of our men, women, and children and held them as slaves (called thralls in Old Norse).

In 1787, The Society for the Purpose of Effecting the Abolition of the Slave Trade was founded. It organised nationwide efforts to highlight the evils of the Atlantic slave trade. Evangelical Anglican, Thomas Clarkson, developed the protype for campaigns to influence public opinion. Easy to read pamphlets was one tool in his armoury. In 1791, there was a boycott of sugar produced by slaves. Hundreds of thousands of people signed petitions calling for the abolition of the slave trade.

Thanks, in part, to action by Christian bodies, particularly Quakers, and vocal campaigners such as Yorkshireman, William Wilberforce M.P., the

British slave trade ended with his Slave Trade Act of 1807. The reformers saw slavery as a betrayal of British values, since it conflicted with their view that the British Empire was based on liberty and free labour, different from previous empires.

The 1807 Act made it illegal for any British subject or ship to trade in enslaved people. For the next sixty years, the Foreign Office tried to persuade other nations to prohibit the trade, whilst the dominant Royal Navy enforced the British ban, by shutting down the slave routes. Targeted were European nations with colonies in the Americas (e.g., France, Portugal, and Spain) as well as those imperialists in Africa and the far east (e.g., Germany, Belgium, and Netherlands). Whilst the triangular trade between Europe, Africa and the Americas was curtailed, slaves in the British colonies were not given their freedom until the Slave Emancipation Act of 1833.

From the 1740s to the 1820s, sugar was Britain's most valuable import, mainly from the West Indies. Profits were so high that it was called 'white gold'. By 1750, there were 150 sugar refineries operating in Britain. And all levels of society became consumers. Eating habits changed and sweet victuals containing refined sugar were consumed in ever increasing quantities. It became a "tropical drug food" and high "octane fuel" for working people. In 1800, annual per capita consumption was eighteen pounds.

We now know that bacteria such as *Streptococcus mutants* and *Lactobacillus acidophilus* colonize the dental surface forming plaque biofilms. They reach a certain stage of maturity after about twelve hours. They convert easily fermentable sugars, such as sucrose, into a sticky, acidic film that dissolves minerals from tooth enamel, leaving microscopic holes. These grow into visible cavities and tooth decay. Prompted by the huge increase in the consumption of sugar in the eighteenth century, a more effective method for teeth-care was required.

William Addis was born in 1734 in the East End of London. He became a rag trader and stationer; pulping rags to make paper. In the early 1770s, he spent a spell in Newgate prison, allegedly for inciting a riot in London's East End district of Spitalfields.

Addis was a clever man who wanted to keep himself presentable. In the western world, twigs or rags were used to apply a gentle abrasive such as coal dust, soot, salt, or brick dust to clean teeth. Addis felt that a more effective tooth brushing tool was feasible and gained inspiration by closely examining a broom. Whilst in prison, Addis experimented with chicken and cattle bones, carved into the shape of a handle and flat head, through which holes were bored with a nail. Horse, badger, pig, or boar hair were tied in bundles and these were drawn through the holes and held in place by wire or glue.

William Addis was released from Newgate prison in 1780. He is credited with developing the prototype for the first mass-produced toothbrush in the western world and hence became the "father" of dental hygiene for the wider British public. Georgian Londoners had awful teeth and bad breath. Some died from gum infections which led to the dissemination of pathogenic organisms to distant body parts. The only option for a bad tooth was a barbaric extraction by a barber surgeon. Any device to avoid the agony of toothache was worthy of investigation.

Serviceable 'mouth brooms' (toothbrushes) were manufactured commercially from his premises in Whitechapel, East London. Ox and bullock thigh bones were bought from local butchers; boiled; cleaned up and cut lengthways to make strips. The strips were cut to one of four sizes: gents, ladies, child's, and Tom Thumb. The insertion of bundles of bristles was performed by women living in Spitalfields and Whitechapel, labouring on a piecework basis. Gender marketing offered tougher hogs' hair to men, and softer, badger hair to women and children.

Modern medical research has shown that brushing teeth properly can prevent cavities, gingivitis (gum inflammation) and periodontitis (gum disease). Failure to brush correctly and frequently can also lead to the calcification of salivary minerals, forming tartar. Brushing at least twice per day is currently recommended—the second one before bed. The first should not occur within one hour of eating, when acid levels are at their highest, so the abrasive action of the brush, in an acidic environment, could have a deleterious effect on tooth enamel. Since smaller amounts of protective saliva are produced at night than during the day, it is important to remove food remnants from teeth before bed so that bacteria cannot feast overnight and cause demineralisation.

Refinements to the toothbrush were made through successive generations of the Addis family. Food rationing during WWII, meant that bones were boiled down to make soups and stews, making them unavailable for toothbrush manufacture. Luckily for the Addis Company, injection moulding technology appeared. Mouldable plastics such as polyethylene or polypropylene are used. The mould forms the entire handle, including the small holes, called cores, into which the bristles are inserted.

In 1940, natural animal bristles were replaced by synthetic Nylon-6 fibres. Nylon bristles were considered a sanitary improvement over natural bristles, since they were less likely to provide a breeding ground for bacteria and fungus. This heralded the arrival of the modern, entirely synthetic toothbrush.

Hard plastic, however, can last hundreds of years, and discarded tooth-brushes are appearing in the oceans. To reduce their environmental impact, some manufacturers are switching back to sustainable, biodegradable materials (e.g., bamboo handles + animal bristles). Alternatively, completely recyclable materials can be used, or, the handle can be retained, and replaceable heads adopted.

The Addis family business adopted the name "Wisdom Toothbrushes". Then, in 1996, after ten generations and 215 years of Addis involvement, the family sold its interests in the company to a management buy-out team for about £15 m. The factory in Haverhill, Suffolk which used to manufacture about fifty million toothbrushes annually is now closed and Wisdom toothbrushes are made outside the UK.

It took an eccentric, resourceful man with conviction to develop the toothbrush for the masses. Ingenuity and innovation have shaped the toothbrush, making it a personal article and one of life's essential oral-care items. The basic design has changed little since Addis' first concept in the 1770s.

9.4 Edward Jenner FRS (1749–1823)—Immunologist; Pioneer of a 'Safe' Smallpox Vaccine

9.4.1 Background

Jenner was born in Berkeley, Gloucestershire. He was the fourth son and eighth of nine children born to the vicar of Berkeley, and his wife. In 1754, when Edward was only five years old, both his parents died. Edward came under the guardianship of his elder brother, Stephen, who organised his education, whilst his aunt provided the necessary home environment.

Edward started school at the age of eight. He was apprenticed to a country surgeon when he was fifteen, acquiring a sound knowledge of medical and surgical practice. In 1770, he moved to London to study medicine at St. George's Hospital. Here, he completed his medical training under the eminent surgeon and experimentalist, John Hunter.

He was also employed by Sir Joseph Banks, Endeavour's chief botanist, to arrange and prepare valuable zoological specimens which Banks had brought back from Captain Cook's first voyage, in 1771. This stimulated Edward's interest in zoology, particularly ornithology, and specifically the habits of the cuckoo. His growing knowledge of animal biology may have motivated his curiosity in animal-to-human disease transmission (zoonoses).

In 1773, he returned to Berkeley where he established himself as the local practitioner and surgeon.

In 1788, Jenner was elected Fellow of the Royal Society, not for his work on smallpox immunology, but for his studies on the behaviour of a hatched cuckoo in a nest.

After twenty years of general practice, together with the recommendation of two lifetime friends, and the payment of a fee, he was awarded the degree of Doctor of Medicine (MD) from the University of St. Andrews, in 1792. He became a consultant and moved to Cheltenham in 1795.

9.4.2   History of the Smallpox Vaccine

Thousands of years ago (ca. third century BCE), possibly in Egypt, the smallpox (variola) virus emerged and began causing illness and deaths in human populations. The name is derived from the Latin word 'varius' meaning "spotted" or varus, meaning "pimple". It refers to the raised bumps on the face and body of the patient. Jenner called it the "speckled monster". Virologists have speculated that it evolved from an African rodent poxvirus ten millennia ago, but it required a threshold population of humans before endemic smallpox could be established.

Smallpox is a highly contagious disease that has killed millions. There are two closely related strains of smallpox—Variola major and Variola minor, the former being fatal in about 30% of cases. Children were particularly susceptible. Survivors suffered blindness and disfigurement. In the eighteenth century, smallpox is estimated to have claimed sixty million deaths in Europe.

Overseas, one of the first methods for controlling the spread of smallpox was via the use of 'engrafting' or 'variolation'—a process by which material from a smallpox sore was given to people who had never had smallpox—in modern parlance, "elective infection". There is evidence that variolation may have been employed in China as early as 1000 CE, followed by Africa in the early 1700s, and the Americas in the 1720s. In Massachusetts, an African slave, Onesimus, made his owner—the Reverend Cotton Mather—aware that the practice of variolation conferred immunity from smallpox. Thereafter, slaves bearing the variolation scar were deemed to be more valuable.

Lady Mary Wortley Montagu, a friend of Caroline, Princess of Wales, was once a 'court beauty' but contracted smallpox in her 20s—leaving her with a scarred face and no eyebrows. In 1717, she was living in Turkey, with her son and husband who was the British Ambassador to the Ottoman Empire. Charles Maitland, an Aberdonian, was surgeon to the Ambassador.

At the behest of Mary, Maitland engrafted her son. This involved a subcutaneous instillation of live smallpox virus into the non-immune recipient. People inoculated in this way then developed milder symptoms of smallpox and fewer people died from variolation (0.5–2%) than those who acquired smallpox naturally (20–30%).

In 1721, after their return from Turkey, there was a smallpox epidemic in London and Maitland was invited to engraft Mary's daughter. She was the first individual to be engrafted in England. In the following year, Maitland engrafted six people in Aberdeenshire, but one died, making the procedure unpopular in the county. Since the technique relies on lymph from an infected person, if that person suffers from Variola major, recipients are likely to suffer a higher mortality. At first, doctors sneered at the technique, but it was endorsed by the Royal College of Physicians in 1755.

Catherine the Great, Empress of Russia, had also seen first-hand the ravages of smallpox. In 1768, it was again sweeping across Europe. Scientifically enlightened Catherine wanted to protect her subjects from it, so she invited Hertfordshire physician, Thomas Dimsdale to St. Petersburg. Dimsdale was another expert on variolation. To convince her subjects of its benefits, she unselfishly ordered him to inoculate, not just herself, but her son, as well as some members of her Court.

In England, it had been noted that, whilst dairymen and dairymaids working with cows easily got cowpox, they did not get smallpox.

In 1796, Jenner took pus from a blister on a milkmaid with cowpox and transferred it onto a cut made on the arm of the 8-year-old son of his gardener, whereupon the boy became infected with cowpox. Eight weeks later, Jenner exposed the boy to smallpox and found him to be immune.

Jenner introduced the term "virus". The Latin word for cow is *vacca* and the cowpox virus is *vaccinia*. Jenner decided to call the new method of protection, against smallpox, "*vaccination*". 'Live' vaccinia virus is a pox virus like smallpox but less virulent and, therefore, safer to use. Putting a small amount of disease-causing cowpox virus into a vaccinee stimulates the production of antibodies which are effective, not only against cowpox, but also against the closely related smallpox virus.

Between 1798 and 1800, he published three works explaining his medical thinking on the vaccine. The first of these evidence-based manuscripts was the "*Inquiry into the Causes and Effects of the Variola Vaccinae*". Ultimately, it was translated into Latin, German, French, Italian, Dutch and Spanish.

Jenner found that lymph taken from cowpox pustules could be dried, kept active for up to three months and sent long distances. Having read about his discovery, doctors throughout the world became interested in his innovative technique and Jenner sent some of them samples of his vaccine so they could inoculate locally. Unfortunately, immunisation was in its infancy and naturally produced cowpox material varied in quality. Clumsy procedures could lead to secondary infections and it was not known that vaccination failed to give life-long immunity.

From 1840, a succession of laws made smallpox vaccinations free. Then, in 1853, thirty years after the death of Jenner, smallpox vaccination became mandatory, in Britain, for infants up to three months old. This was the first governmental intervention into people's health.

In 1850, average life expectancy was slightly above 40 years, whilst 15% of children died before their first birthday. In 1867, vaccination was made mandatory for all children under fourteen years of age. After the smallpox epidemic of 1871, there was a rush for vaccination.

Sadly, the benefits of vaccination were not universally recognised. For an increasingly literate population, pamphlets were produced with a title such as, "Vaccination; its fallacies and evils". The Vaccination Act 1853 met with immediate resistance from those who demanded the right to control what entered their bodies and that of their children. They were those who were offended by the idea that "harmful matter, derived from a cow, should be injected into the human body". Some people were concerned that their children would adopt 'bovine' characteristics. It was deemed to be unhealthy and unchristian.

Part of the population became ever more distrustful of doctors and government. A sceptical public became aggressive leading to riots in some towns. Leicester became the centre of non-compliance. In 1869, the Leicester Anti-Vaccination League was formed. In 1877, the city's medical examiner made it compulsory to report cases of smallpox. To control the spread of disease, patients were isolated, families quarantined, and belongings disinfected or burned. In Leicester, in 1884, prosecutions for non-vaccination grew to about 3,000. These court cases provided incendiary publicity for the anti-vax movement, and a mass protest followed in 1885. In 1892, only 3% of the children of Leicester were inoculated. Since the vaccination acts were not working universally, there followed a softening of legislation. In 1898, a new Vaccination Act allowed conscientious objectors to opt out on moral grounds.

Despite Jenner's best efforts, 300 million people died globally from this virulent disease in the twentieth century. More sophisticated smallpox vaccines have been developed. Coupled with an international vaccination programme, one of the most feared diseases—smallpox—prevalent from ancient times, has been eradicated worldwide. Currently, the variola virus has no known animal host, so elimination of the disease meant total elimination from humans.

Jenner may not have been the first physician to experiment with vaccination. However, his work represented the first scientific attempt to control an infectious disease by the deliberate use of vaccination. His relentless promotion and dedicated research on the subject changed the way medicine was practised in the eighteenth century. Later, Jenner was hailed as the father of immunology and it is said that his work saved more lives than the work of any other human being.

Jenner lived most of his adult life in a house in Church Lane, Berkeley GL13 9BN, now a museum. Dr. Jenner's House and garden can be visited between April and October.

9.4.3 Vaccine Developments

Vaccination is one of the great success stories of public health. In the nineteenth and twentieth centuries, following Jenner's model, new vaccines were developed to fight a variety of diseases caused by viruses and bacteria. For instance, in the UK, the 6-in-1 vaccine protects against diphtheria, hepatitis B, haemophilus influenzae type B, poliomyelitis, tetanus and whooping cough (pertussis). The MMR vaccine protects against measles, mumps, and rubella. World-wide vaccination now averts an estimated 2–3 million deaths every year. It is noteworthy that the Jenner Institute has been involved with the development of one of several COVID-19 vaccines.

Extremely high immunisation coverage can lead to the complete blocking of transmission of many vaccine-preventable-diseases (VPDs). Early immunisation of infants and completion of the full schedule of vaccinations into, and through, adulthood contributes to reducing the incidence of VPDs, and minimises the social and economic burdens of those diseases on communities. Until the virus SARS-CoV-2 appeared, Jenner's legacy was a world where we were not unduly fearful of horrific infectious diseases. That is why the development of a reliable vaccine to protect us against the COVID-19 disease has been so critical.

Some members of society are complacent about vaccinations, others are mistrusting of vaccination, even to the point of hostile opposition. Some argue that the government is imposing too many restrictions on their civil liberties. The term "vaccine hesitancy" is used to describe a range of sentiments. The hardcore and vocal antivax minority spread fear, using pseudoscience, particularly on social media. For instance, there has been an "infodemic" of false and misleading information about Covid-19.

Inevitably, the safety bar for vaccines is higher than for therapeutic drugs because the latter are given to ill patients, whilst the former are given to healthy people, often children. Side effects are common and can be wide ranging but are normally mild, so risks are miniscule. Some are serious, so it is always a societal, ethical, and political decision about the trade-off between benefit and risk. The science of pharmacovigilance is in place to monitor any adverse effects, and to protect the credibility of immunisation systems—it is vital that we do not jeopardise the public's faith in tried-and-tested vaccines.

9.4.4 The Fight Against Coronavirus Disease 2019. SARS-CoV-2 Vaccines

The vast networks of human cells and tissues are constantly on the lookout for invaders. Once an enemy is spotted, a complex, multi-pronged attack is mounted. The goal of vaccination is to reduce or eliminate infectious diseases that can be fatal or have long-lasting effects. In different ways, vaccines mimic the real infection, developing immunity in us, without causing disease. In 2019, citizens of the World lacked any immunity to SARS Coronavirus-2, the causative agent / pathogen of the disease COVID-19. Since the virus knows no borders, it spread very quickly across the globe, temporarily shutting down entire societies. It has killed more people in one year than any other infectious disease has done for over a century.

Viruses are not alive and do not grow. They cannot reproduce themselves unaided. When not inside a host cell, they are inert. Severe acute respiratory syndrome (SARS) coronavirus-2 (SARS-CoV-2) is an RNA virus which has its genetic blueprint encoded in ribonucleic acid (RNA) and not DNA. This viral RNA is sneaky—it has crown-shaped glycoprotein spikes on its surface that it uses to achieve entry into human cells. It then causes the host's machinery for protein synthesis to mistake it for RNA produced by its own DNA.

As a parasite, the virus hijacks our cells, turning them into mini factories to make new copies of the virus which then spread to other parts of the body. In some cases, the infected person exhibits few symptoms and feels reasonably well, whilst in others, the person gets extremely sick, leading to hospitalisation, intensive care, and possible death. The presence of asymptomatic carriers / transmitters made the task of contact tracing and quarantining exceptionally difficult. Long COVID is a secondary disease.

By way of liquid droplets, aerosols, perspiration and other body fluids, the virus is spread from one person to the next. The most important vector is virus-laden particles in the air, especially in poorly ventilated homes, transport situations, work environments and entertainment venues. Infection occurs through the mucous membranes of the nose and upper airways. Transmission was so effective in 2019-21 that a pandemic resulted, exacting an unbearable loss of lives.

At breakneck speed, researchers have tried five main ways of turning the corona virus into a vaccine, to trick the body into thinking it has an infection. The five routes chosen were (i) chemically inactivated virus with spike protein intact (Chinese CDC / Sinopharm / Sinovac); (ii) the spike protein itself with a molecular clamp to stop it unfolding, combined with an adjuvant to improve the immune response (Queensland University); (iii) the spike gene contained in a common cold virus - crippled so it can't replicate (Oxford University / AstraZeneca; Covishield; Johnson & Johnson / Janssen; CanSino and Sputnik V); (iv) the spike gene delivered as messenger RNA in a protective coating of fat (US National Institute of Health / Moderna; Pfizer / BioNTech) and (v) the spike protein gene cloned into a baculovirus that infects insects, facilitating a synthetically-produced protein that is harvested, coupled with adjuvant and assembled into nanoparticles (Novavax).

A conventional vaccine usually contains weakened or inactivated disease-causing organisms or proteins made by the pathogen. When introduced into the human body, it presents antibody generators (*abbrev* antigens) which are substances that mimic the infectious agent—sparking an immune antibody response. The body is hereby primed to respond more rapidly and effectively when exposed to a threatening infectious agent in the future. As a result of vaccination, neutralising antibodies are readily produced that block the pathogen's attempts to invade cells.

AstraZeneca (AZ) adopted the so-called adenoviral-vector approach for combatting COVID-19 disease. The AZ vaccine, AZD1222, was developed at the Jenner Institute and Oxford Vaccination Group, at the University of Oxford. This is where Florey and Chain isolated the penicillin antibiotic (see Sect. 9.7) in 1940. The antigen's operational instructions are packaged for delivery in an adenovirus that is known to cause colds in chimpanzees and other simians, rather than humans. This cold virus itself has first had a specific gene removed in a way that renders it replication-deficient in humans. The gene sequence for the vaccine antigen (the spike protein), genetically manipulated for a stronger immune response in humans, is then inserted in its place, making it a "recombinant" virus.

Optimising the microenvironment to preserve a vaccine's potency, producing sufficient seed-stock, and then mass producing it in an ultra-pure form, for clinical trials and ultimately global immunisation, normally is a slow process. In the case of AZD1222, by performing these tasks in parallel, rather than sequentially, the lab-to-jab time was reduced to one year. The AZ virus can be shipped and stored at normal refrigeration (4 °C) and is relatively cheap to buy since it has been manufactured on a not-for-profit basis. However, it is responsible for an extremely rare platelet clotting reaction, perhaps caused by an auto-immune misfiring response. Despite its vaccine having numerous advantages, AstraZeneca has become a household name for the wrong reasons. It became a political football.

The traditional vaccine approaches to protection against diseases such as smallpox and measles etc, may not have been as effective against rapidly evolving human pathogens such as influenza, or emerging disease threats, like Ebola or Zika viruses. Some believe that new and novel RNA vaccines may have a greater impact in these areas due to their enhanced adaptability, flexibility and short development timelines.

Messenger RNA (see Sect. 13.3.5) vaccines (mRNA) adopt an entirely different approach from the viral vector route. Instead of smuggling the spike protein gene into the human body via another virus, mRNA is encased in fat. Since such vaccines do not require a live or inactivated virus as carrier, they are non-infectious. The Pfizer-BioNTech and Moderna vaccines are synthetically constructed and contain a segment of the pathogen's specific genetic code. It is then 'bubble-wrapped' in a synthetic fat-coated nanoparticle to protect it from disintegration and demolition when it enters the human body. Since mRNA is chemically unstable, it must be stored between −60 and −80 °C before preparation for vaccination.

The mRNA is made from a genetic template, in the laboratory, taking about one week to generate an experimental batch of mRNA vaccine. Unlike the conventional approach, the mRNA technique does not introduce an antigen payload to the body. Rather it carries a copy of the recipe for making the antigen *in vivo*. This triggers production of the viral spike protein which is critical for entry into human cells. Since this is foreign to our bodies, the human cells then respond by making antibodies to neutralise the protein.

When the mRNA is injected into the arm muscle, the invasive mRNA is captured by guarding sentinels called dendritic cells. Protected by a fat-coating, the mRNA flows through the outer membrane of the dendritic cells to the cytoplasm. The strands of RNA seek ribosomes and perch on them. The ribosomes decipher (translate) the coded instructions and churn out new SAR-CoV-2 spike protein. Thus, the spike protein is made, just as it would, had the host vaccinee been infected with the actual virus. Thankfully, the mRNA does not enter the cell's nucleus where the host's hereditary information is stored, so it cannot integrate itself into the host's DNA.

The dendritic cells leave the site of injection and head for lymph nodes in the armpits. Here, they orchestrate an immune response. They alert and 'educate' an assemblage of T-cells and B-cells to launch a coordinated, two-pronged attack on the newly formed antigens.

B-lymphocytes spot the antigen and begin to secrete highly specific antibodies that lock onto the antigen, neutralising it and marking it for death by roaming cells such as killer T-cells. After 7 to 21 days, the B-cells move to the bone marrow where antibodies are released into the blood stream. From there, they enter the nose, throat, lungs and other tissues. Memory T-cells also seed the tissues, colonising our bodies, ready to disarm the actual virus.

As a result, the body is primed to respond to the infectious agent, were it to appear in the future. Overall, it is an exercise in training the body to recognise the spike protein that surrounds the outer surface of the coronavirus. An immunological memory is laid down, so the immune system knows how to swiftly defeat the pathogen, if it encounters the coronavirus again, thereby preventing serious infection.

This gives the body a head start in combatting this specific viral infection and mitigates the symptoms. Both traditional and novel vaccines are proving to be most effective in the battle against COVID-19. Acknowledging that the original SARS-CoV-2 virus is already producing more transmissible and / or infectious mutants, the next step is to produce a mRNA against that part of the virus which is least likely to undergo structural or functional mutations.

Nasal sprays are also being developed because they are easier to deliver and bring the antigens closer to the cervical lymph nodes in the neck—glands closest to the nasal tissues containing immune cells for fighting infection. This gives more immediate and localised protection precisely at the site where Covid infection normally occurs.

The novel mRNA technology has the potential to produce mRNA against any target. For instance, it may be possible to produce a universal influenza virus that does not have to be modified year-on-year. Other areas of interest are vaccines against viruses that cause cancers, as well as parasites responsible for malaria. Work on the use of RNA vaccines to treat non-infectious conditions such as allergies is at an early stage.

9.5 Florence Nightingale (1820–1910)—Applied Statistician; Founder of Modern Nursing; Champion of Strict Sanitary Practices

Florence's father—a rich banker—and his new wife moved to Italy after their marriage in 1818. Florence was named after the city of her birth in 1820.

The family moved back to England in 1821 and Florence was raised in one of two family homes, in Embley Park house, Wellow, Hampshire and Lea Hurst—a summer retreat in Derbyshire. She was brought up by governesses and her father taught her history, mathematics, Italian, classical literature, and philosophy. She displayed an extraordinary ability for collecting and analysing data which she would use to great effect later in life.

She grew up mixing with rich children. Social convention at the time decreed that a daughter of an upper middle-class banker should develop social skills which often meant following frivolous pursuits and domestic routines. Such women did not attend university and did not pursue a professional career, particularly not in nursing, deemed not to be 'respectable'. She was expected to marry 'well' but rejected several proposals of marriage.

Florence sought greater meaning to her life. In a way, like Isaac Newton (see Sect. 6.3.3) Florence's epiphany occurred under a tree—a giant cedar tree in the grounds of the 4,000-acre family estate of Embley Park. Local legend says that, at the age of sixteen, when seated beneath the tree, she heard a calling from God asking her to do important work.

In her twenties, she began making home visits to help poor people in the villages around her home near Romsey, Hampshire. She did volunteer work in hospitals. However, her parents continued to reign in her independent spirit and thwarted her attempts to train as a nurse at Salisbury Hospital. A diary entry in 1849 said she wanted three needs to be fulfilled: intellectual, passionate, and moral.

Convinced that God wanted her to become a nurse, she received some training at the Deaconess Institute for the poor and sick in Kaiserwerth-am-Rhein, Germany, publishing her observations anonymously in 1851. In 1853, when she was thirty-three, she took an unpaid job as Superintendent at a small private hospital in Harley Street, London for invalid gentlewomen, where she quickly made dramatic improvements to hospital procedures.

In 1854, fearful of Russian expansion into the Danube region, Great Britain became involved in a war between the Ottoman Empire and the Russian Empire on the Crimean Peninsula which projects into the Black Sea. There were very heavy casualties. Sidney Herbert, who was Minister of War, and friend of Florence, asked her to take a group of nurses she had carefully selected, to the British Army Hospital at Scutari, near Constantinople, to nurse wounded soldiers.

The conditions she found there were appalling. The hospital was a focus of disease. Wounded men often arrived with contagious diseases like typhus, cholera, and dysentery. It is said that more men died from these diseases than from their injuries in battle. Using her administrative skills, she organised vital supplies and made improvements to bedding and general cleanliness. She faced resistance from military officers who did not approve of her activities.

In March 1855, a visiting Sanitary Commission headed by J G Jennings (see Sect. 5.5) ordered the removal of open toilets, the flushing out of the sewers and improvements to ventilation and fresh-water provision, all of which helped to reduce the mortality rate.

At night, carrying a lamp, she tendered to the wounded and dying. She became immensely popular with the soldiers and they described her as a 'ministering angel'. She was immortalized as the "Lady with the Lamp".

She left what she originally called "The Kingdom of Hell" and returned to England in 1856 as a national heroine. Sadly, she returned a chronic invalid, thought to be due to a bacterial infection, brucellosis, picked up in Crimea. From 1857, much of her analysis and campaigning was done from a couch or bed.

She had a royal audience with Queen Victoria, with whom she discussed how military hospitals could be improved. Whilst in Crimea, she had kept meticulous records about the deaths of soldiers and their causes. Being very gifted with statistical analysis and graphical presentation (e.g., pie charts and polar chart/circular histograms/coxcombs) she was able to use novel method of communicating to the Royal Commission on the Health of the Army (1858), showing that most of the men had died from preventable disease, poor nutrition and/or unsatisfactory hospital conditions. Her 'Rose Diagram' entitled "*Diagram of the Causes of Mortality in the Army of the East*" was so remarkably persuasive that the British Establishment had to take notice.

Her choice of this specific graphical representation was no coincidence—it reinforced her argument that conditions at Scutari before the visit of the Sanitation Commission were dire but improved radically afterwards. One of her major achievements resulting from her Crimean experience was that the field of public health was catapulted to national attention. Ultimately, ameliorative measures were taken which significantly reduced unnecessary deaths during peacetime, as well as wartime, benefitting the whole of society.

She began her next campaign by turning her attention to nursing and sanitary concepts in hospitals. In 1860, she set up a training school for nurses as part of St. Thomas' infirmary in London, introducing new training standards and founded on classless professionalism. It established St. Thomas' as the home of modern nursing. Nurses were to be clean, medically trained and well disciplined. She wrote "*Notes on Nurses*" and "*Notes on Hospitals*".

She became an enthusiastic supporter for the design of healthful buildings for the care of the sick—a place for healing, cure, and recovery. She endorsed the so-called 'pavilion-hospitals'. The guiding principle was that hospitals should do the patient no harm. This was to be achieved via optimal bed layout; removal of vitiated air and its replacement with fresh air; sunlight; sanitary conditions and caring nursing. The "HALO" principle was adopted—incorporating hygiene, air, light and order. In 1869, the iconic General Infirmary of Leeds adopted this principle. It was a masterpiece of design; a state-of-the-art infirmary and became the gold standard across the world.

Her strictures on hospital care were part of a much broader attempt to formulate a policy on public health through the adoption of better sanitation. She lobbied Parliament to legislate for compulsory sanitation in private houses. She worked with reformer, Edwin Chadwick (see Sect. 5.7.2). Chadwick had adopted John Wesley's dictum: "Cleanliness is next to Godliness".

Chadwick was largely instrumental in the Public Health Acts 1848 and 1875, the latter consolidating and amending previous acts. They required enforcement powers to be devolved to local authorities. These authorities had to provide clean water; dispose of all sewage and refuse and ensure that only safe food was sold. It gave them powers to ensure that existing homes were connected to the main sewerage system, whilst the Act forbad the building of new homes without such connection. It meant that minimum standards of sanitation had to be included in every new house built.

Exploiting her social networks and, having the attention of the Cabinet and Queen Victoria, Florence helped to bring about a seismic shift in the UK's sanitation and public health programmes. These reforms set the framework for the next fifty years of public health. Some historians believe that the provision of clean water, drainage and devolved enforcement played a crucial role in increasing life expectancy at birth by twenty years between the mid-1870s and mid-1930s.

Florence Nightingale believed in compassion; commitment to patient care; diligent and thoughtful hospital administration. She had adopted a holistic approach to health. She was awarded the Royal Red Cross and was the first woman to be made a member of the Royal Statistical Society, in 1858. She was also the first female to receive the Order of Merit, in 1907. Her legacy is the fine, noble, nursing profession we all rely on today. The Florence Nightingale Museum can be found in the grounds of St. Thomas' Hospital, London (SE1 7EW).

She disliked unnecessary attention. Before her death, she had been offered, but declined, to be buried in Westminster Abbey, alongside some of Britain's greatest subjects. She never had any wish to be a celebrity. Although a highly influential figure, she chose to be buried next to her parents, in the graveyard of St. Margaret's church, near her childhood home of Embley Park.

9.6   Joseph Lister (1827–1912)—Surgeon, Scientist, and Pioneer of Antiseptic Surgery

Lister was born in Upton Park, now the home of West Ham Football Club. His natural curiosity was encouraged by his prosperous father—Joseph Jackson Lister—who was a Quaker, wine merchant and amateur scientist. Lister junior used microscopes assembled by his father to examine small specimens. He dissected small creatures and knew from an early age that he wanted to be a surgeon.

The only university to accept Quakers was University College, London. In 1844, he began by studying botany and then went on to register for a medical degree. After graduating in 1852, he worked at University College as a house surgeon. Because of his exceptional performance, he entered the Royal College of Surgeons when he was 26. He moved to Edinburgh, Scotland and was eventually appointed Regius Professor of Surgery, at the University of Glasgow, in 1860.

In 1861, he observed that 45 to 50% of amputation patients died from sepsis. He was concerned that some patients underwent a procedure successfully, only to die from a post-operative infection known as 'ward fever'. Many surgeons believed that bad air ('miasma') emanating from a wound was the cause of further infection.

However, Lister had observed that some wounds healed when they were cleaned, and damaged tissue removed. He had also noted that patients with simple fractures had better recovery rates than those with compound fractures. He concluded that infection was coming externally as problems occurred mainly to those with open wounds where the protective skin barrier was damaged. As the son of a wine merchant, he knew that wine went sour because the fermentation process was not done properly.

Many scientists believed that simple life-forms were spontaneously generated. In 1860, the French Academy offered a prize of Fr2,500 to anyone providing convincing experimental proof for, or against, spontaneous generation of life. By 1862, Louis Pasteur—a French chemist and microbiologist—showed that no microbes ever grew in a nutrient broth that had been heat sterilized, providing the air above the broth was also sterilized. However, if unsterilized air was introduced, then microbes began growing. Extending his thinking to human disease and infection, Pasteur concluded that diseases may be caused by microscopic organisms suspended in air. He recommended exposure to heat or chemicals to destroy germs.

Traditionally, some surgeons wore dirty aprons; did not necessarily wash their hands before operations and used surgical instruments that were often unclean. Recovering patients might be placed on bed linen that was stained with someone else's blood and other bodily fluids.

In 1864, Lister learned that staff at the Carlisle sewage works had used carbolic acid—a derivative of coal tar—to render odourless, land that had been irrigated with sewage waste. Furthermore, the treatment had destroyed entozoa (internal parasites) that had infected cattle grazing there. Moreover, the livestock showed no ill-effects.

In 1857, Dr. F. Crace-Calvert is credited with inventing the first commercially practicable method for producing carbolic acid (phenol, C_6H_5OH). From his factory in Manchester, he promoted the sanitary and medical uses of carbolic acid. Lister was able to obtain a crystalline sample of carbolic acid from Crace-Calvert.

In 1865, Lister, inspired by some of Louis Pasteur's research, reasoned that the way to stop post-operative infections was to prevent germs entering the wound. His first trial was with 11-years-old James Greenlee who had an open tibia fracture. Lister wrapped the wound with a dressing dipped in 5% carbolic acid solution. For James, this action changed routine amputation to uneventful healing and healthy discharge. Antiseptic surgery was born.

Lister next experimented with the washing of hands, followed by surgical instruments. He also sprayed the acid in operating rooms. As a result, he reduced mortality rates from 45–50% to 15% over a 4-year period of study.

He had proven that micro-organisms were transferred through contaminated air, hands, clothing, and / or surgical instruments. The carbolic acid appeared to be acting as an antiseptic. His findings were published in the Lancet in 1867.

His introduction of antiseptic procedures dramatically decreased the death rate from childbirth and surgery. By 1879, his approach had gained widespread acceptance around the world. Listerian principles of surgery were adopted by surgeons in many other countries. He paved the way for more advanced, but safer medical procedures, and more effective antiseptics. In 1879, Listerine mouthwash was named after him for his work on antisepsis.

He was Queen Victoria's surgeon for many years and, having successfully treated her for an abscess in her armpit, he was made Baron Lister of Lyme Regis in 1897—the first person to be so honoured for services to medicine. (N.B. Lyme Regis is a small town in Dorset where, many years before, he and his brother had purchased a seaside house). He became President of the Royal Society and, in 1902, one of the first holders of the Order of Merit.

His principle that bacteria must never gain entry to an open wound remains the basis of surgery today. Lister may be known as the 'father of antiseptic surgery' but, by stimulating a range of new principles of cleanliness, in the operating theatre, he is responsible for an incalculable number of lives being saved through evolving approaches to sterile surgery.

9.7 Alexander Fleming FRS, FRSE, FRCS (1881–1955)—Bacteriologist; Discoverer of the 'Miracle Drug', Penicillin

Alexander was the son of a farmer born at Lockfield Farm, Darvel, Ayrshire in Scotland. Fleming's father remarried when he was fifty-nine years of age, having four surviving children from his first marriage. He and his second wife had four more children, Alexander being the third. His father died when Alexander was seven. He attended local primary schools. At the age of eleven, Alexander's academic potential was recognised, and he was awarded a scholarship to board at Kilmarnock Academy.

At the age of thirteen, he moved to London to live with his brother, a physician. He completed his secondary education at Regent Street Polytechnic where he studied business and commerce. With this background, he joined a shipping company and spent four years as a shipping clerk. In 1901, he decided that he wanted to become a doctor, like his brother, so he enrolled at St. Mary's Medical School, London University. His studies were funded by a scholarship, together with a legacy from his uncle. He qualified with distinction in 1906.

At first, he planned to become a surgeon but a temporary position in the laboratory of the Inoculation Department, at St. Mary's, convinced him that his future lay in the new field of bacteriology. Therefore, he retrained and graduated, in 1908, with a degree in bacteriology.

During World War I, Fleming had a commission in the Royal Army Medical Corp and worked as a practising bacteriologist, studying wound infections. He demonstrated that the direct use of strong antiseptics (e.g., carbolic acid, boric acid, and hydrogen peroxide) on deep wounds did more harm than good. Fleming's findings suggested that antiseptics were only useful in treating superficial wound. In the theatre of war, he showed that an irrigation by saline solution was often better for treating deep wounds.

In 1919, fifty-two years after Lister's publication in the Lancet (see Sect. 9.6), Fleming stated that "Antiseptics will only exercise a beneficial effect in a septic wound, if they possess the property of stimulating or conserving the natural defensive mechanisms of the body against infection".

From then onwards, Fleming was on the lookout for more 'benign killers' of bacteria. In 1922, when he was forty-one years of age, he took secretions from inside of the nose of a patient suffering from a head cold. He cultured the secretion and, in the secretion, discovered a new bacterium he called *Micrococcus lysodeikicus,* now called *M luteus.*

At the time, Fleming also had a cold and a drop of his nasal mucous fell into the culture plate of this secretion he was examining. The bacteria in the area where his mucal drop had fallen were killed almost instantly. The substance responsible was lysozyme, an enzyme present in body fluids such as saliva, mucous and tears that has a mild antiseptic effect on non-harmful bacteria. It destroys bacteria by breaking down cell walls (lysis). This was a significant contribution to the understanding of how the body fights infection. Unfortunately, lysozyme has no effect on most pathogenic bacteria.

In 1928, an uncovered Petri dish, sitting near an open window, had become contaminated with mould spores. When Fleming returned from a summer holiday, and, whilst tidying up his laboratory, a green mould was found growing in the Petri dish in which Fleming had been cultivating *Staphylococcus aureus. S. aureus* causes skin infections, sometimes pneumonia, endocarditis, and osteomyelitis. The unannounced mould had created a bacteria-free circle around itself.

Fleming experimented further and found that the culture fluid prevented the growth of staphylococci, even when diluted 1000 times. It is noteworthy that Lister's preferred antiseptic, phenol, loses its inhibitory power when it is diluted more than 300 times.

Fleming named the active substance 'penicillin' after the mould *Penicillium notatum.* About 200 species of Penicillium have been described and they exist as blue or green moulds. *P. notatum* is mostly found indoors and especially in damp buildings.

Between 1929 and 1930, Dr. C.G. Paine used a crude filtrate produced by Fleming's mould to treat eye infections at the Sheffield Royal Infirmary.

Unfortunately, Fleming was unable to isolate penicillin in sufficient quantities to make clinical trials on humans possible. It would take over ten years for significant progress in extraction procedures to be made. Luckily, in 1940, a multi-skilled team of scientists at Oxford University—led by Howard Florey and his co-worker, Ernest Chain, were able to optimise the composition of the broth in which the mould was grown and then to improve the extraction and purification process.

Penicillin was found to be non-toxic to mammals. Unlike antiseptics, such as phenol, it did not destroy white blood cells (leukocytes) involved with protecting the body against infectious diseases. The first human trials with a limited amount of penicillin took place at the Ratcliffe Infirmary, Oxford.

With England at war, further research shifted to USA in 1941, especially because penicillin was deemed to have military importance. A new species of Penicillium—*P. crysogeum*—and its mutants produce much larger amounts of penicillin. By upscaling the process, using aerated steel fermenters, greatly improved manufacturing yields were achieved. By D-Day, in 1944, there was sufficient penicillin available to treat bacterial infections suffered by the troops. It became known as the 'miracle or wonder drug'.

Penicillin can be taken orally or via injection. It kills bacteria by interfering with their ability to synthesize cell walls. It is highly effective against Gram-positive bacteria which are responsible for diseases such as scarlet fever, pneumonia, tuberculosis, meningitis, and diphtheria.

It is less suitable for Gram-negative bacteria which have a protective layer, preventing penicillin from attacking. Gram-negative bacteria are responsible for food-borne disease, urinary tract, respiratory, sexually transmitted, and blood stream infections. Penicillin does not work against colds, influenza, or other viral infections.

At the end of his Nobel lecture, in 1945, Fleming said, "The ignorant man may easily under-dose himself and, by exposing the microbes to non-lethal quantities of the drug, make them resistant" Indeed, antibiotic resistance and 'superbugs' are real issues, particularly with 'over-prescribing' and especially when a high proportion of antibiotics are given to healthy animal livestock to improve their yields but creating yet more opportunities for bacteria to evolve resistance.

Fleming was elected Professor at St. Mary's Medical School, in 1928; Fellow of the Royal Society in 1943 and Emeritus Professor of Bacteriology at University of London, in 1948. He was made Knight Bachelor in 1944. In 1945, Fleming, Florey and Chain shared the Nobel Prize in Physiology or Medicine for the discovery of penicillin and its curative effect in various infectious diseases.

Over the years, Fleming received many other awards. He was recipient of over thirty honorary degrees and was an honorary member of many of the medical and scientific societies across the world. He became an ambassador for science. The Alexander Fleming Museum can be visited at St. Mary's Hospital, W2 1NY.

Over the years, the use of the antibiotic agent, penicillin, has saved millions of lives.

9.8   Patrick Christopher Steptoe CBE, FRS (1913–1988)—Obstetrician, Gynaecologist and Assisted Reproductive Technologist; 'Joint Father' of First Test-Tube Baby

Patrick was the seventh of a family of ten children. His father was Registrar of Births, Deaths and Marriages in Witney, Oxfordshire. His mother worked for the Mothers' Union and Infant Welfare Clinic and was an advocate for women's rights.

He was educated at the Grammar School, Witney (now the Henry Box Comprehensive School). He entered King's College, London and qualified as a doctor from St. George's Hospital Medical School, University of London, in 1939, with degrees of MRCS and LRCP.

He was already a member of the Royal Navy Volunteer Reserve and joined the Royal Navy at the outbreak of WW2, serving as a surgeon. He rose to the rank of Lieutenant Commander. Whilst serving in the Mediterranean, his ship—HMS Hereward—was sunk during the battle of Crete, in 1941. He was captured and remained a prisoner of war, in Italy, until 1943, when he was released as part of a prisoner exchange.

After the war, he became a specialist in obstetrics and gynaecology, with a special interest in infertility. In 1951, he became a Consultant at Oldham General Hospital and it was here that he pursued his interests in laparoscopy and ovulation. Steptoe discovered that it was possible to harvest (aspirate) ripe human eggs (oocytes) from ovarian follicles under direct vision, using a laparoscope—a long, thin telescopic instrument inserted into the abdomen through a small incision.

In parallel, Robert Edwards (see Sect. 9.9) developed a technique for fertilising human eggs in the laboratory. His human culture media facilitated the fertilization of eggs outside the living 'organism', inside a glass culture dish. However, Edwards had difficulty in finding an adequate supply of human oocytes. In 1967–8, Edwards contacted Steptoe to express a possibility of collaboration.

Their pioneering work with infertile couples began, firstly at the Oldham District and General Hospital. A few years later, they moved the work to a cottage hospital, in Royton, near Oldham. The hospital had been built with a legacy from Dr. John Kershaw, a local GP.

Oldham is a former milling and mining town that, at its peak in the nineteenth century, produced more spun cotton than France and Germany combined. In the 1960s, it was a place of faded industrial glory. It seems surprising that ground-breaking medical research, to accomplish fertilisation outside the womb, should take place here.

For about a decade, Steptoe and Edwards worked to unravel the complex secrets of the infinitely complex and delicate process that governs the first steps of life.

Gradually, they overcame one set back after another. They and other staff at Kershaw's Cottage Hospital invested thousands of unpaid, investigatory, laborious hours to prefect the revolutionary treatment. By combining their two skills, Steptoe and Edwards were able to produce mature eggs, at the optimum time, to improve the chances of successful fertilisation and embryo development. Edwards later wrote that their collaboration was "a perfect match".

Edwards and Steptoe, together with their nurse-technician, Jean Marian Purdy, carried out a pioneering artificial conception which resulted in the birth of the first baby to be conceived by in vitro fertilisation (IVF).

Steptoe was one of the 'founding fathers' of the British Fertility Society, in 1974, serving, first as its Chairman, and then President, until his death in 1988. Along with Edwards, Steptoe was awarded the honour of CBE in 1988. Before his death from prostate cancer, he was elected Fellow of the Royal Society, an honour afforded to few clinicians.

9.9 Robert Geoffrey Edwards FRS, CBE, MAE (1925–2013) Physiologist and Assisted Reproductive Technologist; 'Joint Father' of First Test-Tube Baby

Robert was born into a working-class family in Batley, Yorkshire. His father was frequently away from home, working on the railways, maintaining the track in the Blea Moor tunnel. This is the longest tunnel on the Settle to Carlisle railway line, near Whernside, in the Yorkshire Dales. Robert's Mancunian mother was a machinist at a local mill. When he was five years old, he and his two brothers and parents moved to a council house in Gorton, Manchester.

All three boys were academically gifted and gained scholarships to the Central Grammar School for Boys, on Whitworth Street, Manchester which claims fellow Nobel prize winner, James Chadwick, as an earlier pupil (see Sect. 8.5).

Since Manchester was a major producer of aircraft frames and aeroengines, it was a target for German bombers, particularly during late 1940. Therefore, for safety reasons, Robert and brother, Harry, were sent to live with the Bonnick family who owned Broadrake farm in the Yorkshire Dales. The boy's father was working nearby so they could walk to visit him when he was free. Robert laboured on the farm and developed a lifetime friendship with the Bonnick family, as well as an enduring affection for the Dales, going back there whenever he could.

He became a life-long egalitarian with an enduring curiosity about agriculture and natural history, especially the reproductive patterns displayed by the Dales' sheep, pigs, and cattle.

After leaving school in 1943, Edwards was conscripted for war service in the army. When he was demobilised in 1948, he took a BSc degree in agricultural sciences at the University College of North Wales, in Bangor. Disappointed with the course content, he transferred, in his final year, to a zoology course, gaining a simple pass degree, in biology, in 1951, aged twenty-six.

Having learnt that his friend and fellow Bangor graduate—John Slee—had been offered a postgraduate diploma course in animal genetics at Edinburgh University, Robert also applied and was accepted. Receipt of his diploma was followed by a 3-year Ph.D. opportunity.

In 1955, he obtained a PhD in physiology from the Institute of Animal Genetics at Edinburgh University, and this was followed by two years of postdoctoral research. At Edinburgh, he worked on mouse oocytes and embryos and artificial insemination. He realized that to further understand the developmental biology of the mouse, required engagement in an interdisciplinary mix, not just of experts in embryology and reproduction, but also genetics.

Eggs and embryos are not as abundant and freely available as spermatozoa. Overcoming this problem led Edwards to a discovery that proved to be of significance for his later IVF work. He experimented with a variety of exogenous hormone treatments that induced female mice to ovulate. Edwards, and his wife-to-be, Ruth Fowler, worked on controlled induction in the mouse and showed that super-ovulation was possible.

After a year as a research fellow at the California Institute of Technology (1957–8) he joined the Experimental Biology Division of the National Institute of Medical Research (NIMR) at Mill Hill, in 1958. Here, his interests changed from animal to biomedical research, particularly human oocyte development, working in the evenings and at weekends on egg maturation.

He did some of his earliest work on in vitro maturation of human oocytes from biopsied ovarian samples provided by Molly Rose, a gynaecologist, at the Edgware General Hospital, near Mill Hill.

In 1962, Edward's contract at Mill Hill came to an end and he moved to the Biochemistry Department at Glasgow University. He and Robin Cole succeeded in producing embryonic stem cells that were capable of proliferation for over one hundred generations and of differentiating into a variety of cell types. This was futuristic research.

He moved to Cambridge, in 1963, as Ford Foundation Research Fellow in the Department of Physiology. The Foundation funded research into basic reproductive mechanisms, with a view to developing new methods of fertility control. Initially, Edwards was unimpressed by the apparent exclusivity of the university students, but he soon appreciated the beauty of his surroundings and the ambience of scientific excellence.

In 1965, Edwards began a long-standing contact with Howard and Georgeanna Jones at Johns Hopkins University in Baltimore. The Jones were to create the first American IVF baby in 1982.

Fertilization occurs when a sperm enters and fuses with an egg cell (ovum) to form a single cell zygote. The zygote divides repeatedly, first to a solid ball of cells (morula), then to a hollow ball of cells called a blastocyst. After normal (in vivo) fertilization, the blastocyst attaches to the lining of the uterus between 5 and 8 days after fertilization.

Attempts to fertilize mature oocytes consistently failed because spermatozoa failed to 'capacitate' to achieve 'fertilization competence'. This, they do naturally in the uterus. A research student, in the Physiology Department, Barry Bavister, discovered that raising the alkalinity (pH) of the medium in which hamster sperms were suspended enhanced their fertilizing capacity. Bavister developed a culture medium to nourish hamster sperms which were to be used to fertilize hamster eggs.

In 1968, applying the same pH adjustment to the medium in which human, in vitro, matured, oocytes were mixed with sperm, allowed fertilization to take place, for the first time, outside the body.

Also, in 1968, Edwards recruited Jean Purdy who had trained as a nurse at Addenbrookes Hospital, in Cambridge. Whilst Edwards and Steptoe (see Sect. 9.8) were an unlikely partnership, Purdy was able to smooth the bumps in their relationship. The trio of Edwards, Steptoe and Purdy formed a committed working team that lasted seventeen years until Purdy's untimely death in 1985, followed by Steptoe's passing in 1988.

In 1969, the Ford Foundation were so impressed by Edwards' results that they established an endowment fund for a readership, and he was appointed Reader in Physiology, Cambridge University.

By 1970, the team achieved controlled hormonal induction of oocyte maturation in vivo *(Latin for 'within the living')*; the laparoscopic recovery of these eggs from mature follicles, and early development of the eggs when fertilized to the blastocyst stage, in vitro *(Latin for 'within the glass')*.

Although the embryos obtained looked normal morphologically, both in the living state and, as fixed and stained preparations, the question remained, was it safe to implant these in the genital tract of women volunteers?

During early trials, patients were given small doses of exogenous, priming hormones to produce more than one ripe egg to raise the chance of a successful pregnancy.

The hormonal (endocrine) conditions for inducing ovulation—without adversely affecting implantation—took several years to resolve. Edwards concluded that the hormones given to women to induce ovulation, may be making the uterus inhospitable to fertilized eggs, inhibiting implantation. In 1977, it was decided to focus on natural cycles, monitoring urinary oestrogens and luteinizing hormone (LH). An acute rise in LH from the pituitary gland triggers ovulation. This approach led to the first 'test tube baby'.

Edwards found that embryos grew well in a Petri dish for up to three days, but he was eager to reach the five-day stage when the embryo was ready to implant into the womb. In November 1977, they successfully implanted a blastocyst into the womb of Lesley Brown who, with her husband, John, had been trying to conceive a child for fifteen years.

Louise Brown was delivered by caesarean section, in Oldham General Hospital, on 28 July 1978, and became known as the first test-tube baby. Louise was a normal, fit, and healthy baby. In January 2007, Louise gave birth to her first child, a boy, who had been conceived naturally.

Edwards and Steptoe had faced long-term hostility and opposition to their radical approach to reproductive medicine, not only from churches, governments, and the media, but also from some members of the medico-scientific community, including Robert Winston, James Watson, and Max Perutz. Watson questioned the merit of 'tampering with human procreation'. Edwards and Steptoe suffered marginalisation by their professional peers.

Over-population and family planning were viewed as dominant concerns, whilst the infertile were generally ignored. Their research was judged to be scientific experimentation with human beings and not experimental treatments for infertility. New forms of eugenics, human cloning, as well as other social and ethical upheavals were foreseen.

Public disquiet was particularly intense in the USA, where there was concern about what IVF technology would mean for our relationship with nature, God, and each other. The Vatican called Louise's birth "an event that can have grave consequences for humanity because it divorced the conjugal sexual act from procreation".

Following the birth of Louise, Steptoe was quoted as saying: "I am not a wizard or a Frankenstein. All I want to do is to help women whose child-producing mechanisms are slightly faulty". He knew that their infertility was usually incurable. Their research merely gave nature a helping nudge in the right direction. Steptoe had a magnificent obsession that he would crack the problem, and nothing would stop him. Although their characters and personal styles were different, any disparities were shrouded by a mutual respect for the pioneering skills of the other, their joint endeavours overcoming the plight of the irrevocably infertile.

Edwards' humanist, ethical sympathies, as well as his antipathy to religious dogma were to be tested, so he deliberately engaged with theologians, ethicists, and lawyers. He pursued a programme of public education about the issues raised, to challenge and develop bio-ethical thought and disclosure about assisted reproductive technology.

Moral attitudes to IVF began to change and their medical approach to IVF spread across the world giving hope to women who would otherwise remain childless, ending the misery of millions of barren couples. Worldwide, about one in ten couples are infertile and, until IVF, doctors could do little to help.

Infertile women accept the interventionist approach, even though it can be painful, expensive with a variable success rate. According to the *H*uman *F*ertilisation and *E*mbryology Authority (HFEA), in 2017, the birth rate *per* *e*mbryo *t*ransferred (PET) is highest for women under 35. In this group, for those using their own eggs, the figure is 30%.

After the birth of a second IVF child—a boy called Alastair Macdonald—the two men thought that the NHS or the Medical Research Council might fund a national IVF centre. However, their request was turned down on the grounds that the treatment was too experimental. In 1980, they raised venture capital to set up the Bourne Hall Clinic, near Cambridge, the world's first dedicated IVF clinic, where my first and second grandsons were 'conceived'. With Purdy as Technical Director, the clinic advanced their research, and it became a training centre for future fertilization specialists. Before Steptoe died in 1988, Edwards was able to tell him that 1,000 babies had been conceived at their clinic.

Ethical concerns have been addressed by an act of Parliament—the Human Fertilisation and Embryology Act 1990 which made the UK the first country to impose strict regulations on the practice of assisted reproductive technology (ART). The Human Fertilization and Embryology Authority adjudicates, and licenses all work on human embryology, whether for IVF or for scientific study.

Over the years, the whole process of IVF treatment—cycle monitoring; follicular stimulation; oocyte recovery and embryo transfer—has been greatly simplified and streamlined.

In 1984, Edwards was elected Fellow of the Royal Society and, in the following year, he was appointed to a chair in Human Reproduction at Cambridge. In 1985, Edwards and Jean Cohen helped found the European Society for Human Reproduction and Embryology (ESHRE) in Bonn. In 1988, Edwards was granted the title Commander of the British Empire (CBE).

In 2010, Edwards was awarded the Nobel Prize in Physiology or Medicine for the development of in vitro fertilisation. Since the Prize is not awarded posthumously, neither Steptoe nor Purdy was eligible for consideration. At the time of the award, more than 4.5 million babies had been born by IVF. Edwards was made Knight Bachelor, in 2011, for services to human reproductive biology. Edwards was a Member of the Academia Europaea (MAE).

He was a modest, unassuming man whose tenacity of purpose is attributed to his Yorkshire origins, fuelled by his working-class background. He had a passion for science, not just the theory, but its social application and benefits to mankind.

Edwards did not receive the pomp and circumstance of a ceremonial funeral, but he and Steptoe will be remembered forever by an ever increasing eight million couples who have benefitted from their pioneering work.

In 1980, Edwards and Steptoe published the first edition of "A Matter of Life. The Story of IVF—a Medical Breakthrough". In 2012, a second edition was printed. It had been co-edited by Robert Edwards' widow, Ruth, together with Patrick's son, Andrew.

10

Electricity, Magnetism and Light and Their Inter-relationship. Electrolysis and Electrochemistry. Foundations for Both the 'Mechanised Age' (Powered by Electricity) and Radio Broadcasting

As an answer to those who are in the habit of saying to every new fact, "What is its use?" Doctor Franklin says to such, "What is the use of an infant?". The answer of the experimentalist would be, "Endeavour to make it useful." (Michael Faraday, from 5th Lecture in 1816, The Life and Letters of Faraday (1870), Vol. 1, 218, by Bence Jones)

10.1 Précis

The ancient Greeks knew that when amber is rubbed with wool or fur, it will attract light objects such as feathers or bits of straw. The word 'electric' (from the Latin word, *electrum* meaning amber, and the modern Latin word *electricus*) was first used by Gilbert in 1600. Nowadays, we use the term "triboelectric effect" to describe the electrification of dissimilar materials which are brought together and then separated. For example, when hair is rubbed with an inflated balloon, electrons from the hair migrate to the rubber latex wall of the balloon, leaving behind positively charged strands of hair which repel one another.

With the invention of the Voltaic pile, it became possible to produce electricity continuously. This propelled us into our modern world, without which it would be darker, colder, and quieter.

Humphry Davy was the first of a new generation of electricians who used electricity to establish the composition of chemical substances and produce several pure elements for the first time.

© The Author(s), under exclusive license to Springer Nature
Switzerland AG 2022
J. Bailey, *Inventive Geniuses Who Changed the World*,
https://doi.org/10.1007/978-3-030-81381-9_10

Of the 118 chemical elements currently appearing in the periodic table, a majority (20%) of them were discovered, co-discovered and/or isolated by British scientists. Davy alone was implicated in isolating, or discovering the rudimentary nature of, up to ten of these elements, and W. Ramsey in five.

Ninety-two of these elements are found naturally but only eight of them were involved in the formation of 98% of the rocks constituting the Earth's crust. Fifty-six of these 92 elements make up at least 0.1 mg of a typical human, with both light and heavy elements playing some role in the body's biological processes.

Davy effectively established the new scientific field of electrochemistry. He deduced that, for metal salts, chemical bonds are electrical in nature and that an electric current, involving the movement of electrons, could stimulate the making and breaking of bonds, resulting in chemical reaction. Electro-synthesis is a 'clean' process requiring no heat application or added reagents.

Davy is also remembered for his design of the Davy lamp to prevent fires and explosions, from methane-laden air in coalmines, causing injury and deaths to many coalminers. Davy was the first man to be knighted for service to science since Sir Isaac Newton. He was the first to be awarded a baronetcy. He conferred popularity, and even glamour, on the discipline of chemistry.

Michael Faraday came from a poor family; had limited education but became one of the greatest scientists in history. He was a man of relentless wonder and curiosity. He lived in the age of steam power, but he laid the foundations for the 'Age of Electrical Mechanisation'. The practical applications of Michael Faraday's discoveries have transformed the world. The versatility of his achievements, in diverse branches of science, was truly outstanding, and all mankind has benefitted from his findings. He deserves his place in the pantheon of great scientists.

Almost anything electrical uses the scientific principles that Faraday established, relying on the interplay between electric current, magnetic field, and mechanical motion. He was a talented experimentalist who constructed a device that exploited the interaction between electricity and magnetism, converting electrical energy into mechanical energy/continuous circular motion.

One of his greatest breakthroughs involved electromagnetic induction, whereby a magnetic field of excited electrons, momentarily produced an electric current. Next, he combined magnetism with mechanical motion to generate electricity continuously. In his first arrangement, an in–out motion of a magnet produced an alternating current that changed direction. In the second, he produced a direct current by spinning an electrically conductive disc in a permanent magnetic field.

His rudimentary laboratory devices and apparatus were the springboards for electric motors, DC generators, AC alternators, transformers, and miniature batteries. We have harnessed electricity to illuminate and power our modern world.

With W. Whenwell, he introduced the nomenclature of electrochemical terms. His two laws of electrolysis laid the foundation for other modern industries involving electroplating and the production of some chemicals.

He demonstrated that a hollow electrical (Faraday) cage can offer protection from an induced charge. He was the first scientist to make a link between electromagnetism and light, showing that an external magnetic field could cause the plane of light polarisation to rotate.

He demonstrated that cooling results from the evaporation of a gas previously compressed to a liquid, this forming the basis of today's refrigerators and freezers.

James Clerk Maxwell was a child prodigy and gifted mathematician. His many areas of scientific interest included astronomy (especially Saturn's ring) and optics. He worked on colour vision, determining a colour equation which gave quantitative measurements of the ability of the eye to match real colours. He co-produced the first ever colour photograph.

He played a key role in the development of statistical mechanics, paving the way for quantum mechanics. For his kinetic theory, he applied methods of probability and statistics to describe the speed distribution of an assembly of gaseous particles and how this would change as the temperature was raised.

Theoretical physicist, Maxwell, adopted a mathematical approach to some of Faraday's empirical findings. He succeeded in unifying three realms of physics, namely electricity, magnetism, and light. The basic rules by which light behaves, electric current flows and magnetism functions can be expressed in Maxwell's equations. His equations have a reach that extend to the extremities of the Universe.

His unified 'field theory' became a cornerstone in physics. Maxwell was a bridge between the mechanical world of Newtonian physics and the theory of fields as espoused by Einstein and others. He was possibly the greatest theoretical physics in the nineteenth century, later known as 'Scotland's Einstein'.

In the 1870s, his notions pointed to the existence of an 'electromagnetic spectrum', suggesting that 'nature's storehouse' might contain other types of radiation with frequencies both higher and lower than the visible spectrum. This speculation was vindicated during the next thirty years with the discovery of radio waves (1886) X-rays (1895) and γ-radiation (1900).

The discovery of these other forms of electromagnetic radiation have had far-reaching social impacts, setting the stage for modern lifestyles, information, and communication technologies, as well as medical applications, via X-ray machines and gamma rays.

The warm radiance of sunshine; a rainbow; the colourful beauty of Michelangelo's frescos; the soreness of sunburn; the sound and sight of radio and TV transmissions; the incandescent light bulb; the friendly telephone conversation; the hot meal taken from a microwave oven; the X-ray revealing a broken bone, all are brought to us by electromagnetic radiation, resulting from electrons, either accelerating along a metal conductor or descending from excited atomic orbits.

10.2 Humphry Davy 1st Baronet, PRS, MRIA, FGS (1778–1829)—Electro-Chemist and Inventor of the Davy Lamp

Humphry was born in Penzance, Cornwall. His father was a carpenter. At the age of six, he attended a grammar school in Penzance. At the age of nine, his father inherited a small farm and the family moved four miles to Varfell, near Ludgvan. During term-time, Davy boarded with his godfather, John Tonkin, who later became his guardian. John Tonkin was the town's apothecary, and his storeroom became a source of chemicals for Davy's pranks. Apparently, he was full of mischief with a penchant for explosions. He was a born chemist.

He entertained his school friends by writing poetry. Over the years, as a poet, he wrote over 160 poems. In later life he befriended poets of the Romantic period: Samuel Taylor Coleridge, William Wordsworth, and Lord Byron.

In 1793, Davy left the Penzance school and Tonkin paid for him to finish his secondary education at the Truro Grammar School. When Davy was sixteen, his father died, leaving significant debts. He found an apprenticeship with John Bingham Borlase, a surgeon-apothecary in Penzance. He began performing basic chemical research. A refugee priest taught him French and, in 1797, he read the works of Lavoisier (especially his "Elements of Chemistry") that influenced him greatly.

Whilst still an apprentice, Davy met the Sheriff of Cornwall, Davies Gilbert, who granted him access to his private library and introduced him to notable scientists, including Thomas Beddoes, a prominent physician who sought treatments for tuberculosis. The latter had established the Pneumatic Institution in Bristol—an experimental hospital to study the therapeutic effects of factitious airs and gases (including laughing gas, nitrous oxide, N_2O) on the treatment of certain diseases.

Davy joined the Pneumatic Institution as Chemical Superintendent and became an expert on the physiological effects of some of these gases. He and some of his friends became addicted to the pleasurable sensation arising from the inhalation of nitrous oxide, which became a recreational drug. Later it was used for its anaesthetic and pain-reducing effects in surgery and dentistry.

The Royal Institution was founded in 1799 and Davy moved there in 1801 as Director of the Laboratory and Assistant Lecturer in Chemistry. Here, despite an ungainly appearance, Davy grew a reputation as a dramatic and compelling lecturer, able to deliver spectacular chemical demonstrations, attended by huge audiences. Scientific culture became a fashionable part of high society and this enhanced the prestige of science. However, some members of the London aristocracy were not immediately receptive to Davy's seemingly rough, provincial ways. Nonetheless, Davy became a scientific luminary and was elected Fellow of the Royal Society in 1803.

Davy helped propel us to a new, electrical age. As a young researcher at the Pneumatic Institution, Davy became aware of the work of Count Alessandro Volta on what became known as a voltaic pile, an early type of electric battery that produced a steady source of electric current. In 1800, Volta built a column (pile) comprising pairs of dissimilar metals, alternating with a cardboard disk soaked in an electrolyte (e.g., saltwater or dilute acid).

Volta listed metals in order of their so-called electromotive force (now called standard electrode potential). For two of the metal-pairs he used—zinc/copper and zinc/silver—the potentials at 25 °C are now recorded as: $Zn^{2+}/Zn = -0.76$ V; $Cu^{2+}/Cu = +0.34$ V and $Ag^+/Ag = +0.80$ V.

In 1808, in the cellar of the Royal Institution, Davy built, what was then the world's largest battery. It was composed of over 800 connected Voltaic piles. He attached two carbon filaments to his giant battery. When these were brought together, continuously flowing electricity created a carbon arc light. The world stage was now set for inventors such as Thomas Edison and Joseph Swan to develop incandescent lights by holding filaments in an evacuated glass bulb and passing currents through them to make them white hot.

Volta's work triggered a further burst of discovery. Volta's pile became a new tool for decomposing compounds into their constituent elements. For instance, in 1800, William Nicholson and Anthony Carlisle used a zinc/silver pile to generate a sustained stream of electric charge into water, splitting it into its component parts. Hydrogen arose from a brass wire attached to a zinc plate, whilst the brass wire attached to the silver plate tarnished due to oxidation by the oxygen liberated there. When platinum wire—resistant to oxidation—was used, the ratio of gases collected was 2:1 hydrogen to oxygen.

Davy studied the effect of electricity on chemical reactions. His first, electrolytic conduction experiments on alkali solutions produced hydrogen and oxygen because water can be both oxidised and reduced:

$$2H_2O \rightarrow O_2(g) + 4H^+(aq) + 4e^-$$
$$4H_2O + 4e^- \rightarrow 2H_2(g) + 4OH^-(aq)$$

Therefore, he refined his experiments to include the electrolysis of slightly moistened, molten salts for which the ions are more mobile than solid salts. Earlier attempts, by numerous scientists, to separate caustic potash into its component elements had been fruitless. In 1807, Davy isolated metallic globules from electrolysed molten potash (potassium hydroxide, KOH) melted, at 370 °C. These globules burst into flames when thrown into water. In air, it burnt with a pink flash. He named the new element 'potassium'. Some months later, he isolated sodium from caustic soda (sodium hydroxide, NaOH) by a similar method. Where the metal is represented as 'M', the following reactions occur at the two electrodes (Table 10.1).

In 1808, he went on to isolate four of the six alkaline earth metals (now known to be in Group IIA of the periodic table). Other scientists had worked previously on minerals of these alkaline earth metals but had never isolated the pure metal. For instance, in 1755, J. Black recognised magnesium as an element. Barium salts were discovered by K. W. Scheel in 1774. In 1790, A. Crawford and W. Cruickshank recognised a new mineral, "strontianite" (strontium carbonate) as different from barium minerals such as witherite and baryte (barium sulphate).

In 1808, Davy isolated magnesium from magnesia (magnesium oxide, MgO); calcium from quicklime (calcium oxide, CaO); strontium from strontium chloride. This, he achieved by mixing the moistened substrate with mercuric oxide. Using a platinum wire (as anode) in a puddle of mercury (as cathode), he produced enough amalgam at the mercury electrode so that he could distil away the mercury and isolate the new metal. In the case

Table 10.1 Electrolysis of molten alkali metal hydroxides

Negative electrode	$4M^+ + 4e^- \rightarrow 4M(s)$	Reduction
Positive electrode	$4OH^- \rightarrow O_2(g) + 2H_2O + 4e^-$	Oxidation

of barium, he electrolysed molten baryte, in a manner like his isolation of potassium and sodium.

Davy studied the forces involved in these separations, inventing the new scientific field of electrochemistry. He deduced that, for metal salts, chemical bonds are electrical in nature. He concluded that an electric current could stimulate chemical reaction, whilst a chemical (redox) reaction could generate a current.

When the source of electricity to two electrodes, in an electrolytic cell, is a direct current, each ion in the molten salt or solution is attracted to the electrode of opposite charge. The positive metal ions move towards the negative electrode (the cathode), whilst the negative ions move towards the positive electrode (anode). For the origin of this nomenclature, (see Sect. 10.3.9).

At the negative electrode—where there is an excess of electrons—positive, metal ions pick up electrons and are reduced to native metal. At the positive electrode (where there is a deficiency of electrons) the negative ions deposit electrons and are, therefore, oxidised. (N.B. When talking about galvanic cells, the labels are reversed and a source of electricity results from a spontaneous oxidation–reduction reaction taking place in a solution).

Thus, during the process of electrolytic conduction, there are electrons flowing through the external electrical circuit and positive and negative ions flowing in opposite directions in the liquid. The process of electrolysis provides sufficient energy to cause an otherwise, non-spontaneous oxidation–reduction reaction to occur, whilst electrical neutrality is maintained.

In 1808, on his thirtieth birthday, during a lecture at the Royal Institution, Davy announced the discovery of a new element which he called boracium (now named boron). This he achieved by heating boric acid and potassium in a copper tube. In France, in the same year, Gay-Lussac and Thénard made a similar discovery.

Scheel, in 1774, had reacted manganese dioxide with hydrochloric acid (aqueous HCl, then known as muriatic acid). Scheel had liberated a green gas (chlorine) but thought that a compound of oxygen had been produced (oxymuriatic acid) with an unknown element. In 1811, Davy's efforts to remove oxygen from this green gas were fruitless. He concluded that the gas was an independent element, giving it the name 'chlorine', derived from the Greek word *chloros*, meaning pale green or yellow green.

Whilst having a youthful association with political radicals, Davy eventually became part of the Establishment. In 1812, Davy was knighted and married a wealthy widow. Now financially comfortable, he was able to retire from routine work and devote his whole time to the pursuit of discovery. He and his wife purchased 23 Grosvenor Street in London's exclusive Mayfair, in 1815.

Unfortunately, in 1812, when experimenting with nitrogen trichloride, an explosion occurred which caused him temporary injury. Therefore, he hired a book-binding apprentice, Michael Faraday, as his amanuensis. Like Davy, Faraday had no university training but had a great appetite for research. Faraday went on to establish an even more prestigious reputation than Davy.

Davy's research was recognised throughout Europe and, in 1813, he embarked on a two-year European tour. He visited Paris, even though Britain was at war with France, and was awarded the *Prix Napoleon*.

Whilst in Paris, he was given a sample of lustrous, violet crystals by A-M. Ampère that fellow Frenchman, B. Courtois, had produced from seaweed in 1811. Davy characterised it, recognised that it had properties like chlorine and named it *iodine*, the Greek word for violet-coloured.

In 1815, Davy was approached by James Wilkinson—founder of the Society for Preventing Accidents in Coalmines—who told him of the dangers that miners faced from firedamp (mostly methane gas). The gas is often present in mines and could be sparked by the candles they had in their helmets to illuminate their work. The resulting fires and explosions caused many injuries and deaths.

Davy's objective was to design a lamp that segregated its flame from the outward atmosphere so that it could not impact with, and ignite, the methane-heavy air in the mine. The resulting Davy lamp had an iron gauze cylinder enclosing the lamp's flame. It allowed light to travel through it but absorbed the heat of the candle's flame. It became widely used before being superseded by more sophisticated lamps.

In 1817, Arfvedson discovered lithium in petalite ore but was unable to isolate it. In 1818, Davy and W.T. Brande independently isolated small quantities of elemental lithium by the electrolysis of lithium salts.

In summary, Davy was involved in the characterisation and nomenclature of boron, as well as two of the halogens (namely, chlorine and iodine). He was the discoverer, co-discoverer and/or isolator of three of the six alkali metals, as well as four of the six alkali earth metals. Fortuitously, these seven metals had relatively high reduction potentials (between -2.38 and -3.04 V) facilitating their isolation by electrolysis.

Davy was a charismatic speaker, a brilliant communicator, a showman who could entertain Londoners with his latest discoveries. He was fearless in the face of some extremely dangerous experiments, particularly with molten caustic substances. To some members of the aristocracy, he was a pretentious social climber. However, in 1819, Davy was awarded a baronetcy, by the Prince Regent, for his outstanding contribution to science. This was the first time that this honour had been conferred on a man of science.

Between 1820 and 1827, Davy was President of the Royal Society. At the time, the Society was transitioning from a club for gentlemen from the political and social elite who were interested in natural philosophy, to an academy representing specialized scientists.

His list of titles included: Baronet (Bart) of Grosvenor Street; President of the Royal Society (PRS); Member of the Royal Irish Academy (MRIA) and Fellow of the Geological Society (FGS).

In 1826, Davy suffered his first stroke. In 1829, during one of his numerous European tours, Davy suffered another stroke. On his journey home, he died, aged 50, in a hotel room, in Geneva, the city where he was buried. His widow, Jane, organised a memorial tablet for him in Westminster Abbey.

At the end of his life, a total of fifty-five elements had been identified by a host of researchers, and the world had a new science—chemistry.

10.3 Michael Faraday FRS (1791–1867)—Natural Philosopher. Expert in Electrochemistry and Electromagnetism, Pioneering a Mechanised World Powered by Electricity

10.3.1 Family Background, Education, and Early Employment

Michael Faraday was not born to be a man of influence. He came from a poor family but became one of the greatest scientists in history. He was born, in 1791, in Newington Butts, about two miles southeast of Westminster Abbey, in London. He was the third of four children, born to a blacksmith in poor health who could barely support his family. Before marriage, Michael's mother had been a servant. His parents had moved from Cumbria, just before he was born. In the mid-1790s, the family moved into rooms in Manchester Square, on the western edge of London.

He learnt the rudiments of reading, writing and arithmetic at a common day school which he attended until he was twelve years of age. He also attended a Sunday school for the Christian literalist sect called the Sandemanians which greatly influenced the way he approached and interpreted nature. Members believe that the accumulation of wealth is unscriptural and improper, whilst the acceptance of worldly honours is wrong. Faraday was appointed Deacon (in 1832) and Elder (in 1840).

To supplement the family income, he started working as an errand boy for George Riebau who was a bookbinder and bookseller who owned a bookshop at No. 2 (now numbered 48) Blandford Street, Marylebone, a short distance from Faraday's home in Manchester Square. His hard work impressed his employer, and he was promoted to apprentice bookbinder. In his spare time, he read a variety of books in the bookshop or brought in for binding, including the electrical treatises in *Encyclopaedia Britannica*, as well as Jane Marcet's *Conversations on Chemistry*.

In 1812, a customer of the bookshop gave him a ticket to hear Humphrey Davy (see Sect. 10.2) lecturing at the Royal Institution. This was a turning point in his life. When Davy was injured in an explosion, Faraday worked for Davy, taking notes and, later, at the age of twenty-one, he became his chemistry assistant. Davy took Faraday, as his secretary and valet, on an 18-month tour of Europe, where he met notable scientists, including Ampère and Volta.

What Faraday lacked in formal scientific training, he made up for by his exceptional talent as an experimentalist. In 1821, aged twenty-nine, he was promoted to be Superintendent of House and Laboratory of the Royal Institution (RI). In 1824, he was elected to the Royal Society. One year later, he became Director of the RI's Laboratory and then, in 1833, its first Fullerian Professor of Chemistry.

Having begun his career as a chemist, Faraday became particularly interested in three facets of physics, namely electric current, magnetic field, and mechanical motion. It is the interplay between these three that is a recurrent theme in his research.

10.3.2   Magnetism; Magnetic Field Lines; Magnetic Force Between Poles

In 1600, Dr. W. Gilbert founded the scientific study of magnetism. He worked on auto-magnetized samples of the mineral magnetite (iron oxide; known as lodestone). In his famous treatise, *De Magnete*, he told how iron shavings (filings) assumed certain mysterious lines when sprinkled on a piece of card sitting on a natural magnet.

Iron filings will arrange themselves into semi-circular arcs, emanating from the ends—the north and south poles—of a magnet. The greatest concentrations of filings are at the poles (see www.rigb.org/our-history/iconic-objects-list/faradays-iron-filings-lines of force, field theory, paramagnetic). Faraday called these 'magnetic lines of force' (nowadays, magnetic field lines).

The space around a magnet is not 'empty', devoid of properties. Instead, it is 'modified space' with specific, invisible properties, and called a 'magnetic field'.

We now know that each piece of iron filing becomes a temporary magnet pointing along the magnetic field line, the north end of one piece behind the south end of another. Field lines converge where the magnetic force is strong and spread out where it is weak.

There is no single source of magnetic field lines—they are continuous loops with no start or end. It is only accepted convention that shows them travelling from north to south pole. The number of flux lines passing through a given space represents the flux density or the strength of the field. It should be noted that a magnetic field is smooth and continuous and does not exist of discrete lines as is shown in Fig. 10.1.

All magnets have two poles (are dipolar). No magnetic monopoles are known to exist. Pairs of magnetic poles attract and repel each other, much

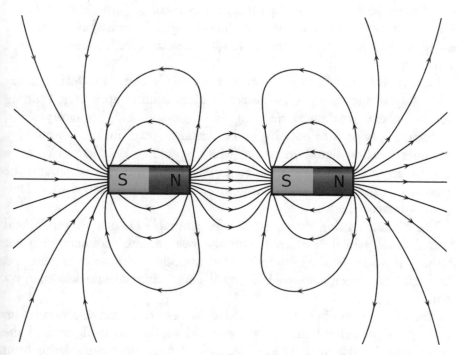

Fig. 10.1 Magnetic field lines. *Source* By Geek3—Own work, CC BY-SA 3.0, Attribution-ShareAlike 3.0 Unported https://creativecommons.org/licenses/by-sa/3.0. https://commons.wikimedia.org/w/index.php?curid=10515628.   https://commons.wik imedia.org/wiki/File:VFPt_cylindrical_magnets_attracting.svg#/media/

as electric positive and negative charges do. In 1851, Faraday placed a small, permanent magnet between two larger magnets. A pole of the middle magnet was attracted to the opposite pole of both outer magnets, north to south, and south to north. They were pulled together by a 'magnetic force'.

10.3.3   Creation of Magnetism from Electricity. Electromagnetic Rotation (1821). Mechanical Energy Derived from Electrical Energy. Principle of an Electric Motor

In 1820, Danish scientist H. C. Ørsted was credited with the discovery of electromagnetism. He ran an electric current through a wire whereupon a nearby compass needle was temporarily deflected, indicating that a magnetic force was at play—an electric effect produced a magnetic effect.

A-M Ampère had shown that an electric current produces magnetic lines of force that caused two wires to be attracted to, or repelled from, each other, depending on whether the currents were flowing in the same or opposite directions. In response to Ørsted's observation, Ampère argued that magnetic force was a circular one, producing a cylinder of magnetism around a wire carrying an electric current. No such circular force had been observed previously.

Faraday was the first to understand what this implied. In 1821, Faraday saw a way of harnessing these forces in a mechanical apparatus, utilising electromagnetic rotation. Exploiting his ingenuity and laboratory skills, he built a device incorporating a bar magnet fixed in the centre of a container holding liquid mercury (which conducts electricity). When a current-carrying wire was suspended in the mercury, it rotated around the stationary magnet (see www.rigb.org/our-history/iconic-objects-list/faradays-motor—mercury bath).

The current passing through the wire created a circular magnetic field around itself, which interacted with the field of the stationary magnet, pushing the wire in a continuous circular motion whilst the current was flowing. This device converted electrical energy into mechanical energy / motion.

Producing motion from the interaction between electricity and magnetism is the principle behind the electric motor. Faraday had demonstrated the earliest ever electric motor. The first surviving Faraday apparatus, dating from 1822, can be seen in the Basement Laboratory, at the Royal Institution. By the 1830s, the performance of practical electromagnetic engines was being studied by J Joule, amongst others.

10.3.4   Liquefaction of Gases in 1823. Refrigeration. 21st Century Liquid Air Battery

In 1823, Faraday showed that, by compression, mechanical pumps could change a gas (e.g., ammonia) at room temperature, into a liquid. When the gas was evaporated, cooling occurred. The resulting gas could be collected and compressed by a pump into a liquid again. This cycling principle of gas compression/evaporation led to today's refrigerators and freezers.

Chemical batteries can only store a relatively small amount of electricity. Currently, spare (green) energy is being used to compress atmospheric air into a liquid, so it becomes a cryogenic energy store. When demand for domestic electricity is high, the liquid air can be released to power a turbine.

10.3.5   Chemistry Firsts; Syntheses of Chlorinated Hydrocarbons and His Discovery of Benzene in 1825

Faraday synthesized the first known compounds of carbon and chlorine – hexachloroethane (C_2Cl_6), as well as the product of its thermal decomposition—tetrachloroethylene (C_2Cl_4; now a dry-cleaning solvent).

The Portable Gas Company produced 'illuminating gas' for lamps in private and public buildings. This was achieved by dropping whale or fish oil into a hot furnace and compressing the resulting gas. In 1825, Faraday distilled the oily deposit formed when the gas was pressurised. He succeeded in separating a new compound that had equal numbers of carbon and hydrogen atoms and a boiling point of 80 °C. With an empirical formula "CH", he named it 'carbureted hydrogen' or 'bicarburet of hydrogen' (later named benzene, C_6H_6). Benzene is an important substrate for numerous other substances.

10.3.6   Creation of Electricity from Magnetism. Faraday's Induction Ring (1831). Electromagnetic Induction. The Electric Transformer

It is noteworthy that the magnetic field of Ørsted's current-carrying, straight wire can be intensified by coiling it, to form a 'solenoid', giving a stronger magnet. For a coil of wire of N-turns, the effect is N-times greater than if it were a straight wire.

Volta and Faraday's former employer and patron, Davy, had created electric currents with primitive batteries in which a current was produced by chemical reactions within the voltaic pile. Several scientists, including Faraday, had demonstrated that electricity can produce magnetism (see Sect. 10.3.3). Faraday persisted longer than most to demonstrate that the reverse is possible and, in 1831, he built a device that could create electricity from magnetism.

He did this by using a soft iron ring, a 100-plate voltaic cell, two lengths of copper wire and a galvanometer (a newly devised instrument for detecting current). The two wires were wrapped round opposing sides of the iron ring to form Coil A and Coil B embedded in cotton wadding for insulation. The coils had 2 mm spacing between each winding which were separated by insulating thread.

When Coil A was attached to the battery, the needle of the galvanometer attached to Coil B immediately deflected. A current had been produced in the secondary coil by one in the primary coil. The effect (called mutual induction) faded when the current was left on. Thus, the induction was only operational whilst the magnetic effect was changing.

Astonishingly, as soon as the current was switched off, deflection occurred in the opposite direction. Somehow, turning off the current also created an induced current in the secondary circuit that was equal and opposite to the original current. The effect was reproduced when the current was switched on and off.

When electrons flow along the windings of Coil A, magnetism is induced in the iron ring. A magnetic field of excited electrons is created, transiently producing an electric current in Coil B, which is inside the magnetic field. This is one of Faraday's greatest discoveries—electromagnetic induction— whereupon a momentary current was generated in Coil B every time the battery was connected or disconnected from Coil A.

Faraday's induction ring experiment demonstrated that, with a coil of wire and a magnet, electricity can be produced. His apparatus can be seen in the museum of the Royal Institution (see https://www.rigb.org/our-history/ iconic-objects-list/faraday-ring—ring-coil apparatus, transformer, induction ring).

When two differing coils of wire are wound on a metal core, an 'electric transformer' results. Using such an apparatus, it is possible to 'transform' the electrical voltage of the current from Coil A to Coil B by deliberately having differing numbers of windings on the two coils. For instance, if Coil B has twice the number of windings of Coil A, then the voltage is reduced by half. By judicious choice of primary and secondary coils it is possible to engineer step-up or step-down transformers.

10.3.7 Combining Magnetism with Mechanical Motion to Generate Electricity Continuously in the Form of an Alternating Current (1831). Magneto-Electric Induction. The 1st Alternator. Faraday's 1st and 2nd Law of Electromagnetic Induction

Using Faraday's induction ring, only a transient effect was produced, so he looked for a way of generating electricity continuously.

In the autumn of 1831, Faraday showed something quite startling, namely that when an electrical conductor and magnetic field are moved relative to each other, they generate a voltage and, in a complete electrical circuit, a current. The magnitude of the current is dependent upon the number of lines of force "cut" by the conductor in unit time. The voltage is not generated solely at one location—it is distributed around the circuit. Such a 'distributed voltage' is called an 'electromotive force'.

This observation formed the basis of his first law of electromagnetic induction. This states that "Whenever a conductor cuts magnetic flux, an electromotive force is induced in that conductor". By the same token, "whenever a magnetic flux linked with a circuit changes, an electromotive force is always induced in it".

In summary, electromagnetic induction is the process by which current can be **induced** to flow due to a changing magnetic field. Either moving a wire through a magnetic field, or (equivalently) changing the strength of the magnetic field over time can cause a current to flow. Faraday was on the track of generating electricity mechanically rather than chemically, as in a battery.

Faraday made a major adaptation to his induction ring. He took a simple permanent bar magnet, together with a hollow coil of insulated copper wire, the ends of which were attached to a galvanometer. By thrusting the rod-shaped magnet quickly into the stationary coil resulted in deflection of the needle. Reversing the process caused the needle to deflect in the opposite direction (see www.rigb.org/our-history/iconic-objects-list/faraday-generator).

If the magnet is moved back and forth repeatedly, the current keeps changing direction. In other words, an alternating current (AC) is produced where the flow of current keeps reversing its direction. As the magnet moves, lines of magnetic force repeatedly intersect with the wire. This excites electrons in the wire, generating an electric current. This primitive device is the forerunner of AC generators (now known as alternators) (Fig. 10.2).

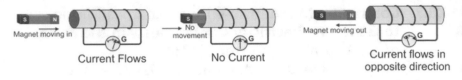

Fig. 10.2 Effect of passing a magnet through a stationary coil

The current produced by electromagnetic induction is greater when (i) the magnet or coil moves faster; (ii) the coil has more turns and/or (iii) the magnet is stronger.

Voltage is induced whether a magnetic field of a magnet moves near a stationary conductor, or the conductor moves in a stationary magnetic field. The induced voltage in a circuit is proportional to the rate at which the magnetic field changes through that circuit. This forms the basis of Faraday's second law of electromagnetic induction.

These principles are exploited in the modern 'Faraday flashlight' in which a magnet goes back and forth, through a coil of wire, creating an electric current that is stored in a capacitor which lights the bulb.

It is impractical to generate large amounts of electricity by passing a magnet in and out of a coil of wire. Instead, generators induce a current by either spinning a coil of wire inside a magnetic field, or by spinning a magnet inside a coil of wire (see Sect. 10.3.8). One simple example of a generator is the bicycle dynamo.

An electric transformer can be used to increase or decrease the voltage of alternating currents. An alternating voltage in a primary coil results in a changing (alternating) magnetic field in the metal core. The secondary coil senses this changing magnetic field and so has an alternating voltage induced in it.

The National Grid transmits electricity, from power stations, over long distances, at extremely high voltages. At sub-stations, 'step-down transformers' reduce voltages to a level that are safer for customers to use.

10.3.8 Combining Magnetism with Continuous Mechanical Motion to Generate Electricity and a Direct Current in 1831. Magneto-Electric Induction. Faraday's Spinning Disc and the 1st Electric Generator. Electrical Energy Derived from Kinetic Energy. Electric Power from Magnetic Power

Later in 1831, Faraday modified François Arago's disc design. He mounted an electrically conductive, copper disc on a brass axis so that it could freely rotate between two poles of a powerful, permanent magnet. He connected the disc to a galvanometer by attaching one wire to its centre and another touching its rim.

When the disc was rotated, the galvanometer registered a continuous current that had to be travelling in a radial direction through the spinning disc, 'cutting' through a magnetic field. Reversing the direction of spin caused the needle to be deflected in the opposite direction.

The machine, later called a Faraday disc, produced a continuous, direct current (DC). It was, therefore, a simple dynamo-electric machine, the first DC electric generator. By today's standards, Faraday's generator was crude, but he had discovered the first method to generate direct current by means of motion in a magnetic field. Electric generators use magneto-electric induction to transform mechanical energy (the work done in moving the magnet or coil, to change the magnetic flux) to electrical energy (in the form of current in the coil).

Much of our electric power is produced using this principle. Turning a generator (turbine) to cause continuous rotation in a magnetic field is achieved by high-pressure steam (from coal, oil, gas, geothermal or nuclear) in a power plant, or by a wind turbine.

Faraday lived in a world of steam power but laid the foundations for the age of electricity. This would become a powerful new technology. Faraday's primitive electric motor, transformer, alternator, and generator were the springboards for major technological advances that have become the very foundation of our modern world, bringing tangible benefits to us all. We have harnessed electricity to illuminate and power our modern world.

10.3.9   Theory of Electrolytic Decomposition in 1832. Nomenclature of Electrochemical Terms

Davy (see Sect. 10.2), in 1807, had isolated sodium and potassium by passing direct electric currents through their molten hydroxides. Faraday began his scientific work as Davy's laboratory assistant.

In the early 1830s, Faraday returned to his earlier research on the electrolytic decomposition of molten salts and aqueous salt solutions, by electricity. Cooperating with William Whenwell, Faraday introduced the nomenclature of electro-chemical terms, namely: '*electrode, ion, cation, anion, cathode and anode*'. A cation is attracted towards the electrode which is negatively-charged—the cathode. An anion moves towards the positive terminal—the anode.

If two electrodes are placed in a vessel containing molten sodium chloride, and an electric potential is applied, then metallic sodium is produced at the cathode and chlorine gas at the anode. Such electric decomposition of a substance is called *electrolysis*.

During the nineteenth century, it became evident that electric charge had a natural unit that could not be further sub-divided. In 1891, Stoney proposed to name it '*electron*'.

At the time of Faraday's research, J. J. Thomson had not proven experimentally, the existence of electrons (see Sect. 8.3.2). Thomson's research was conducted in 1897. We now know that *electrolytic conductivity* occurs because, in its molten form, sodium ions (Na^+) are sufficiently mobile to move to the cathode and collect electrons, whilst the chloride ions (Cl^-) migrate to the anode and give up their 'extra' electron.

Electrochemistry is the science that has produced lithium-ion batteries and metal hydride batteries capable of powering ubiquitous mobile instruments.

10.3.10   Faraday's Laws of Electrolysis (1834); Electrochemical Equivalent; Faraday Constant

In 1834, Faraday summarized his results on electrolytic decomposition in his two fundamental *Laws of Electrolysis*. His first law states "That the mass (m) of a substance liberated at an electrode, during electrolysis, is proportional to the quantity of electricity passed". The quantity of electricity is measured in coulombs (equivalent to the number of amperes {I} in a period {t} in seconds):

$$\text{Mass of substance liberated} \propto I \times t$$

$$m = e \times I \times t$$

The value of e for any element is an important quantity known as the *electrochemical equivalent (ECE)*.

Faraday's second law states that, "When the same quantity of electricity is passed through solutions of different electrolytes, the masses of the substances liberated at the electrodes are in ratio of their electrochemical equivalents.

The electrochemical equivalent for silver is 107.880 and it can be shown that 96,487 coulombs are required to liberate this amount of silver. The "*Faraday constant*" is the quantity of electricity (96,487 coulombs) required to liberate one electrochemical equivalent of any monovalent ion from an electrolytic solution. The *faraday* is a dimensionless unit equal to 6.02×10^{23} electric charge carriers. This is equal to one mole, also known as Avogadro's constant.

The SI unit of electrical capacitance is the "*farad*" (F), named in Faraday's honour.

10.3.11 Invention (in 1836) of Faraday Cage to Offer Protection from an Induced Charge

In 1836, Faraday's "ice pail experiment" showed that, if a hollow, electrical conductor (bucket) is charged, all the induced charge sits on the outside of the conductor. This means that the extra charge does not appear on the inside of a metal enclosure.

A Faraday cage or shield prevents electromagnetic fields getting into, or escaping from, the cage. For example, the metal body of an aircraft protects its passengers from any lightning strike. The construct of a microwave oven stops microwaves escaping. Sensitive electrical components can be placed inside a Faraday cage to prevent interference from external electrical activity.

10.3.12 Faraday Effect or Rotation; Optical Phenomenon Resulting from an Interaction Between Light and Magnets or Electric Currents (1845)

In September 1845, prompted by W Thomson (see Sect. 7.4.2) Faraday experimented with a beam of polarised light passing through an optical glass, of high refractive index, in a uniform magnetic field. He showed that, when the magnetic field was aligned in the direction of the light beam, the plane of polarisation (plane of vibration) was caused to rotate by an angle proportional to the field intensity. He further observed that, if the light was reflected backwards, the direction of optical rotation was reversed.

Hence, the direction of rotation of the plane of polarization depended solely on the polarity of the lines of force. Moreover, the rotation can be reversed by either changing the field direction or the light direction. In 1846, he was awarded the Royal Society's Rumford medal for his discovery of this optical phenomena developed by the action of magnets and electric currents on light passing through optical glass.

Faraday was the first scientist to study an interactive link between electromagnetism and light. This link would be described fully by J.C. Maxwell, in 1864, when he proposed that light is an electromagnetic wave (see Sect. 10.4.5).

10.3.13   Diamagnetism (1845)

Again in 1845, Faraday demonstrated that all substances are diamagnetic to varying degrees, some weak, others stronger. Diamagnetism competes against the direction of the magnetic field, so a diamagnetic substance exhibits a weak repulsion from a magnetic field. This can make things levitate.

10.3.14   Governmental Posts. Bow Creek Experimental Lighthouse. South Foreland Lighthouse. The World's 1st Practical Provision of Electric Power

Because of his moral code, Faraday refused to assist the British government in the production of poisonous gases for the Crimean War. However, for thirty years, he served as scientific advisor to the English and Welsh lighthouse authority, the Corporation of Trinity House.

Englishman F. H. Holmes was a pioneer of electric lighting. In 1857, he invented a magneto-electric machine powered by a steam engine. The carbon arc lamp it powered was tested under Faraday's supervision in the Bow Creek experimental lighthouse, at Trinity Buoy Wharf, on the River Thames. When it was installed in the South Foreland lighthouse at Dover cliffs, it became the first example in the world where electricity was generated for the practical provision of power.

10.3.15 Later Life, Legacy, Distinctions Accepted and Declined

In 1840, Faraday's suffered a major nervous breakdown, and for four years, he avoided research activities. In 1848, Prince Albert, Queen Victoria's consort, gave the Faraday family a comfortable house—37 Hampton Court—near the royal palace, in recognition of his contribution to science.

Faraday twice received the Copley Medal—thought to be the oldest scientific prize in the world, and the highest award from the Council of the Royal Society. He was awarded it in 1832 for his research into magnetoelectricity, and again in 1838 for his discovery of electrical induction.

Faraday was an unpretentious man of strong convictions. He declined a knighthood and requests to become president of both the Royal Society and the Royal Institution. Before his death, he was offered burial in Westminster Abbey where lie the most illustrious UK citizens, but he declined. At his own request, he is buried in an un-consecrated plot, beneath a simple headstone, with his wife, Sarah, a fellow Sandemanian, in the dissenters' section of Highgate cemetery. They died as they lived, simply, reverently, and unsung. In the twentieth century, a memorial floor plaque dedicated to him, was placed near Isaac Newton's burial spot in Westminster Abbey.

Faraday's magnetic laboratory from the 1850s is replicated in the Royal Institution's Faraday Museum (W1S 4BS).

He was praised by E. Rutherford, father of nuclear physics, who said of Faraday, "*When we consider the magnitude and extent of his discoveries and their influence on the progress of science and industry, there is no honour too great to pay to the memory of Faraday, one of the greatest discoverers of all time*".

On his study wall, Albert Einstein had pictures of his pantheon of great scientists. They included Isaac Newton, James Clark Maxwell, and Michael Faraday.

Faraday was a self-taught man who became a supreme communicator and helped popularise science and engineering. His "Christmas lecture" format continues today. He was a man devoted to discovery through experimentation. The experiments suggested the theories, and the theories guided the experiments. The ability to work in both the theoretical and experimental realms, simultaneously and creatively is a rare gift.

He specialised in understanding the physical basis of electromagnetic induction and electrolysis using rudimentary apparatus that demonstrated the potential for large-scale applications that were to be realized later in the nineteenth century and beyond.

Technologists refined his primitive devices to create electric engines, AC alternators, DC generators and transformers that would power the 'Age of Electrical Mechanisation'. His contribution to electrochemistry paved the way to miniature batteries, facilitating portable, electrical devices of the Digital Age. His laws of electrolysis laid the foundation for that field and another modern industry involving chemical-production and electroplating. His work on gas liquefaction / evaporation was the stimulus for the refrigeration industry.

10.4   James Clerk Maxwell FRS, FRSE (1831–1879)—Theoretical Physicist and Mathematician. Colour Perception; Molecular Motion in Gases and Electromagnetic Radiation

10.4.1   Background; Education and Early Career; Saturn's Rings

James was born at 14 India Street, in Edinburgh, the only surviving child to a wealthy lawyer, John Clerk. His parents had married late in life and his mother was forty years old when he was born. Shortly afterwards, the family moved ninety-five miles to a small country estate called Middlebie, situated near Corsock in Dumfries and Galloway. To satisfy some legal requirement, John Clerk added "Maxwell" to the family name to inherit the Middlebie estate from ancestors who were called Maxwell. John supervised the construction of a house which he called "Glenlair". His son was devoted to the house, living there during childhood and adolescence, returning there after his father's death and using it as a refuge until his own passing, in 1879.

James' mother died from abdominal cancer when he was eight and, two years later, he was sent to live, during term-time, with his paternal aunt, Isabella Wedderburn. Maxwell attended Edinburgh Academy which was 0.6 miles from his aunt's home. He was an intensely inquisitive child, an avid reader with a phenomenal memory but not a prodigy. His remarkable intellectual abilities were matched by his social awkwardness for which he was given the nickname "dafty" by fellow scholars.

At the age of fourteen, he displayed geometric imagination for the first time. Working with such basic tools as a length of twine, some pins and pencil, he generalized the definition of an ellipse and succeeded in producing true ovals. This appeared as his first academic paper which was submitted to a meeting of the Royal Society of Edinburgh. Since submissions from minors were forbidden, his paper had to be presented, on his behalf, by Professor J. D. Forbes.

His father wanted James to become a lawyer, but he struggled with the classics at school, and it was obvious that his talents lay in mathematics and problem-solving. He was better suited to philosophy where he could search for the 'truth'. Between 1847 and 1850, he attended Edinburgh University, reading mathematics, natural philosophy, and metaphysics. He then moved to Cambridge University, where his exceptional powers were recognised by his mathematics tutor. In 1854, he achieved the second highest mark amongst those undergraduates who obtained a first-class degree in mathematics. Thus, he joined the group known as the Second Wranglers which includes Lord Kelvin (1845) and J. J. Thomson (1880).

In 1855, Clerk Maxwell was elected to a fellowship at Trinity College, Cambridge but two years later, he decided to terminate his academic career at Cambridge. This was prompted by his father's failing health, whereupon James decided to return to Scotland. He, therefore, applied for the position of Professor of Physics at Aberdeen's Marischal College. On 3rd April 1856, his father died, and James became a Scotch laird and the proprietor of an 1,800-acre estate. A few weeks later he learnt that his appointment to the professorship had been confirmed.

In 1856, at the age of 25, he was elected fellow to the Royal Society of Edinburgh (FRSE).

Galileo had spotted Saturn's rings when looking through a telescope, in 1610, but no one could account for their composition. Maxwell spent two years hypothesising about Saturn's rings, before showing mathematically that gravity would not allow regular solid rings to survive. He postulated that they must comprise a myriad of small, solid particles, orbiting the planet. Who would have predicted that, at the end of the twentieth century, spacecrafts Pioneer and Voyager would transmit photographs of the rings via electro-magnetic micro-waves (see later) to confirm Maxwell's hypothesis?

Maxwell was not gifted at oral exposition and could not always maintain order during lectures. In 1860, the University of Aberdeen was formed by the merger of two colleges, namely, King's College (Aberdeen) and Marischal College and Maxwell was made redundant. In 1861, he was appointed to the Chair of Natural Philosophy and Astronomy at King's College, London. About two centuries after Newton's "*annus mirabilis*" (his 'year of wonders'), the 1860s were possibly the most creative of Maxwell's life.

10.4.2 Colour Spectrum; Colour Triangle; Perception of Colour

In 1668, Newton demonstrated that white light could be split in to seven coloured 'rays' (see Sect. 6.3.2). He invented the first, spinning, colour wheel (top) and investigated how different colour combinations were perceived. In 1801, Thomas Young put forward a wave theory of light, as well as a theory of 3-colour vision.

Maxwell experimented further with a spinning colour wheel. He decided that the primary 'light' colours are red, green, and blue (RGB). When equal parts of the three primary colours are combined, white light results. Conversely,

White light - red - green - blue = no light.

Primary colours are like prime numbers—they are non-divisible and cannot be formed by combination of other colours. The primary 'light' colours are said to be 'additive'. When lights of the primary colours are combined in pairs, then 'secondary' hues are produced. Red + blue give magenta; green + blue → cyan and red + green → yellow. In these paired circumstances, both wavelengths travel to our eyes without any mutual interference. Because they are emissive devices, TVs, computer monitors, and mobile phones operate on additive colours.

Adopting his characteristic mathematical approach, Maxwell developed equations that quantified how much of each primary light colour was necessary to create any secondary colour. He represented this as paired ratios on a colour triangle in which the distance from the corner (vertex) indicated how much of each primary colour must be mixed (Fig. 10.3).

In 1860, when he was twenty-nine years of age, Maxwell was awarded the Rumford Medal by the Royal Society for his research into the theory of colour vision. In 1861, he was elected Fellow of the Royal Society of London.

After floundering in the dark, visible light permits us to see the world. The human eye can see visible light with wavelengths up to 760–800 nm and down to 360–400 nm. It is now accepted that photo-receptor cone cells in the retinas of mammalian eyes are responsible for colour vision. There are three types of cone cells, each able to detect a range of colours (hues) but each group being most sensitive to a specific colour, namely red, green, or blue. Their absorption maxima are 563, 534 and 420 nm. The lens in the human eye absorbs UV light, so any UV photoreceptors would be redundant.

The range of photon energies for visible light, from red to violet, is 1.63–3.26 eV. These energies are comparable with the energy required to excite an

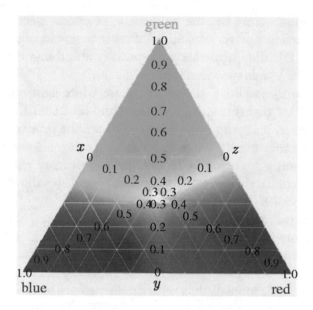

Fig. 10.3 Maxwell's colour triangle

electron in the outer shell of atoms and molecules. A single light photon with energy 1.63 eV is sufficient to stimulate red receptor molecules in the retina which triggers a nerve impulse. The brain integrates signals from the cones, allowing us to 'see' colour.

10.4.3 Colour Theories—Additive (RGB) Versus Subtractive (RYB and CMY). 1st Colour Photograph

Light enters our eyes in two ways: (i) directly from a light source and (ii) indirectly via reflection from an object. H. von Helmholtz pointed to key differences in behaviour between the mixing of coloured rays (see Sect. 10.4.2) and the depositing of coloured pigments on a substrate. For instance, when (i) coloured paints on solid objects or canvas, (ii) dyes in fabrics, or (iii) printer's ink on paper are illuminated with white light, some of the wavelengths are absorbed (subtracted) by the chemical constituents of the pigments so they cannot be reflected to the eye.

A dye is soluble in its carrier, a pigment is insoluble. The colouring agents used are inorganic, natural or synthetic substances. Representative examples include organo-metallic compounds of chromium (yellow → orange) and copper (blue → green), as well as azo dyes (red → yellow). Preferential colour absorption occurs because ambient light excites electrons in transition metal complexes or via conjugated double bonds in organic chemicals (see Appendix D). Blue pigments are especially absorbing, digesting about two thirds of the visible spectrum (red → green).

According to the 'additive' theory of colour, white light results from the mixing of primary (light) colours, red, green and blue (RGB). In traditional painting, however, red, yellow and blue (RYB) are the primary (subtractive) colours because they cannot be made by mixing other pigments. Secondary colours for painting are made by combining two primary colour pigments. Examples of secondary paint colours are orange (red + yellow), green (blue + yellow) and violet/purple (red + blue).

Professional printers, however, discovered that the three optimal primary 'pigment' colours for printers' inks are cyan, magenta and yellow (CMY). When these three are mixed in equal measure, all the colours are subtracted and the colour black (or dark brown) results. For computer printers, the black combination is made more distinct with supplementary black ink (designated 'K') to give a four-colour printing system (CMYK) (Fig. 10.4).

As with painted or dyed surfaces, the colour that a pigmented printed surface displays depends on which components of the visible spectrum are not absorbed, thereby remaining visible. The light reflected or transmitted from coloured objects stimulates the cone cells in our eyes and we perceive colour. In the case of printers' inks, the primary subtractive colours are cyan (red absorbing); magenta (green absorbing) and yellow (blue absorbing), the consequences of which are summarized in Table 10.2.

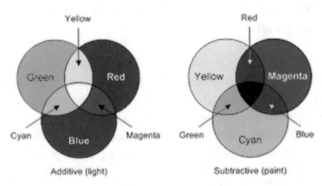

Additive and subtractive color combinations

Fig. 10.4 Comparison of the mixing of coloured rays with the mixing of coloured dyes or pigments in printers' inks

Table 10.2 Tristimulus theory of colour perception from coloured print

Colour of printers' inks (CMY)	Visible light component that this ink absorbs when illuminated with white light	Visible light components not absorbed, but reflected or transmitted and received by photo-receptor cone cells in our eyes	Brain integrates signals from cone cells so the human sees
Cyan	Red	Green + Blue	Cyan
Magenta	Green	Red + Blue	Magenta
Yellow	Blue	Red + Green	Yellow

Possibly because the best additive primary colours are RGB, the best subtractive primary colours are CMY (which absorb RGB).

Six years after his work on primary colours of light, Maxwell decided that he could produce a colour photograph. He reasoned that, if three monochromatic lights could reproduce any perceivable colour, then a colour photograph could be produced with a set of three coloured filters. He employed a professional photographer, Thomas Sutton, to take three separate exposures of a tartan ribbon taken, first through a filter of red light. then through a green filter and finally a blue-violet filter.

At a meeting of the Royal Institution, in 1861, the three developed negatives were simultaneously projected, through separate magic lanterns (projectors), with one of the same three filters, on to a screen. A reasonable, recombined, colour image of the tartan ribbon resulted. This was the first colour photograph.

10.4.4 Kinetic Theory of Gases; Maxwell–Boltzmann Distribution

During the nineteenth century, concepts were developing that atoms and molecules are in continuous motion and that the temperature of a body is a measure of the intensity of this collective motion. The idea that the physical behaviour of gases could be explained by considering the motion of their constituent particles had occurred to several people and was developed into a detailed kinetic theory of gases by Clausius, Maxwell and Boltzmann.

In 1860, Maxwell applied the methods of probability and statistics to describe the speed distribution of an assembly of gaseous particles, in thermal equilibrium, in a closed system. In the late 1860s, Boltzmann broadened Maxwell's concept of a molecular distribution function.

Maxwell assumed that particles have random motion and exchange energy when in collision between themselves and the walls of their container. The result is that the speeds of individual particles will vary enormously. Most particles have a speed close to the mean (c_{mean}), but some particles acquire considerably higher or lower speeds resulting from a series of favourable or unfavourable collisions.

On average, for a given temperature, heavier particles will travel more slowly than lighter particles. Therefore, heavier particles will have a narrower speed distribution.

If the temperature is raised, the probability of particles gaining higher energy increases. Therefore, as the temperature is raised, the number of particles travelling at high speeds increases (Fig. 10.5).

Consequently, the curve flattens to maintain the same total area under each curve (since the total number of particles is unchanging). The speed directly under the peak (c_{peak}) of the curve is known as the most probable speed. Since the curve is asymmetric, with a 'tail' to the right, the mean speed (c_{mean}) is a little to the right of c_{peak}.

The spread of particle speeds in a gas is referred to as the Maxwell-Boltzmann distribution. The distribution of kinetic energy follows the same pattern as the distribution of speeds. Here was the first use of statistical methods to describe the microscopic properties of molecules, whilst offering

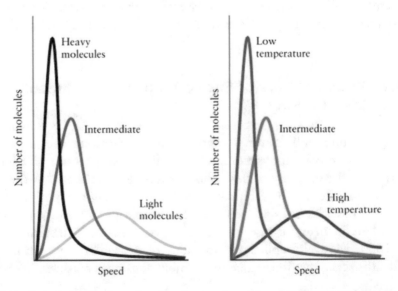

Fig. 10.5 Speed distribution of molecules as a function of their size and temperature

an explanation for the macroscopic behaviour of gases. This became a subtle theoretical tool, now called "statistical mechanics". By introducing probability theory into the physics of exceedingly small particles, Maxwell laid the foundation for quantum theory.

10.4.5   Electromagnetic Radiation; The Wave Model; Maxwell's Equations and the Heaviside Re-statement

In 1865, Maxwell resigned his professorship at King's College, London and returned to the family estate, Middlebie. For the next six years, most of his attention was devoted to experimentation, calculations, and writing. His major objective was to convert Faraday's empirical ideas on electromagnetic induction into a mathematical form. During this time, he wrote much of his ground-breaking "*Treatise on Electricity and Magnetism*" which was published in 1873.

Maxwell had two key mentors—Michael Faraday and William Thomson. Between 1845 and 1855, Faraday had developed his own theories about electro-magnetism. Maxwell attended lectures at the Royal Institution, where he came into regular contact with Faraday. The latter was quite pleased to have Maxwell apply his full panoply of mathematical skills to his experimental results and empirical evidence. Faraday's law of electromagnetic induction became one of Maxwell's four key equations.

For many scientists, magnetism was a conundrum, whilst electricity and magnetism were thought to be unrelated. Maxwell shifted attention away from the magnet per se to the space around it. Eventually, he was able to demonstrate an intimate relationship between electricity and magnetism by bringing together numerous equations, experimental results and observations related to them both, and assembling them into a consistent electro-magnetic theory. He said that "In every branch of knowledge, the progress is proportional to the number of facts on which to build and, therefore, the facility of obtaining data".

He unified existing laws and observations on electricity, magnetism and light emanating from Newton (1704), Coulomb (1785), T. Young (1801), Ørsted (1820), Ampère (1821), Faraday (1831), Gauss (1831), Lenz (1834), Lord Kelvin (1845), Fizeau (1849), Kirchhoff (1854) and Foucault (1862). By analogy, he translated mechanical relationships into electromagnetism. He derived connections between charges at rest (electrostatic charge); charges in motion (electric current); electric and magnetic fields (electro-magnetism).

Maxwell formulated a complete mathematical description based on field theory. The concept of a 'field', although lacking any mechanical parts to make it work, became central to much of classical physics. His 'wave model' of electromagnetic radiation (EMR) was described by a significant number of simultaneous equations and variables. However, lacking any experimental evidence to verify his theory, he was surrounded by sceptical colleagues.

Using a terrestrial approach, Fizeau, in 1849, had determined that visible light travels at 3.15×10^8 m/s. In 1862, Foucault improved Fizeau's rotating mirror technique and arrived at a figure of 2.98×10^8 m/s.

In 1857, W. E. Weber and R. A. H. Kohlrausch, working independently, made an empirical observation that an electromagnetic signal propagates at light velocity along a thin wire of negligible resistivity.

In 1861/2, Maxwell speculated about an elastic medium capable of supporting a transverse wave motion (i.e., perpendicular to the direction of the wave propagation). He calculated the elasticity of a medium that could support such wave motion. From that he was able to calculate the speed of the electromagnetic propagated through the medium. To his amazement, it corresponded closely to the known speed of light prevailing at the time.

He announced his conclusion as follows: *We can scarcely avoid the conclusion that light consists of transverse undulations of the same medium which is the cause of electric and magnetic phenomena"*. Maxwell had brought together under his equations, two branches of physics—electromagnetism and optics, implying they are close relatives. In 1865, Maxwell published his crowning achievement—"*A Dynamical Theory of the Electromagnetic Field*".

Simple mathematical statements can eloquently unite and express a multitude of concepts. That is why mathematics is the language of science. In his *Treatise*, Maxwell relied on twelve key equations. In 1884, Oliver Heaviside reduced the complexity of Maxwell's theory, recasting much of the original cumbersome format, condensing it to four key differential equations as they appear in their present form. These were to prove highly influential in the development of future technologies.

The following conclusions can be drawn from their joint overall analyses, as well as observations by other scientists, both before and after:

- All matter in the universe is composed of small components called atoms (see Sect. 8.6). Every atom is composed of one or more negatively charged electron, together with one or more positively charged protons. Apart from the hydrogen atom which only has one proton and one electron, all other atoms also carry one or more neutrons which are neutral—carrying no charge. If an atom gains or loses an electron, forming an ion, it becomes electrically charged since the number of electrons and protons are no longer equal, whereupon an electric field will form.

- A charged particle is a particle such as an electron or proton with an electric charge.
- The three laws of electric charge are (i) opposite charges attract; (ii) like charges repel and (iii) charged objects can be attracted to neutral objects.
- Stationary charges (electrons or ions) produce only electric fields.
- If an electron starts to wiggle up and down, side to side and / or back to front, in a steady rhythm, an electric field will form waves, moving in every direction.
- For a charge to accelerate, it needs to increase its speed or change its direction.
- If an electrically charged particle starts to move, the electric field will become a flowing electric current and will automatically form a magnetic field around it. The magnetic field will mimic the electric field in its movement but at an angle of 90° to it.
- A magnetic field is only produced once an electric device is switched on and current flows. The higher the current, the greater the strength of the magnetic field.
- Electric charges in uniform motion (constant velocity) produce both electric and magnetic fields. For every imperceptible movement of an electron, the fields will follow it in a wave-like manner.
- A changing magnetic field creates an electric field and current, and a changing electric field gives rise to a magnetic field, so electricity and magnetism are inextricably linked.
- Under the right circumstances, electric and magnetic fields can sustain one another and can transmit waves of electromagnetic (EM) energy (radiation) which spread out, travelling long distances into empty space, or shorter distances through matter.
- Once EM waves have achieved sufficient distance from the moving charge that produced them, they are free to propagate themselves (radiate) without influence of the moving charge.
- The wave front of electromagnetic waves emitted from a point source such as a light bulb is a sphere.
- Waves carry on for ever in space unless something absorbs them. They do not lose energy significantly as mechanical waves do.
- Electromagnetic radiation is a self-perpetuating coupling of an oscillating electric field and an oscillating magnetic field, the two fields being perpendicular to each other, and to the direction of energy transfer and travel

of the electromagnetic wave. For this reason, they are called "transverse" (across) waves.

- These fields are in phase at any point of propagation. Moving in concert, they both reach their maximum and minimum values at the same time and can be illustrated by two transverse sinusoidal waves. Thus, each field vibrates at the same frequency, this being the frequency of the propagating wave. The frequency is the number of wave-length cycles that pass a given point per second. The wavelength is the distance between adjacent wave crests or troughs and defines the type of electromagnetic radiation involved.
- The field strength is at its maximum at the wave crest. The amplitude is the distance between the peak or trough of a wave and its 'still point'. It is a measure of the displacement of the wave from its rest position. For a light wave, it defines the intensity or brightness, from 'intense' to 'dim'. Taller waves carry more energy (Fig. 10.6).
- Since all matter has electrons in orbit round a nucleus, under the right conditions, all matter can generate electromagnetic waves.
- Maxwell derived the speed at which electromagnetic waves are propagated through empty space. He found that it approximated to the speed of light. It would appear, therefore, that visible light is a manifestation of electromagnetic wave propagation.
- If electromagnetic radiation has wave motion, then it must have wavelength and frequency. The formula relating the frequency (f) of radiation and its wavelength (λ) to the speed (c) of light in a vacuum is: $f\lambda = c$.
- The period (T, seconds) it takes for one complete oscillation is related to frequency (f; Hz) of the wave by the equation: $T = 1/f$.

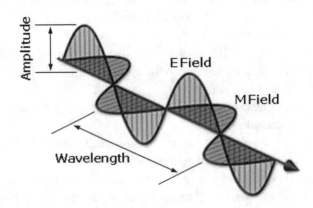

Fig. 10.6 Coupling of an oscillating electric field with an oscillating magnetic field to give electromagnetic radiation

- Maxwell proposed that light is an 'undulation' in the same medium that is the cause of electrical and magnetic phenomena. What we now know is that:
- Unlike longitudinal sound waves (see Sect. 15.2.3), electromagnetic waves do not need a supporting medium and can travel through empty space because, as envisaged, the waves do not consist of particles (but see Sect. 10.4.11 for wave-particle duality). There are consequences when EMR passes through a substance (see below).
- All electromagnetic (EM) waves travel at the same speed in a vacuum, this being the speed of light (viz 299,792,458 m/s) in a vacuum. However, EM waves travel more slowly in a static medium and, if they cross boundaries between different media, their speed changes but their frequency remains constant. For instance, white light travels more slowly through glass than air—the wave crests are closer together, but the light still oscillates at the same number of times per second, staying the same colour.
- Light slows down a little in air because it will interact with the few gaseous particles it encounters. Atoms will temporarily absorb, and then re-radiate the light, releasing it into empty space between particles, where its velocity increases again to 299,792,458 m/s.
- The degree of slow down through different transparent media will depend on the electron density of the medium—the light velocity therein being determined by the absolute refractive index of the medium (Table 10.3).
- When white light passes through a drop of water, the more energetic photons slow down more than the less energetic ones, causing the colours to separate (witness a rainbow).
- Electromagnetic waves travelling through a vacuum are unaffected by any static electric or static magnetic field. Because they have no "rest mass" and no overall charge, they cannot be accelerated, so their speed is constant.
- Electromagnetic waves travel in straight lines (within the limits set by diffraction).

Table 10.3 Velocity of light through various media

Material/medium	Absolute refractive index (η)	Velocity in medium ($c_{med} = 3 \times 10^8/\eta$) Thousand km/s	Velocity in medium ($c_{med} = 3 \times 10^8/\eta$) Thousand miles/s
Vacuum	1	300	186
Water	1.3	231	140
Window glass	1.5	200	124
Diamond	2.4	125	77.5

- Electromagnetic waves can be reflected, refracted, diffracted, and interfere with each other.
- Since the electric and magnetic fields are transverse (perpendicular) to the direction of propagation, and to each other, they are capable of being polarized (see Sect. 10.3.12). For instance, light composed of waves with vibrations at only one angle can be engineered. A polarising filter has all its molecules aligned in the same direction. As a result, only waves with vibrations in the same direction can pass through.
- Light from the Sun's photosphere is largely unpolarised since its atoms emit light in an entirely independent and random manner. Their vibrations could be vertical, horizontal or at any angle in between, so that the light waves are evenly distributed across all angles.
- Electromagnetic waves propagating (radiating) through space carry electromagnetic energy away from the source. Thus, energy can reside in fields as well as bodies.
- When the nucleus of a radioactive element disintegrates, the decay can feature gamma rays (with properties characteristic of EMR) or high speed sub-atomic particles (e.g., α- or β-particles).
- Electromagnetic radiation is all around us and takes many forms.
- Maxwell's classical physics does not account for the spectra of black-body radiators or the photo-electric effect (see Sect. 10.4.10).

Maxwell's model implied that electromagnetic fields could be propagated through empty space, as well as through conductors. During the 1880s, experimental proof materialized to demonstrate that electrical and magnetic phenomena are propagated as transverse waves. Heinrich Hertz achieved this through air; Oliver Lodge through a conducting wire (see Sect. 10.4.7).

10.4.6   Hertzian (Radio) Waves (1887)

Maxwellian theory implied that, as well as light and radiant heat, there were likely to be other forms of electromagnetic waves that propagate through space, according to electro-magnetic laws. He realized that oscillating charges, like those in AC circuits, would produce electric fields. He predicted that these changing fields would propagate from the source, rather like mechanical waves generated on a lake, after a stone lands in it.

In 1887, eight years after the death of Maxell, Heinrich Hertz was the first to generate and detect experimentally, electromagnetic waves that had features both similar to, and different from, light.

Hertz began a series of experiments generating electric sparks in different ways. For instance, he used a Rümkorff induction coil and a Leyden jar capacitor to induce a huge voltage, alternating current, across the spark gap of a di-pole antenna. He was able to produce a regular electric vibration within the copper wires supplying the gap. The violent pulses moved back and forth and comprised rapidly accelerating and decelerating electric charge. Brass spheres served to detect the sparks, the length of each antenna arm being an integral number of quarter wavelengths of the electromagnetic waves generated. In a darkened room, the sparks were visible.

Next, he found it possible to transmit the waves to a receiving, loop antenna with an adjustable micrometre spark gap. These produced surges of electric current within the receiving loop, causing minute sparks to cross the gap. It did not matter whether the two circuits were connected or not, sparks were obtained at the receiver—waves were propagated through a wire or through free space and their presence was manifest by sparks.

At a lecture, delivered in 1889, Hertz said, "We perceive 'electricity' in a thousand places where we had no proof of its existence before. In every flame, in every luminous particle, we see an 'electric' process. Even if a body is not luminous, provided it radiates heat, it is a centre of "electric disturbances". Thus, the domain of 'electricity' extends over the whole of nature".

As Maxwell had predicted, the oscillating electric charge produced electromagnetic waves that radiated through the air, around the wires. Hertz had succeeded in making electric and magnetic fields detach themselves from wires and travel through air, from one device to another, without any connecting wires (i.e., wireless transmission). Hertz did not appreciate the ramifications of what he had done. Marconi and other inventors would do that.

By way of other experiments, Hertz showed that these waves propagated at the same speed as light. Their frequency was about 100 million cycles per second (= 100 MHz). They had other properties like those of light in that they could be polarized and refracted, further verifying Maxwell's theories. Furthermore, Hertz observed that these charged objects lost their charge more readily when illuminated with ultraviolet light, demonstrating a photo-electric effect (see von Lenard, Sect. 10.4.10).

The discovery of 'Hertzian' waves (later termed radio waves) triggered an explosion of related experimentation which resulted in wireless telegraphy, leading to radio broadcasting and, later, its offspring, television.

10.4.7   Oliver Lodge—Almost the Father of the Radio in 1894

In 1888, Oliver Lodge used a pair of 29 m long wires as wave 'guides' connected to strategically placed spark gaps. When two Leyden jars (capacitors) were discharged, intense sparks occurred at the terminal spark gap. This represented an integral number of half-wavelengths of the propagating wave and the place where the incident wave and reflected waves were in phase. Less intense sparks at two interceding spark gaps occurred at half-wavelength distances. Lodge had created a standing electromagnetic wave along a wire in much the same way as a single note and its overtones are produced in a musical instrument.

After Hertz's death in 1894, Lodge's research extended well into the era of practical radio communication. In 1894, before the work of Marconi, he gave a public demonstration of a radio, in Oxford, and filed a patent for a tuneable radio in 1897.

10.4.8   Electromagnetic Spectrum

All the energy arriving in. and departing from, the Earth's climatic system is in the form of electromagnetic waves. Human eyes are photon detectors which are capable of spotting electromagnetic waves with a wavelength between 390 and 700 nm, equivalent to about 0.0035% of the full electromagnetic spectrum. All other electromagnetic waves are invisible to the human eye. In addition to radio waves, these other forms include microwaves, heat in the form of infra-red radiation, ultra-violet light, X-rays, and gamma rays. Electromagnetic waves are produced by the motion of electrically charged particles. Because they radiate from these charged particles, they are also called electromagnetic radiation (EMR).

Collectively, they are classified by their energy, frequency, or wavelength ranges (see Appendix D). Shorter waves carry more energy than longer ones. Essentially, there are seven overlapping sub-ranges (portions) across the spectrum, with gamma-rays at one end and radio waves at the other. Gamma rays have short waves (smaller than atomic nuclei) and higher frequencies, whilst exceedingly long radio waves are the size of buildings.

The seven categories can be further classified as to whether they are ionizing radiation or non-ionizing radiation. Ionizing radiations are short wavelengths, extremely high frequency electromagnetic waves (e.g., X-ray and gamma-rays) which have enough photon energy to ionize atoms or molecules

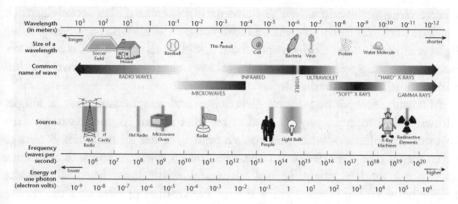

Fig. 10.7 Electromagnetic spectrum. *Source* (1) https://commons.wikimedia.org/wiki/File:Cont_emspec2-es.png#/media/File:Cont_emspec2.jpg. http://son.nasa.gov/tass/content/electrospectrum.htm. Unknown author. Public Domain. NASA usage guidelines: https://www.nasa.gov/multimedia/guidelines/index.html. (2) https://www2.lbl.gov/MicroWorlds/ALSTool/EMSpec/EMSpec2.html; courtesy Lawrence Berkeley National Laboratory

or break chemical bonds (see Appendix D). X-rays pass through the human body, making them useful in medical applications.

On the other hand, non-ionizing radiation (e.g., ultra-violet, visible, infrared, and radio [including TV, microwave, and radar] waves) has photon energies too weak to have these effects. Any effect on chemical systems or living tissue that they do have is caused primarily by heating effects, not from the energy input of one photon, but from the combined energy inputs of many photons. The more photons arriving per second, the greater the potential for damage.

The electromagnetic spectrum is explored broadly in Appendix D and shown diagrammatically in Fig. 10.7.

10.4.9 Summary of Maxwell's Hypotheses. Scientific, Technological, and Social Impacts

As well as radio communication, Maxwell's research into EMR would lead, one day, to numerous other applications, including the development of TV, microwave ovens, radar, satellite communication, mobile phones, infra-red and radio telescopes, and X-ray machines. Microwaves are useful for transmitting telephone signals and for cooking food. The eventual social impacts of his research are profound.

The movement of electrons is implicated in the generation of different types of EMR. For instance, a metal antenna is a device to transmit and receive electromagnetic waves. An alternating current makes loose electrons oscillate back and forth, so creating radio waves, TV, and micro-waves. Visible light is emitted when electrons fall from a high-level atomic orbit to a lower one. X-rays are given off when high-velocity electrons collide with a metal surface, and "braking radiation" is emitted.

Maxwell incorporated light, electricity, and magnetism into a single, unified field theory in physics, and it remains one of its cornerstones. This was not a synthesis of what was known before, rather a fundamental change in concept, that departed from the Newtonian view and was to influence greatly the modern scientific and industrial revolutions. Maxwell was the bridge between the mechanical world of Newtonian physics and the theory of fields, as espoused by Einstein and others.

There were, however, two major shortcomings with Maxwell's wave theory, and these became the bridge between Newtonian physics and the radical proposals in physics in the early decades of the twentieth century, concerned with quantum physics and relativity.

10.4.10   Anomalies with Maxwell's Classical Electromagnetic Wave Theory. Blackbody Radiation; The 'Ultraviolet Catastrophe' and the Photo-Electric Mystery. A 20th Century Explanation Via Quantum Theory

A blackbody is an idealized object that allows all incident radiation to pass into it (so there is no reflected energy) and internally absorbs all the incident radiation (so no energy is transmitted through the body). Such a body is a perfect absorber for all incident radiation and for that reason, is jet black.

All bodies heated above absolute zero emit energy in the form of electromagnetic radiation. According to Wien's Law, as the temperature of a blackbody is increased, the peak wavelength decreases. successively resulting in shifts in the visible spectrum, from dull red, to bright red, orange and yellow, finally giving a white glow, suggesting that the colour blue has joined the other primary colours. This change in colour or frequency distribution is somewhat surprising since one might expect more of the same colour with rising temperature.

In some ways, bodies with surface colours other than black behave like blackbodies when heated. For instance, a blacksmith working with a piece of iron in hot charcoal, at about 1000 K, sees the iron glowing red and it is surrounded by charcoal glowing with the same colour, even though the chemical properties of the two substances are enormously different. Steelmakers who routinely heat materials to a specific temperature have a reasonably standard nomenclature as is shown in Table 10.4.

A second feature of a blackbody is that its radiation is independent of its surface and its chemical composition, being dependent on only one parameter, namely temperature. An idealized blackbody emits the maximum amount of electromagnetic radiation for its absolute temperature. Equations (e.g., Rayleigh-Jeans) to describe the spectrum of perfect radiator, derived from Maxwellian classical theory, produce a reasonable match with experimental data at long wavelengths but do not fit at short wavelengths. In practice, as the temperature is raised, the peak wavelength emitted by a blackbody decreases. However, intensive ultraviolet radiation does not predominate (is suppressed) at high temperatures. This contradiction was called the 'ultraviolet catastrophe'.

Apart from reservations by Thomas Kuhn, most science historians assert that, in 1900, Max Plank arrived at an equation which did account for the observed spectrum. However, it required a new assumption to be added to those of classical physics. He assumed that oscillators emitting radiation from a hot blackbody can only have discrete energies, namely E, 2E, 3E etc. Thereby, any change in energy must be discontinuous because a change of less than one is not allowed. The resonator energy is, to use the modern terminology, "quantized". This was a revolutionary departure from classical physics.

Eventually, Plank's hypothesis led to an acceptance that each resonator can have an energy (E) which is zero or some multiple of a fixed amount which depends on the frequency (f) of oscillation, according to the formula: E = nhf (where 'n' is an integer, 0, 1, 2 etc.; and 'h' is a constant, now known as Plank's constant). For instance, in the case of an infrared frequency of 10^{14} Hz, being emitted by a blackbody, the difference in energy levels is about 0.4 eV which would be unnoticed on a macroscopic or classical scale.

If the thermal energy of the body is randomly distributed, the chance that a high frequency (short wavelength) oscillator will acquire sufficient

Table 10.4 A steelmaker's colour—temperature chart

Temp (°C)	480	600	800	950	1100	1300	1500
Colour	Barely red in dark	Dark red	Cherry red	Orange barely visible in sunlight	Orange—yellow	Light yellow, blinding	Nearly white, blinding

energy to start vibration is much smaller than for the lower frequency oscillators. Effectively, high frequency emissions are suppressed and the 'ultraviolet
dilemma (catastrophe)' is solved. The graph below shows that, at 6000 K,
invisible infrared and visible range emissions predominate over UV emissions
(Fig. 10.8).

The inescapable conclusion is that electromagnetic radiation (EMR) is not
emitted continuously, with one wavelength changing smoothly into another.
Rather, it is emitted discontinuously, in a stream of discrete energy bundles—
'energy quanta', now called 'photons'. In quantum mechanics, EMR consists
of photons, uncharged elementary particles with zero rest mass which are the
"quanta" of the electromotive force. A quantum is the base 'unit' of EMR.
A photon, as an energy carrier, is associated with a probability wave, with a
frequency proportional to the energy carried. A light ray, for instance, consists
of a finite number of energy quanta which are localised in space, which move
without dividing, and which can only be produced and absorbed as complete
units.

EMR emissions, in the visible and ultraviolet regions, can result from excitation of electrons to higher energy orbitals. Excitation can be caused by
heat stimulation (see above) or by bombardment with photons (see [a] in the
schematic Fig. 10.9). When the excited electrons lose the acquired energy and
return to their ground state, the freed energy escapes from the atom in the
form of photons (see [b] in Fig. 10.9 below and Table D.1, Appendix D).

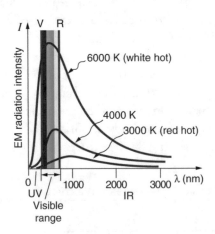

Fig. 10.8 Blackbody radiation showing UV intensity is low even at high temperatures. *Source* https://courses.lumenlearning.com/physics/chapter/29-1-quantization-
of-energy/. Figure 1 https://openstax.org/books/college-physics/pages/1-introduct
ion-to-science-and-the-realm-of-physics-physical-quantities-and-units. TOC, Chap. 29
(Quantum physics), Sect. 29.1 (Quantization of energy), Fig. 29.3. Creative Commons
Attribution License 4.0. https://creativecommons.org/licenses/by/4.0/

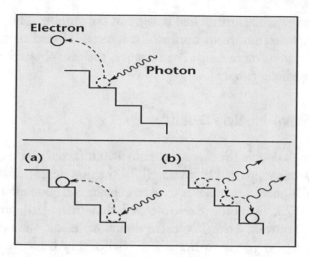

Fig. 10.9 Photon absorption to give excited state or electron ejection. Courtesy Lawrence Berkeley Laboratory

Sometimes, electrons will absorb sufficient energy to escape completely from the experimental material (see top of schematic). This is called the photoelectric effect (Fig. 10.9).

Philipp von Lenard extended the work of Hertz on the photoelectric effect (see Sect. 10.4.6). The photoelectric effect suggests that certain (mass-less) photons have sufficient 'momentum' to eject electrons. Working with high vacuum apparatus, von Lenard showed that, when ultraviolet light falls on a negatively charged metal, electrons are released more readily, and these are propagated through the vacuum. Red light, irrespective of its intensity was incapable of releasing electrons.

According to Maxwell's wave theory, light energy is spread evenly across the wave front, so electrons should be emitted, only if sufficient energy is delivered to the electrons. The ejection should depend only on the intensity of the incident light and not its frequency. This is contrary to observation of the photoelectric effect.

In 1905, Einstein realized that Plank's idea about light behaving as packets of energy could be key to understanding this 'photoelectric mystery'. If the wavelength of the light is short enough, the electrons at the surface of the negatively charged metal receive sufficient energy to break free from the surface and escape. A certain threshold value must be exceeded. Conversely, if the wavelength is too long, the absorbed energy is insufficient to facilitate escape of the electrons. Drenching the metal with excessive amounts of 'impotent' bundles of light energy has no effect.

It is the energy of the quanta that is key, not the number. Thus, weak ultra-violet can emit electrons from zinc, whilst intense infrared cannot. Reactive metals lose electrons more easily. For instance, the ease of ejection is caesium > potassium > zinc > copper.

10.4.11   Wave-Particle Duality

Thomas Young's double slit experiment, in 1805, favoured a wave theory of light and this was supported by Maxwell. In the first decade of the twentieth century, Plank's photon model was used to explain two anomalous behaviours of light. In 1923, Compton demonstrated that (mass-less) photons could collide with electrons, changing their direction, energy and momentum, acting in every way like a particle. The relationship between the photon's momentum (p) and its wavelength (λ) is given by the formula: $p = h/\lambda$. Newton's corpuscular theory was resurrected.

It became accepted that neither wave theory nor quantum theory, in isolation, can fully describe the behaviour of light. Scientists came to the view that Newton's mechanical model could co-exist with a continuous field model which is not mechanically explicable. Paradoxically, there appears to be a wave-particle duality, with the most appropriate model being applied, depending on the phenomenon under scrutiny, as illustrated in Table 10.5.

Because of this parallelism, EMR can be described as a stream of mass-less particles, called photons, each travelling in a wave-like manner, at the speed of

Table 10.5   Wave-particle duality

Phenomenon	Can be explained in terms of waves	Can be explained in terms of particles
Reflection	✓	✓
Refraction	✓	✓
Interference	✓	✗
Diffraction	✓	✗
Polarization	✓	✗
Black body radiation	✗	✓
Photoelectric effect	✗	✓

light. A brighter light source delivers more packets of energy, but the amount of energy each packet contains is the same.

Maxwell's work was admired by Einstein who said, "*Since Maxwell's time, physical reality has been thought of as represented by continuous fields and not capable of any mechanical interpretation. This change in the conception of reality is the most profound and the most fruitful that physics has experienced since the time of Newton*". Hanging on his study wall, Einstein had portraits of his three heroes of science—these were Newton, Faraday and Maxwell.

10.4.12 Maxwell's Later Life and Legacy

In 1871, Maxwell returned to Cambridge to become the 1st Cavendish Professor of Physics. He was placed in charge of designing the Cavendish Laboratory and specifying apparatus. His inputs were practical and clever, serving the needs of the physics department for over a century.

Although one of the giants of physics, Maxwell's contribution is less well-known, and he received fewer awards and decorations than other notable scientists, possibly because his equations were so hard to comprehend, and they needed equally knowledgeable mathematicians such as Boltzmann and Heaviside to simplify them. Moreover, Maxwell was a quiet, modest, self-less man who derived joy in his intellectual activities. He did not attract publicity. Neither did he need the fame and appreciation of others to give him satisfaction in what he had achieved.

His development of statistical physics was a keystone in quantum physics, where the probability, rather than the certainty, of an event behaving in a certain way is recognised. A 'field' although lacking any mechanical parts to make it work, became central to much of modern physics. It is arguable, therefore, that Maxwell is the nineteenth century scientist who had the greatest influence on twentieth century physics, laying the foundation for such disciplines as quantum mechanics and special relativity. He has been described as 'Scotland's Einstein'.

His unified theory of physics, showing electricity, magnetism and light are intimately connected, was revolutionary. His electro-magnetic theory underpins so much of our modern world. His equations have a reach that extends to the extremities of the universe. He set the stage for many great accomplishments in physics, electrical engineering, colour photography, telecommunications, domestic, medical, security and contact-less banking applications etc.

In recognition for his work on magnetism, a 'maxwell' (Mx) was adopted in the CGS system to represent a unit of magnetic flux. However, this was superseded in the SI (MKS) system by the weber (Wb).

Maxwell died in 1879, at the same age as his mother and from the same disease, namely abdominal cancer. His remains were brought back to Galloway and buried besides those of his parents, within the ruins of the Old Kirk which lies in the graveyard of Parton's Parish Church, seven miles from his ancestral home at Glenlair.

The James Clerk Maxwell Foundation acquired his place of birth, at 15 India Street (EH3 6EZ) and displays a growing collection of heritage material.

11

Palaeontology and Evolution

Am not I a fly like thee? Or art not thou a man like me?. (William Blake, Songs of Experience, 1794)
Although Nature needs thousands or millions of years to create a new species, man needs only a few dozen years to destroy one. (Victor Blanchford Scheffer, Spires of Form: Glimpses of Evolution, 1983)

11.1 Précis

Mary Anning grew up in poverty; was self-taught and lived at a time when most people believed in Creationism—a conviction that nature is static and unchanging, and species are immutable. She had a life-time passion for fossil hunting, and by reading and application, she became a world expert in her field, arriving at a level of knowledge exceeding those formally educated in the discipline.

Fossils provide information about the nature of species that existed at specific times in the Earth's history. The spectacular fossils she unearthed shook the scientific world into looking for a new approach to explain why some creatures are extinct, whilst others took their place. She lived on the Jurassic coast, now a World Heritage site and some of the fossils she found are housed in the Natural History Museum, in London.

In 1831, Charles Darwin was a budding naturalist given the opportunity to join a 5-year expedition to see ecologically diverse regions in the Southern

© The Author(s), under exclusive license to Springer Nature
Switzerland AG 2022
Γ. Bailey, *Inventive Geniuses Who Changed the World*,
https://doi.org/10.1007/978-3-030-81381-9_11

Hemisphere. His epoch-making voyage, on HMS Beagle, had a monumental effect on Darwin's view of natural history. He logged similarities amongst species across the world, as well as variations based on specific location and habitat. For instance, on each of the Galápagos Islands, he found a variety of unique species. He found finches with beaks that differed from island to island. The contribution that finches made to Darwin's future thinking may have been exaggerated but they became an emblem of evolution.

Wherever he went, he collected samples of flora, fauna, and fossils. He sent pertinent samples and scientific notes to a colleague in England who acted as a conduit to the scientific community. As a result, when he returned home, he had become a respected, well-known scientist.

As Newton and Clerk Maxwell had done with physics, Darwin developed a grand, unifying theory of biology. His revolutionary new thesis about the origin of living things was contrary to the popular view of divine intervention and would change the course of scientific thought. It led Darwin to believe that contemporary species have transmuted (evolved) from common ancestors. Since life on Earth began, the changing environment has put pressure on individual organisms to adapt or face extinction. Those individuals that adapted favourably to the local conditions preferentially survived and passed their favourable traits to their offspring. The mechanism by which each slight variation, if useful, is perpetuated, is called natural selection. This is one reason why evolution occurs, mainly through a series of short steps.

Survival of the fittest does not mean the strongest, nor the most intelligent, but the one that is most adaptable to change. Darwin did not say that humans are descended from apes, rather he said that chimpanzees, apes, and humans may have a common ancestry because of their many similarities.

Humans have managed to occupy almost every ecological niche. We now look at ourselves and question how we relate to all creatures and how our lifestyles impinge on theirs. *Homo sapiens* may be part of the natural world but has evolved to a point where it can either destroy life or preserve it with all its splendour. We must act decisively to preserve the healthy condition of the world's biosphere and atmosphere. By protecting eco-systems, we preserve biodiversity.

11.2   Mary Anning (1799–1847)—Fossilist and Palaeontologist

Mary was born in Lyme Regis, Dorset. She was one of ten children but only two survived to adulthood. Indeed, almost half the children born in the nineteenth century died before the age of five years. Mary's father was a cabinetmaker, amateur fossil-hunter, and religious dissenter. From the age of eight, Mary attended a Congregationalist Sunday school, where she learnt to read and write.

In the Jurassic period, about 200 million years ago, the region around Lyme Regis was a shallow sea. This was a period when life was abundant. Giant marine reptiles inhabited the sea; dinosaurs ruled the land masses and pterosaurs flew across the skies. Whilst soft-bodied creatures post-mortem rot away, leaving no trace, hard-bodied creatures leave behind their shells, bones, and teeth.

The Blue Lias of Lyme Regis exists in layers of limestone interspersed with soft clay. It contains numerous remains of prehistoric sea creatures that were fossilized after their deaths. As the sea-level dropped, these fossils could be found on the beach and above it, especially in exposed rock cliffs. The geology is such that the fine sediment facilitates recovery of skeletons whilst a specific set of chemical conditions means good preservation, even for pieces of skin.

Mary, her father, and other siblings looked for spiral-shaped shells which were ammonites—a type of mollusc that lived in the Jurassic period. In the late 1700s, Lyme Regis was a popular summer seaside town and Mary's father became a dealer in these fossil "curiosities". They included 'snake stones' (ammonites) 'devil's fingers' (belemnites) and 'vertaberries' (vertebrae).

In 1810, when Mary was eleven, her father died from the combined effect of tuberculosis and a serious fall from the local cliff. Food had become more expensive because of Napoleon Bonaparte's wars to conquer Europe, as well as a colder than normal decade, resulting in lower crop yields. Destitute, the Anning family burnt furniture to keep warm and were under constant threat of the workhouse. They did receive some assistance from the Overseer of the Poor, a role established in 1597 to help the parish poor. Nonetheless, Mary had to work harder to find fossils for sale to supplement their meagre income.

She particularly liked to fossil-hunt after a storm. The wind, rain and waves made the rocks in the local cliffs crumble, making it easier to spot fossils. It is believed that Terry Sullivan's tongue-twister, "she sells seashells by the seashore" was written about her.

In 1812, when she was twelve, her brother found the 4 ft skull of an ichthyosaur (translates to fish-like lizard). Some months later, Mary found some other parts of the skeleton. They sold it to a local collector. The skull of this Temnodontosaurus platyodon is now displayed at the National History Museum in London (SW7 5BD).

In 1821, Mary found three more ichthyosaur skeletons, ranging from 5 to 10 feet in length. They were marine vertebrates with fossils resembling those of terrestrial dinosaurs.

In 1823, she found the first complete Plesiosaurus skeleton. The Plesiosaurus dolichodeirus was another "sea dragon" 10 ft (3 m) long. It had a snake-like neck, short tail, tiny head and four pointed flippers, shaped like paddles. This particular skeleton is the holotype—the specimen used to describe the species.

In 1828, she discovered the anterior sheath containing an intact ink bag of the nominal genus, Belemnosepia. Belemnites are fossilized, squid-like, dart-shaped invertebrates of an extinct group of marine cephalopods, related to the modern cuttlefish. They contained an ink sac and had ten arms, but no tentacles. Ink could be ejected into the sea water to confuse predators.

In the same year, she found the first, (post cranial) skeleton of a pterosaur discovered outside continental Europe. It was a "flying dragon", a strange winged-creature and the first pterodactyl of the Dimorphodon genus. More complete skeletons found later by others revealed that it had a huge head and given the name Dimorphodon macronyx.

Also, in 1828, she found fossilized faeces in the abdomen of ichthyosaurs which were called bezoar stones (later coprolites). When they were closely examined, she found small bones and fish scales, demonstrating that ichthyosaurs were carnivorous, their prey being small fish, squid, and shell-fish.

In 1829, she unearthed a Squaloraja polyspondyla, a cartilaginous fish that appeared to be an evolutionary step between rays and sharks. A year later, the large skulled Plesiosaurus macrocephalus was discovered. It is now housed at the Natural History Museum.

Her findings were key to the development of palaeontology as a scientific discipline. She was in contact with geologist Adam Sedgwick, one of Charles Darwin's tutors and gained the respect of geologists A. de la Beche and W. Buckland.

However, because she was working class, and female, with no formal education, she was not completely accepted by the nineteenth century, British scientific community. For instance, the Geological Society of London was founded in 1807 but its first female fellow was not elected until 1919. As was common with most 'learned' bodies, women were excluded from both membership and attendance of Society lectures, during the nineteenth century, as they were deemed to lack the intellectual rigor to engage in scientific study.

Anning grew up in poverty and was self-taught. Her short life was scarred by hardship, and tragedy. Although female, working class and poor, by reading and application, she arrived at a level of knowledge exceeding those formally educated in the field. To this extent, she was consulted and visited by many eminent scientists of the era.

She died from cancer at the age of forty-seven and is buried in the grave-yard of St Michael's parish church, in Lyme Regis. Four stained-glass panels in the church were commissioned by the Geological Society, in recognition of her charitable work and for furthering the science of geology.

Geology is the study of the Earth's 4.5-billion-year history, its various natural processes, and materials. Palaeontology lies on the boundary between biology and geology since palaeontology focusses on the record of past life, drawing its main source of evidence from fossils and deceased organisms buried in rocks. It uses these to better understand their relationship with each other and their evolution.

The building that was once Mary's home and her fossil shop, until 1826, was demolished in 1889 to make way for the Lyme Regis Museum (DT7 3QA). A 96-mile-long stretch of the Jurassic coast has been awarded the status of a World Heritage Site, famous for its geology and fossil finds. The search for fossils continues. In 2016, another Ichthyosaur skeleton was unearthed and that formed the basis of a BBC 1 documentary, presented by Sir David Attenborough.

Numerous scientists have speculated about how life arose on a rocky planet, third nearest to the Sun. The Earth was formed about 4.54 billion years (Gy) ago; oceans following about 0.13 billion years later. Scientists believe that cellular life began in the seas. Stromatolites (fossilized mats of cyanobacteria—prokaryotic organisms that could photosynthesize but lacked a nucleus) have been found that date back about 3.7 b years.

Such prokaryotic organisms have reproduced themselves by cell division, using RNA for self-replication. Evolution was initially slow because new forms appear rarely by mutation. About 2.8 billion years ago, eukaryotes appeared, the cells of which contain a distinct nucleus containing DNA. Their genetically diverse offspring triggered the multiplication of a dazzling array of organisms with different shapes, sizes, colours and behaviour.

The first fish possibly appeared 500 million years ago. The age of dinosaurs began about 230 million years ago and ended 65 m years ago. Prehistoric creatures are not the same as creatures living today.

Mary Anning was a devoted Christian and lived in a time when many people believed in Creationism—the Biblical account of creation, whereby everything in the universe had been created by God, within six days. The spectacular marine reptiles that she unearthed shook the scientific community into looking for a different explanation for changes in the natural world. Some creatures are extinct, and others took their place. This is the process of evolution and scientists, including Charles Darwin, were compelled to speculate how it might occur. Extinction, however, implied that God's initial creation had been imperfect.

11.3 Charles Robert Darwin FRS, FRGS, FLS, FZS (1809–1882)—Naturalist and Biologist

11.3.1 Background

Charles was born in a grand Georgian house in Shrewsbury, Shropshire. He was a child of wealth and privilege. His father was a prosperous society physician and Charles was the fifth of six children. His mother died when he was only eight years old. After his mother's death, he was sent, as a weekly boarder, to the Anglican Shrewsbury School. Here, he was judged to be a very ordinary boy, rather below the common standard in intellect.

As a growing boy, he had an intense interest in the diverse range of animals and plants that occupied his surroundings. He was particularly fascinated in beetles and, under the encouragement of one of his teachers, he developed a large beetle collection, including some rare species. Later in life, he would use beetles to illustrate different evolutionary phenomena.

In the summer of 1825, Charles, aged sixteen, served as apprentice in his father's medical practice. He then followed in his father's footsteps to the prestigious medical school at Edinburgh University. Unfortunately, the sight of blood made him queasy, and he was perturbed by the rather brutal techniques of surgery practised at the time.

In his second year at medical school, he became more interested in zoology. Whilst in Edinburgh, Darwin had taxidermy lessons from a freed slave, J Edmonstone who lived nearby. This skill would prove indispensable for preserving finches and other creatures during his Beagle voyage. He befriended Edmonstone who would have told him about the exotic flora and fauna living in his birthplace, in the Southern Hemisphere.

In 1827, his father next sent him to Cambridge University to study theology with a view to his becoming a clergyman. He became an undergraduate at Christ's College.

He became friends with geologist, Adam Sedgwick. Also, at the University was the Reverend J.S. Henslow—a popular botany professor, having just resigned as Professor of Mineralogy. In 1828, Darwin secured an invitation to one of Henslow's scientific soirées. Darwin was stimulated by Henslow's progressive teaching methods which relied on field- and garden-work, as well as encouraging students to make observations on their own. Darwin regularly attended Henslow's field trips, and he became known as "The man who walks with Henslow".

Darwin graduated in 1831 with a Bachelor of Arts degree in theology, Euclid, and the classics.

11.3.2 Voyage on HMS Beagle

At the time, R FitzRoy, captain of HMS Beagle, was organising a five-year expedition to the South Seas but involving a circumnavigation of the globe. The Beagle was one of the Royal Navy's survey ships. FitzRoy asked his superiors whether a well-educated and scientific gentleman could accompany him as an unpaid naturalist. Firstly, the position was offered to Professor Henslow, but he declined and recommended his protégé, the 22-year-old Darwin. This was an excellent opportunity for a budding naturalist to see ecologically diverse regions and would ultimately change the course of scientific thought.

Before HMS Beagle set sail in December 1831, Darwin was elected as a Corresponding Member of the Zoological Society of London (ZSL), becoming a member of its Council in 1839. The London Zoo (NW1 4RY) is the world's oldest scientific zoo and Darwin used his visits there to develop his theories.

Whilst sailing, Darwin read Charles Lyell's *Principles of Geology*, published in 1830. Lyell's book was an attack on the common belief that unique catastrophes or supernatural events, such as Noah's flood, shaped the Earth's surface. Lyell made a "uniformitarian" proposal that the formation of the Earth's crust took place through countless small changes occurring over vast periods of time.

Darwin's first stop was the volcanic Cape Verde Islands in the Atlantic Ocean, off the western coast of Africa. Here they found seashells in the cliffs, 45 feet above sea-level, providing clear evidence of land uplift and supporting Lyell's theory of long-acting, gradual, planet-sculpting forces.

For about one month, the Beagle was anchored at Bahia Blanca, 400 miles southwest of Buenos Aires. In a small boat they sailed ten miles from their anchorage to some reddish mudstone cliffs near Punta Alta. Darwin made a major find of fossil bones of nine huge extinct mammals alongside some modern seashells, indicating relatively recent extinction with no signs of sudden change in climate or catastrophe. One of the mammals was a 3-m-long, giant, ground sloth (later named Mylodon darwinii) which would have weighed about 200 kg.

In Chile, Darwin experienced an earthquake. He saw mussel-beds stranded above high-tide, indicating that the land had been raised.

In 1835, on the Galápagos Islands, he found a variety of unique creatures including giant tortoises, marine iguanas and what became known as Darwin's finches. There are nineteen major and minor islands, and these form part of the Republic of Ecuador, 600 miles to the west. The islands emerged from the floor of the ocean; are several million years old; wholly volcanic in origin and have never been connected to the South American mainland. Here was to be Darwin's moment of enlightenment, his epiphany.

Each island seemed to have its own distinct varieties of wildlife. He found finches with beaks that differed from island to island. Wherever he went, Darwin collected samples of flora, fauna, and fossils. He sent scientific notes and pertinent samples to Henslow who acted as a conduit to the scientific community. When Darwin embarked on this epic voyage, he was an unknown graduate. When he returned to Falmouth, England, he was a respected, well-known scientist.

When he started the voyage, Darwin was fresh out of divinity school and was quite religious. However, when he saw slavery first-hand and witnessed the wretched living conditions of the natives of Tierra del Fuego, he began to question his religious beliefs. He began to wonder why God allowed such inhumanities to occur. He supported the emancipation of slaves and the abolition of slavery. He further lost faith when three of his ten children died before reaching eleven years of age. From the early 1850s, he stopped attending church with his family and from that point, he was probably an agnostic.

11.3.3 Developing His Theory of Evolution and the Mechanism of Natural Selection

In 1837, Darwin presented some of his geological findings to the Geological Society of London. He hypothesised that the South American continent is gradually rising from the sea. On the same day, he presented specimens of the mammals and birds that he had collected from the Galápagos Islands. Ornithologist, curator and preserver at the museum of the ZSL—John Gould—was able to establish that Darwin had collected twelve new finch species, forming an entirely new group of finches.

Biographers and writers may have exaggerated the contribution that Galápagos finches made to Darwin's thinking on evolution and natural selection.

For instance, whilst present on the Islands, he did not identify the birds collected as finches (that was done by J. Gould). Neither did he systematically label them by island of origin, nor did he study their diets. However, finches have become the emblem of evolution. Increasingly, biologists were drawn into resolving questions about finches that Darwin left unanswered.

Adaptive radiation is an evolutionary pattern whereby a single-centred form or species diversifies (or speciates) into several related forms. According to this concept, the different descendent taxa are similar, but each has adapted to a particular environmental niche. The Galápagos Islands are living laboratories for both allopatric speciation (in separate geographic areas) and sympatric speciation (same geographic location).

In the case of the Galápagos finches, they differ in body size, as well as the shape and proportions of their beaks and feet. They are all probably descended from a common ancestor that arrived in the Archipelago from central or South America, breeding pairs that were possible blown off course by the prevailing winds. Most finches are generalized eaters, but these have undergone adaptations to allow for specialized feeding, facilitating survival during the dry season or other times when food is scarce. Fourteen Galápagos finches are now recognised, and these are classified in five genera.

Their specific kind of beak equips them to acquire distinct food sources from unexploited ecological niches. For instance, four finches of the Geospiza genus are seed eaters. So-called cactus finches boast long, pointed beaks for picking seeds from cactus fruits. Their two Geospiza relatives—the ground-foraging finches—have short, stout, thicker beaks serving best for the harvesting and crushing of seeds found on the ground.

The five finches of the Camarhynchus and Cactospiza genera are insect-eaters. Their bills vary because they eat different types and sizes of insects, and they capture their prey in different ways. In the genus Certhidea, the two Warbler Finches catch their prey in flight, spearing insects with beaks that are sharp and slender. A vegetarian finch—with a heavy bill for grasping and wrenching buds from branches—is to be found in the fifth genus, Platyspiza. A selection of nine, of the total of fourteen finches, is shown in Fig. 11.1.

After his return from South America, Darwin developed a life-long illness that left him severely debilitated or bed-ridden for long periods. In January 1839, Darwin and his first cousin, Emma Wedgewood were married. They had ten children. Whenever his children fell ill, he feared that they might have inherited weaknesses from in-breeding due to the close family ties he shared with Emma, his cousin. Despite his misgivings, most of their surviving children and many of their descendants went on to have long and distinguished careers.

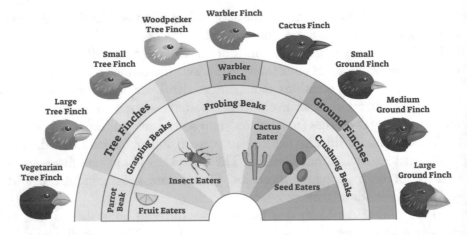

Fig. 11.1 Galápagos finches—adaptive radiation results in optimal beak morphology for local food sources

The Beagle epoch-making voyage had a monumental effect on Darwin's view of natural history. Through his observations and studies of birds, plants and fossils, Darwin noticed similarities among species all over the world, as well as variations based on specific locations and habitats.

Over several years, he began to develop a revolutionary new theory—built on hard scientific evidence—about the origin of living-beings which was contrary to the popular view. This led him to believe that the species he knew had gradually evolved (his word *transmuted*) from common ancestors. He could have published a theory of evolution by natural selection in the 1840s but did not.

He started to question the idea of religious providence. He recognised that his theory conflicted with Creationism and the religious view that all of nature is born of God, that it is static and unchanging. Perhaps concerned about public and ecclesiastic condemnation of his theory of evolution, he delayed any public announcement for as long as possible. However, in 1858, his Scottish, geologist friend and mentor, Charles Lyell, urged him to publish his theory to establish precedence.

In that year, Darwin received a paper from Welshman, A.R. Wallace who asked for his opinion on it. The paper was entitled "*On the Tendency of Varieties to Depart Indefinitely from the Original Type*". Wallace's theory was based on species adapting to environmental pressures to survive, resulting in their divergence (or evolution).

Darwin's theory had a different emphasis. He looked at the pressure of competition between the same or similar species. It was decided to read the respective theories of Darwin and Wallace at the annual meeting of the

Linnaean Society of London, in 1858. The Society is the world's oldest, active biological society and Darwin had been made a Fellow (FLS) in 1854. Unfortunately, his baby son, Charles Waring, had recently died from scarlet fever and Darwin was too distraught to attend the reading. Wallace, on the other hand, was still conducting field work on the Malay Archipelago.

In 1859, after twenty years of painstaking study, Darwin published his seminal work. The original title was *"The Origin of the Species by Means of Natural Selection or the Preservation of Favoured Races in the Struggle for Life"*. In his sixth and final edition, in 1872, the title was simplified to *"The Origin of the Species"*, but this did not mean the origin of life. It was in this edition that Darwin used the word "evolution" for the first time. It is a beautifully written work, full of argument and observation and free from much jargon.

Despite its title, Darwin did not give much thought to how life started on Earth. He did say that "Probably all the organic forms which have ever lived on this Earth have descended from some primordial form, into which life was first breathed". He believed that all complex life evolved from simpler ones. He did not hypothesise extensively about the building blocks of life, only how and why species change over time. His book contains a large amount of evidence to support his ideas.

Darwin's theories were beginning to make sense of the natural world and he started to formulate mechanisms for the multiplicity of natural forms. In 1868, he proposed his "pangenesis" theory. coining the concept of "gemmules"—"particles" of inheritance between parents and offspring. This unleashed the search for a deeper understanding of the mechanism of inheritance. It became a story of scientific enquiry that spanned a century, beginning with Mendel in the 1850s (see Sect. 13.3.3) and culminating in the 1950s with Crick (see Sect. 13.3.2). Eventually, the concept of gemmules was replaced by that of "genes".

Darwin may be remembered for saying, "In the long history of humankind (and animal kind too) those who learned to collaborate and improve most effectively have prevailed". Since life began on Earth, the changing environment has put pressure on individual organisms to adapt or die out.

Just as Copernicus, Galileo and Newton had challenged the orthodox, Earth-at-the-centre-of-the-universe view, so Darwin was equally challenging about the Establishment's view of living organisms being attributed to the will of a divine Creator. He saw the living world as a system of matter in motion, governed by natural laws. Here was another head-to-head confrontation between religion and science, with Darwin at its fore.

There is a competition for resources between individuals, as well as populations of organisms, in the setting in which they inhabit. Individuals and populations with adaptations favourable to their local environment will live long enough to pass those desirable traits to their offspring. Favourable traits become more common in successive generations. Conversely, unfavourable traits become less common and will eventually disappear from the gene pool. Over the millennia, these adaptations accumulated and created the biodiversity we witness today, as well as much more that has since died out through mass extinctions.

In his 1837 'sketch of the tree of life', the tips of branches show the species alive then, as well as those extinct. Many a limb and branch has dropped off the tree and decayed. Those fallen branches of various sizes may represent whole orders, families and genera which have no living representative and are known to us only in a fossil state (see Mary Anning's fossils, Sect. 11.2).

In Chap. 3 of his 1859 publication, he said, "I have called this principle, by which each slight variation, if useful, is preserved, by the term 'natural selection'". Natural selection is an inevitable outcome of three principles, namely: (i) most characteristics are inherited; (ii) more offspring are produced than are capable of surviving and (iii) offspring with more favourable characteristics will survive and have more offspring than those individuals with less favourable traits. 'Evolution' and 'natural selection' are not synonymous. Natural selection is just one mechanism by which evolution occurs.

11.3.4 Cataclysmic Events Leading to Mass Extinctions and an Evolutionary Reset

Life appeared on Earth about four billion years ago. In that time, the diversity of living things has varied wildly, as new species appeared, and others became extinct. Ecology is the study of the interaction between organisms and their environment and other organisms. Evolution is the study of life, how it arose, and how and why it changes through time. Ecology shapes evolution. Diversity is the propeller of evolution.

Many of these changes have been gradual as life forms responded to ever-changing environmental conditions. Several, however, were so sudden and cataclysmic that they led to abrupt mass extinctions and irrevocable alterations to entire ecosystems . It is surmised that, in the last 500 million years, there have been at least five such calamitous extinction events. These were caused by asteroid impact or extensive volcanic eruptions releasing massive amounts of heat, spewing ash, and ejecting copious quantities of sulphur dioxide and other greenhouse gases into the atmosphere. High levels of free radicals in the atmosphere may have destroyed the protective ozone layer allowing more harmful cosmic rays to penetrate. Coupled with dwindling food availability, these adverse conditions caused the extinction of many terrestrial creatures.

These catastrophic events led to global warming (melting ice caps) or global cooling (as phytoplankton absorbed carbon dioxide). Enhanced weathering of rocks created the first soils for root-stabilised plants. Plants that formed symbiotic relationships with bacteria—legumes—appeared and they enriched the soil with nutrients, enabling flowering plants to thrive.

However, leaching of soil rich in nutrients delivered toxic chemicals to sea water leading to ocean acidification. Anoxia and the asphyxiation of fish resulted in the extinction of many marine creatures. On these five occasions, there were many losers and some winners. For ecosystems, a massive reset event occurred, putting them on an entirely new evolutionary pathway. New life forms emerged—the World's flora and fauna changed dramatically.

The most researched event of mass extinction is that caused by the impact of a 12 km asteroid, in shallow waters, near the small town of Chicxulub Puerto, on Mexico's Yucatán Peninsula. Sixty-six million years ago, the asteroid struck land with such unimaginable force that it triggered an unstoppable chain of events—vaporising the crater's rocks causing a blast wave; a tsunami across the Gulf of Mexico; a global firestorm consuming plant vegetation, and a superheated cloud of rock debris that circulated the globe, blocking out sunlight.

It changed the World beyond recognition. As many as 75% of plant and animal species were sent spiralling into extinction. Animals in the vicinity of the impact were killed instantly. It caused perpetual night-time for over a year, plunging temperatures below freezing. Food sources dwindled, so terrestrial and marine animals remote from the impact would eventually die.

As the atmosphere cleared, dormant animals came to life; the giant beasts (dinosaurs) were gone but smaller mammals survived and multiplied. Evolution eventually led to human beings who learnt to contemplate their own existence.

11.3.5   Survival of the Fittest

A contemporary of Darwin—Herbert Spencer—coined the phrase "survival of the fittest" and extended Darwin's biological concept to the realms of sociology, ethics and economics. Darwin adopted the phrase in his fifth edition. The word 'fittest' is not used in the context of 'strongest' or 'best physical condition'. Rather, it means 'most suited to the immediate environment'. Hence, it is not the strongest of the species that survives, nor the most intelligent—it is the one that is most adaptable to change.

As with most geological changes, Darwin believed that biological changes do not usually occur suddenly. He said, "Natural selection acts solely by accumulating slight, successive, favourable variations; it can produce no great or sudden modification; it can act only by very short steps". The Galápagos finches are one of the fastest evolving vertebrates because their appearance and behaviour adapted relatively quickly to the isolated environment.

In 1871, Darwin published "*Descent of Man, and Selection in Relation to Sex*" some of the contents of which were misinterpreted by certain parts of society. Darwin never said that humans are members of the ape family and descendants of apes. But he did see connections between higher primates (simians/anthropoids).

He thought that chimpanzees, apes, and humans must have a common ancestor because of the great similarities between us and the differences from other species. Collectively, we are tail-less members of the superfamily, Hominoidea. Humans are not descended from apes or chimpanzees; they simply belong to a different branch of the same family tree. It is generally accepted that chimpanzees are our closest living relative, sharing about 98% of DNA.

When Darwin's work on the theory of evolution appeared, certain parts of the Church considered it to be deeply blasphemous and attacked him vociferously. Eventually, his theory of evolution and the process of natural selection became known as 'Darwinism'. Over the years, scientific evidence accumulated that provided further evidence for his theory.

11.3.6   Similarities and Differences Between Species

Starting with Darwin's beloved insects, approximately seven out of every eight living animal species are Arthropods. They have had a long tenure on Earth, probably because of their small size and capacity for rapid change.

Though unbelievably diverse, they have roughly the same body plan. They are invertebrate animals having an exoskeleton with segmented bodies divided into three regions, namely a head (with a pair of antennae and compound eyes), thorax and abdomen, with pairs of appendages. The subphylum, Hexapoda, have three pairs of segmented legs and may have wings attached to the thorax. Ants, cockroaches, butterflies and flies are examples of Hexapoda. Fleas are wingless ones.

The basic anatomy of a hexapod insect below shows a developed digestive system (yellow–orange); a respiratory system (blue); a circulatory system (red) and nervous system (purple) (Fig. 11.2).

The limbs of Arthropods are jointed. Externally, the forelegs of frogs, rabbits, lizards, and birds look quite different because they have adapted to function in different environments. However, the forelimbs of all verte-brates have a common structure—one bone (humerus) followed by two bones (radius and ulna), then sets of little bones (carpals, metacarpals and phalanges) arriving at the fingers, toes, flipper or wing, at the end.

A similar structure is seen in a 370-million-year old fossil of Eusthenopteron, an extinct lobe-finned fish. Such homologies suggest a common ancestry for all these animals. It also implies that the patterning of the hands of terrestrial vertebrates—the tetrapods—was developed in aquatic creatures which grew muscles in their fins. Pentadactyl (5-fingered) limbs facilitate locomotion on land (Fig. 11.3).

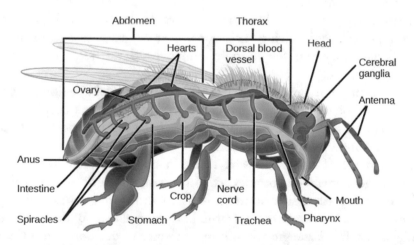

Fig. 11.2 Hexapod insect body plan. *Source* https://courses.lumenlearning.com/wm-biology2/chapters/subphylums-of-arthropoda. Figure 1. CC licensed content shared previously. https://creativecommons.org/licenses/by/4.0/. Provided by: OpenStax. Located at: http://cnx.org/contents/185cbf87-c72e-48f5-b51e-f14f21b5eabd@10.8. Biology; Unit 5, Biological diversity; Chap. 28. Invertebrates; Sect. 28.4 Superphylum Ecdysozoa; Subsection Arthropods. License: CC BY: Attribution. License Terms: Access for free at https://openstax.org/books/biology-2e/pages/1-introduction. Biology 2e. Chapter—Biological Diversity. Section 28, Invertebrates. Section 28.6, Superphylum Ecdysozoa: Arthropods

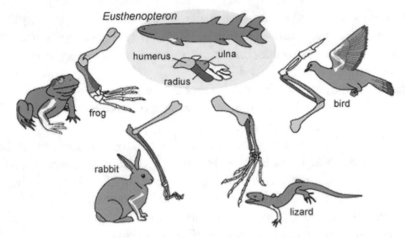

Fig. 11.3 Similarities between the pentadactyl forelimbs of different species of tetrapod. *Source* Lines of evidence: The science of evolution: Homologies (forelimbs of tetrapods). Understanding Evolution. 2021. University of California Museum of Paleontology. Accessed on 29 March 2021. https://evolution.berkeley.edu/evolibrary/article/0_0_0/lines_04

Like many primates, Homo sapiens have opposable thumbs, but humans can stretch their thumb across the hand to the little finger and the ring finger next to it. They can flex those fingers towards the base of the thumb to create a powerful grip giving exceptional dexterity to hold and manipulate tools. Because they walk fully upright, human hands are free to conduct such activities.

Humans may not have the heaviest brain in the animal kingdom, but their extraordinary brain power allows them to be skilful, to think deeply, to solve problems, and be creative in diverse ways. They are the most inquisitive, imaginative and inventive of all animals.

When tribes of hominids were living in open grassland habitats, they hunted large animals—a complex, collective task which drove the evolution of larger brains. Alongside hunting, foraging, cultivating crops and rearing domesticated animals, humans learnt how to make food nutrients more readily available for the growth and maintenance of our bones, muscles and organs. They did it by pounding raw food material with a rock; grinding it to a powder; letting it partially rot; cooking and roasting it. This increased the nutrient and energy supply to our bodies facilitating the deposition of more optimally connected neurons in our brain. This allowed us to evolve bigger, smarter and more energy-demanding brains.

Fossil records, though incomplete for most species, provide information about the nature of species that existed at approximate times of the Earth's history. Fossils signal the existence of past species that are related to existing species.

Some of the best studied fossils are those of the horse lineage. Scientists have constructed a branching family tree for horses, beginning with their now extinct relative—the short, forest-living 'dawn horse' (genus Hyracotherium/Eohippus). Hoofed mammals first appeared in the Eocene epoch. Successive generations learnt to adapt to a grassland environment in the late Eocene, Oligocene and Miocene periods. Here, they were more exposed to predators and their survival depended on running faster. Like other grazing herbivores, the horse relatives underwent typical adaptations for eating— strong, high-crowned teeth suited to grinding grasses, and a relatively long digestive tract to digest cellulose.

Whilst not a direct line of descent, five transitional stages are shown over 50 million years, from the 4-toed foot of Hyracotherium/Eohippus to a single, middle toe (hoof) of the present, tall, Equus that appeared in temperate climates in the Pleistocene period (Fig. 11.4).

11.3.7 Artificial Selection

Nature is no longer the only arbiter of what is a favourable trait and what is not. In artificial selection, humans are the selective agents, whereupon 'desirable' characteristics are arrived at by human intervention. Examples include the colour of flowers, breeds of dogs etc. Animal breeders 'improve' domestic animals by selecting animals with what they judge to be the 'best' characteristics and qualities.

About 2,500 years ago, common uncultivated mustard (Brassica oleracea) was solely a wild plant found growing on seaside cliffs. Then, humans became selective agents. Many different varieties of cruciferous plants were created by selective breeding to maximise certain attributes. For instance, wild mustard was bred to grow more leaves (kale); enlarged stems (kohlrabi); enlarged leaf buds (cabbage); flowering structures (broccoli and cauliflower) or numerous small heads (Brussels sprouts). They have a distinctive bitter flavour due to the presence of glucosinolates. The family of Cruciferae vegetables are rich in bioactive chemicals (Fig. 11.5).

11.3.8 Commonalities Between Plants and Animals

All organisms are composed of cells filled with water that contains genetic material, proteins, lipids, carbohydrates, salts and other minor substances.

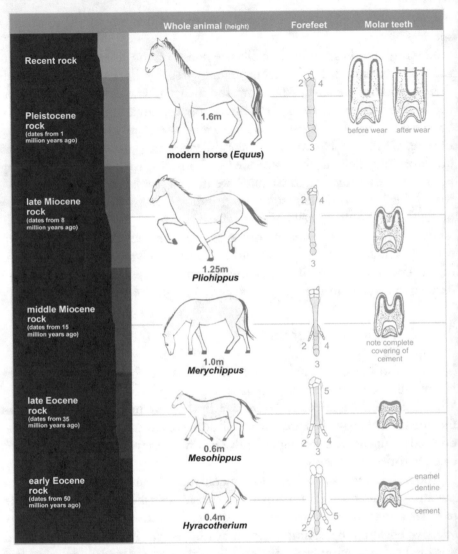

Fig. 11.4 Five transitional stages of horse lineage through the geological ages. *Source* By Mcy jerry at the English-language Wikipedia, CC BY-SA 3.0, http://creati vecommons.org/licenses/by-sa/3.0/. See https://commons.wikimedia.org/w/index.php? curid=496577/.%20.File:Horseevolution.png

Carbohydrates and lipids are sources of fuel whilst proteins are building blocks, messengers and enzymes.

The cells of plants and animals are remarkably alike, suggesting that life shares a common ancestor. There are few structural difference between plant and animal cells. Plant cells have a rigid cell wall, but animal cells have a membrane. Plant cells have a single large vacuole, animal cells have several

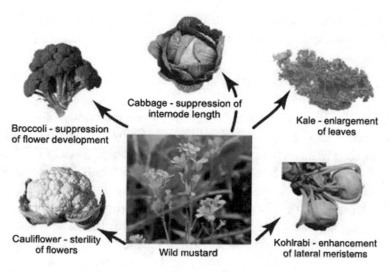

Fig. 11.5 Plant breeding outcomes of wild mustard. *Source* Evolution 101, Artificial selection. Understanding Evolution. University of California Museum of Paleontology. 19 April 2021. https://evolution.berkeley.edu/evolibrary/article/0_0_0/evo_30

small ones. Plant cells may have chloroplasts, but animal cells have more mitochondria (Fig. 11.6).

At the molecular level, DNA and RNA possess a simple 4-base code (see Sect. 13.3.4) that provides the recipe for all living things. Parts of the genetic

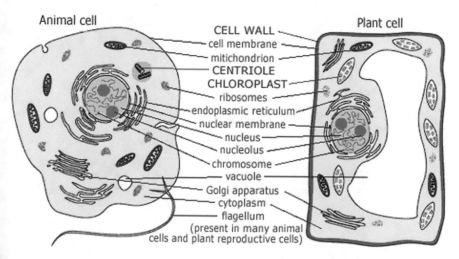

Fig. 11.6 Structural similarities and differences between animal and plant cells. *Source* Lines of evidence: the science of evolution.: homologies: cellular/molecular evidence. Understanding evolution. University of California Museum of Paleontology. 29 March 2021. https://evolution.berkeley.edu/evolibrary/article/0_0_0/lines_08

code for all living organisms are similar. The DNA code is a homology that links all life on Earth to a common ancestor.

11.3.9 Darwin's Accolades

Darwin received numerous accolades during his illustrious career. He was elected Fellow of the Royal Society (FRS, 1839); Fellow of the Linnean Society (FLS, 1854); Fellow of the Royal Geographical Society (FRGS) and Fellow of the Zoological Society of London (FZSL, 1839). The Royal Society Council awarded him the prestigious Copley Medal in 1864.

By building a wealth of scientific evidence, Darwin was able to put forward a grand unifying theory of biology. By attributing the diversity of life to natural causes, rather than to supernatural creation, Darwin gave biology a sound scientific basis. He had strong, steely convictions, able to articulate radical intellectual concepts in a kind, gentle and courteous manner. When he died, in 1882, he was laid to rest in England's most revered church—Westminster Abbey—near other notable British scientists such as Isaac Newton, John Herschel, Ernest Rutherford, J. J. Thomson, and Lord Kelvin.

12

X-ray Crystallography of Biomolecules

In physical science a first essential step in the direction of learning any subject is to find principles of numerical reckoning and practicable methods for measuring some quality connected with it. I often say that when you can measure what you are speaking about, and express it in numbers, you know something about it; but when you cannot measure it, when you cannot express it in numbers, your knowledge is of a meagre and unsatisfactory kind; it may be the beginning of knowledge, but you have scarcely in your thoughts advanced to the stage of science, whatever the matter may be. (Baron William Thomson Kelvin, from a series of six lectures to the Institution of Civil Engineers, London. (3 May 1883), on 'Electrical Units of Measurement', published in Popular Lectures and Addresses (1889), Vol. 1, 73–74)

12.1 Précis

Father and son, William Henry Bragg and (William) Lawrence Bragg developed experimental methods and mathematical formulae that tell us how atoms are spatially configured in crystals of simple substances, as well as more complex macromolecules of living cells. In 1915, the Braggs were awarded the Nobel Prize in Physics for their work on X-ray crystallography, Lawrence being the youngest laureate at the time, being only 25 years of age.

William Bragg Senior was one of the motivators for Dorothy Hodgkin to use X-ray crystallography to examine the structures of biologically active substances. In part, this revolutionized modern medicine and improved health expectations. By advancing novel techniques of X-ray crystallography,

© The Author(s), under exclusive license to Springer Nature Switzerland AG 2022
R. Bailey, *Inventive Geniuses Who Changed the World*, https://doi.org/10.1007/978-3-030-81381-9_12

she was able to elucidate the structures of numerous compounds, the most noteworthy of which were cholesterol, penicillin, vitamin B_{12} and insulin. Once Sanger had revealed the chemical structure of insulin, this led to its laboratory synthesis and improved treatments for diabetes—an autoimmune condition that was a major economic and health care burden. Hodgkin is the only British woman to have been awarded a Nobel prize (in 1964).

Lawrence Bragg inspired co-workers John Cowdery Kendrew and Max Perutz to use X-ray crystallography to determine the molecular structures of myoglobin and haemoglobin, physiologically important substances in binding molecular oxygen in animals. Myoglobin was the first protein to have its atomic structure determined by X-ray crystallography. For their research, Kendrew and Perutz shared the Nobel Prize in Chemistry, in 1962.

In the UK, with about 1% of the Earth's population, 'Bioscience Britain' has been responsible for major advances in identifying the circulation of blood; smallpox vaccination; evolution; penicillin; X-ray crystallography of biologically active substances; amino acid and nucleotide sequencing; in-vitro fertilisation; structure of, and fingerprinting by, DNA; cloning and genome sequencing.

12.2 William Henry Bragg OM, KBE, PRS (1862–1942) and William Lawrence Bragg CH, OBE, MC, FRS (1890–1971)—X-ray Crystallographers

William Bragg Senior was born at Westward, near Wigton, Cumberland, the son of a merchant navy officer and farmer. When he was seven years old, his mother died, and he was raised for six years by his domineering uncle, in Market Harborough, in Leicestershire. He was educated at the local grammar school before his father elected to send him to board at King William's College on the Isle of Man. In 1881, William was elected a minor scholar at Trinity College, Cambridge, where he studied mathematics, graduating with a 1st class honours degree in 1885. During part of 1885, he studied physics in the Cavendish Laboratory.

At the end of that year, when he was twenty-three years old, he was elected to the Professorship of Mathematics and Physics in the University of Adelaide. Whilst in Australia, his son, William Lawrence, was born in 1890.

Recognising certain deficiencies in the equipment in his teaching laboratory, Bragg Senior apprenticed himself to a firm of instrument makers to assemble necessary apparatus. Soon after W. C. Röntgen's discovery of X-rays in 1895, Bragg and his assistant, A.L. Rogers, constructed their own X-ray tube. With their primitive equipment, they obtained an X-ray image of William junior's broken elbow.

In 1909, the family returned to Great Britain and William snr. was appointed Professor of Physics at Leeds University (1909–1915). William jnr. entered Trinity College, Cambridge, taking a 1st class honours degree in the Natural Science Tripos, in 1912. At the age of twenty-three, he began his collaborative work with his father which was to lead to a Nobel prize.

It is impossible to observe anything smaller than the wavelength of the electromagnetic waves being used for the observation. In 1912, Max von Laue and his students, Friedrich and Knipping, discovered that diffraction patterns occur when X-rays pass through crystals. It implied that, in this situation, X-rays must be waves, analogous to light travelling through an array of slits in a transmission grating, but of much shorter wavelength, namely 0.01–10 nm (see Appendix D). They concluded that the wavelength of 'hard' X-rays are comparable to the size of atoms (0.1–0.2 nm). However, the diffraction pattern they obtained was too complicated for them to establish the structure of the crystal from which it came. William and son, Lawrence, were inspired to pursue this line of research.

With his training in instrument making, William snr. designed the Bragg ionization spectrometer—the prototype of all modern X-ray diffractometers and neutron spectrometers, with which he made the first exact measurements of X-ray wavelengths and crystal data. His instrument can be seen in the Faraday Museum (https://www.rigb.org/visit-us/faraday-museum—ionisation spectrometer, 1912) at the Royal Institution (London W1S 4BS). X-ray tubes produce X-rays that penetrate matter and have wavelengths smaller than the sizes of many atoms and molecules.

Scientists now know that crystals are like a three-dimensional grid or endless lattice of regularly repeating units. Lawrence visualized this lattice as being constructed from a series of sheets of atoms or ions, laid one on top of the other, with each sheet behaving like a mirror. The intricate pattern of photographed dots that were produced by passing X-rays through a single crystal results from a complex series of interactions in which some X-rays are reflecting off the first layer of atoms encountered, some from the second, some from the third and so on, as illustrated below (Fig. 12.1).

Superposition of diffracted waves results in a series of maxima and minima. Maxima are formed in directions where rays scattered from consecutive layers are in phase.

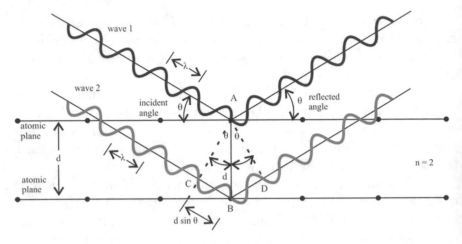

Fig. 12.1 X-rays impinging on chemical crystals

For each maximum, Lawrence was able to formulate a relationship between the wavelength (λ) of the monochromatic X-rays, the angle of incidence (Θ) and the distance (d) between the atomic layers / planes inside the crystal. This became a powerful tool for determining the precise location of individual atoms. The Bragg equation:

$$n\lambda = 2d \sin \Theta$$

where n is a whole number, is basic to X-ray diffraction. Between 1913 and 1914, father and son had effectively founded a new branch of science and were rewarded, in 1915, with a Nobel Prize in Physics for their services in the analysis of crystal structure by means of X-rays. By then, the Braggs had determined the crystalline structures of sodium chloride, zinc sulphide (zinc blende) and diamond.

During WWI, William snr. was put in charge of research connected with the location of submarines. He was awarded a CBE in 1917 and was knighted in 1920. The Order of merit followed in 1931. Having been a Fellow since 1907, he was elected President of the Royal Society in 1935 and served in this role until 1940.

When Bragg jnr. received the Nobel prize in 1915, he was only 25 years of age—the youngest-ever Nobel laureate. During WW1, he served as an army officer in France, where he endeavoured to use sound to determine the location of enemy guns.

In 1919, Lawrence succeeded Ernest Rutherford, firstly as Professor of Physics at the University of Manchester, and then at the Cavendish Laboratory. At the latter, he established the Unit for the Study of Molecular Structure of Biological Systems which was funded by the Medical Research Council. This is where Crick, Watson and Kendrew carried out their groundbreaking studies on the structure of proteins and DNA.

Lawrence was elected a Fellow of the Royal Society in 1921. He was knighted in 1941 and made a Companion of Honour in 1967.

12.3 Dorothy May Crowfoot Hodgkin OM, FRS, HonFRSC (1910–1994)—Biochemist and Crystallographer

Dorothy was born in Cairo, Egypt to parents John and Grace (Molly) Crowfoot. John was a colonial civil servant working in the Egyptian Education Service. After his retirement, he devoted his time to archaeology. Both parents had a healthy respect for intellectual enquiry. Grace taught Dorothy a variety of subjects and encouraged her scientific ambitions.

During the First World War, with Dorothy's parents in the Middle East, she was sent with her three sisters to live with her paternal grandparents and nanny at Worthing, in Sussex. She was given a chemistry set at the age of eight or nine.

In 1920, the family decided to make their English home in Geldeston, near Beccles, in Sussex, where her paternal grandfather had been raised. During a settling in period, Dorothy attended classes organised by the Parents' National Educational Union (PNEU), at Beccles rectory. The PNEU provided support, and resources for home schooling. Here she made aqueous solutions of alum and copper sulphate; allowed them to evaporate and watched as the crystals gradually appeared. Crystallography captured her imagination for life. Her mother encouraged her to pursue this passionate curiosity in crystals. A family friend—the chemist, Dr. A. F. Joseph—further stimulated her interest by giving her a box of reagents and minerals which allowed her to set up a home laboratory.

From 1921 to 1927, she attended the state-funded, Sir John Leman Grammar School, a co-educational establishment. Here, she had to struggle to be allowed to study chemistry with the boys. Her soft spoken, gentle, modest demeanour hid a steely, patient resolve which was to propel her to an outstanding career. In 1927, she obtained the highest marks of any girl candidate in the School Leaving Certificate, set by the Oxford Local Examinations Board. She gained six distinctions, not including chemistry.

With a supportive female chemistry teacher, Miss Deeley, she was encouraged to sit the Oxford University entrance examination. Women had only been awarded degrees at the University since 1920. The number of women was capped at 1 in 4 students. Therefore, she faced many hurdles, including a requirement for two science subjects, better mathematics, and physics, as well as Latin. Since she had a year in hand, she achieved the pre-conditions via one-to-one tuition.

Between school and university, in 1928, she spent some time with her parents, excavating in Jerash, Jordan, where she became fascinated with floor mosaics in Byzantine churches. She recorded their patterns and symmetries, and her draughtsmanship would be helpful in her future career as a crystallographer.

At university, she was inspired by reading William Bragg's 1915 Christmas lecture in which he revealed that X-ray crystallography provided the means to determine the three-dimensional spatial arrangement of atoms in pure crystals. In her final year as an undergraduate, at the University of Oxford, she was one of the first to study the structure of a simple organo-metallic compound (viz thallium alkyl halide) using X-ray crystallography.

Because of her exceptional intelligence, she became the third woman to get a first in chemistry, at Oxford. Thereafter, she studied for a PhD, in Cambridge, under J. D. Bernal who was especially interested in the relationship between physics and biology. She conducted research on steroids, vitamin B_1 and the protein, pepsin, a digestive enzyme. She was awarded a PhD in 1937.

In 1936, she was invited back to Oxford, as a research fellow, where she spent the rest of her illustrious career. However, it was not until 1946 that she was made a permanent member of the university staff. She had a collaborative approach to scientific investigation, encouraging multi-disciplinary teams, supporting her students and co-workers.

In 1937, she married T. L. Hodgkin, a Marxist historian of Africa. In later life, Dorothy became a campaigner for peace and social justice. She was a socialist and pacifist, putting her at odds with one of her former students, Prime Minister, Margaret Thatcher (née Roberts).

By virtue of their uniform spacing, atoms of a pure crystal cause an interference pattern when subject to an incident beam of X-rays (see Sect. 12.2). If a single crystal is rotated, any diffracted, 'Bragg maxima' waves can be captured as dots on an encircling photographic film, sensitive to X-rays

Where there are many overlapping atoms, a mass of diffraction images, taken at different angles, allow individual slices through a molecule to be examined and drawn, facilitating determination of the crystal structure.

If a powdered sample is used, it will contain atomic planes in all directions, so it does not have to be rotated. Planes for which the Bragg maximum applies will create a cone of scattered X-rays which strike the photographic film in a slightly curved line which can then be analysed and interpreted.

In the case of penicillin, Dorothy obtained 2-D contours of electron density and, with the help of her sister, Betty, these contour sheets were drawn on Perspex and stacked to visualise the model in 3-D. Finally, Dorothy made a cork-and-wire (ball-and-stick) model.

Extensive mathematical calculations, and astute analysis are required to determine the three-dimensional structure of biomolecules. Before the arrival of computers with sufficient power, Dorothy used Beevers-Lipson strips as an aid to determine the atomic structure. Her most noteworthy analyses were penicillin (in 1945) vitamin B_{12} (1955) and insulin (1969)—a crowning achievement.

In 1939, Howard Florey (see Sect. 9.7) also at Oxford, had isolated crystalline penicillin, known to be active against bacteria. The objective was to mass produce penicillin, rather than relying on the growth of mould, making it available for soldiers in WW2. Dorothy was asked to solve the structure of the natural material, and this would help with the synthesis of future antibiotic derivatives. By 1945, she had succeeded, and her work was rewarded by her election to the Royal Society, in 1947.

Dorothy continued to work on steroids, long after her doctoral studies. Cholesterol is a crystalline alcohol, called a sterol, which belongs to the class of compounds called steroids. The latter include, not only cholesterol (the chief animal sterol), but also sex hormones (e.g., testosterone) and adrenal cortical hormones (e.g., cortisone). In 1945, she and student, C.H. Carlisle, published the first structure of a steroid—cholesteryl iodide—using the heavy atom, iodine, and Patterson maps to work out the stereochemistry of this representative compound. It confirmed the 3D structure of the sterol skeleton to be comprised of three six-membered carbon rings and one five-membered ring.

In the mid-1950s, she made extensive use of computers to carry out complex computations, to determine the structure of vitamin B_{12}. The sample had been supplied by E. L. Smith, of Glaxo Laboratories, who isolated it from liver tissue. Todd's chemical studies on the degradation products of B_{12} (see Sect. 13.2.2) were helpful to Hodgkin. Arguably, B_{12} (a.k.a.

cobalamin) is the largest and has the most complex structure of all vitamins. It contains the biochemically rare element, cobalt, and has a formula $C_{63}H_{88}CoN_{14}O_{14}P$ and molecular weight of 1,355.4 g/mol.

Pernicious anaemia is an auto-immune disorder causing the destruction of an intrinsic factor that facilitates the absorption of vitamin B_{12}. Without this vitamin the body is unable to produce adequate amounts of healthy red blood cells. Compared to animal tissue, plants are a poor source of this vitamin. Knowing the structure of B_{12}, it was possible to synthesize it, making it available as a supplement for medical treatment.

By 1955, using electrophoresis and chromatographic techniques, Sanger had determined the amino acid sequence of the hormone, insulin (see Sect. 13.4.2). Insulin is a complex protein, composed of hundreds of atoms bonded together. It controls the concentration of sugar in the blood. Periodically, for thirty-five years, Hodgkin had attempted to determine the crystal structure of insulin. Using higher speed computers, and working with an international team, her patience eventually paid off and they were rewarded with the elucidation of its rudimentary structure in 1969.

Hodgkin received many national and international awards for her work. She was the sole person nominated for the Nobel Prize for Chemistry, in 1964, for her determinations, by X-ray techniques, of the structures of important biochemical substances. She is the only British female Nobel laureate in any of the three scientific categories. In the following year, 1965, she was made a member of the Order of Merit, being the second woman (after Florence Nightingale [see Sect. 9.5]) to be so honoured.

Dorothy was a perfectionist with a caring and sharing mentality. Her pioneering work on biologically active substances revolutionized modern medicine and improved health expectations. In 1996, she was featured on a postage stamp, celebrating 'Twentieth Century Women of Achievement'.

12.4   John Cowdery Kendrew CBE, FRS (1917–1997)—Molecular Biologist and Crystallographer

Kendrew was born in Oxford. His father was Reader in Climatology at the University of Oxford, his mother an art historian. He was educated at Dragon School, Oxford (1923–1930) and Clifton College, Bristol (1930–1936). In 1939, he graduated from the University of Cambridge and then started a doctorate on reaction kinetics. WW2 intervened and Kendrew worked on radar and was a scientific adviser to the Air Ministry. During missions overseas, he met crystallographer, J D Bernal, and physical chemist, L Pauling, who persuaded him that protein crystallography would be a challenging, but worthwhile field to enter.

Therefore, in 1945, when Kendrew returned to Cambridge, he began a collaboration with Max Perutz, under the direction of Sir Lawrence Bragg (see Sect. 12.2), at the Cavendish Laboratory. The three joined the Medical Research Council (MRC) Laboratory of Molecular Biology in 1947. They continued working on the structure analysis of haemoglobin. This work gained Kendrew a Ph.D. in 1949.

Max Perutz was an Austrian. He used X-rays to map the structure of haemoglobin—the protein that allows blood to transport energy-giving oxygen long distances from the lungs to the tissues and carries carbon dioxide away from the tissues to the lungs (see Sect. 9.2). Perutz completed his study in 1959, achieving a resolution of 5.5 Å.

In the interim, Kendrew's research shifted to the X-ray analysis of sperm whale myoglobin—a protein one quarter the size of haemoglobin. Myoglobin is a monomeric protein found in highly oxidative muscle fibres (such as cardiac and striated muscle cells) where it combines reversibly with molecular oxygen, thereby behaving as a local storage reservoir for oxygen. Myoglobin effectively facilitates oxygen diffusion, down an oxygen gradient, to mitochondria where oxygen is consumed in respiration.

Diving animals (e.g., whales) require greater oxygen reserves. Therefore, their dark red tissues contain larger amounts of myoglobin, making it more accommodating for crystallographic research. Kendrew was able to isolate large myoglobin crystals with clean X-ray diffraction patterns.

The chemical components of myoglobin (and haemoglobin) are detailed next. Porphyrins are a group of heterocyclic organic compounds based on four pyrrolic rings, linked by four methine (=CH–) bridges to form a macrocycle. Being highly conjugated, its iron derivatives absorb strongly in the red-blue region of the electromagnetic spectrum. Indeed, the name 'porphyrin' is derived from a Greek word meaning 'purple'. They are substituted forms of the parent porphin molecule ($C_{20}H_{14}N_4$). Porphin has twelve positions available for substitution, usually at the periphery.

A porphyrin molecule is large enough to accommodate a metal ion in the interior of the ring. Its tunability and versatility is probably why Nature has chosen the porphyrin family to be one of its work horse macrocycles. Numerous pigmented metal complexes of porphyrins occur naturally. Iron is the chelating metal found in haemoglobin and myoglobin, whilst magnesium is found in chlorophyll, and cobalt in vitamin B_{12} (see Sect. 12.3).

Porphyrins can make four linkages to a metal ion (such as ferrous ion [Fe^{2+}]) using each of the nitrogen atoms, in the four pyrrole rings (see A to D below) as electron-pair donors.

The most important metallo-porphyrins are the iron complexes called haems or hemes. Haem is a nearly planar, iron porphyrin (Fe-protoporphyrin-IX, FePPIX) and the non-protein precursor of haem-containing globular proteins, haemoglobin and myoglobin. Haem biosynthesis takes place in bone marrow and liver cells. Its ability to reversibly bind molecular oxygen has life-giving physiological importance. Haem is responsible for oxygen storage in the muscles, via myoglobin, as well as oxygen transport in the blood via its hetero-tetrameric blood relation, haemoglobin (Fig. 12.2).

The principal secondary structure of the protein in myoglobin is a coiled shape (alpha helix) in which an amino acid is bonded to another amino acid, four residues ahead of it in the single polypeptide chain. Myoglobin belongs to the globin family of small globular proteins, being about 150 amino acids in length, characterized by eight α-helical segments, connected by loops.

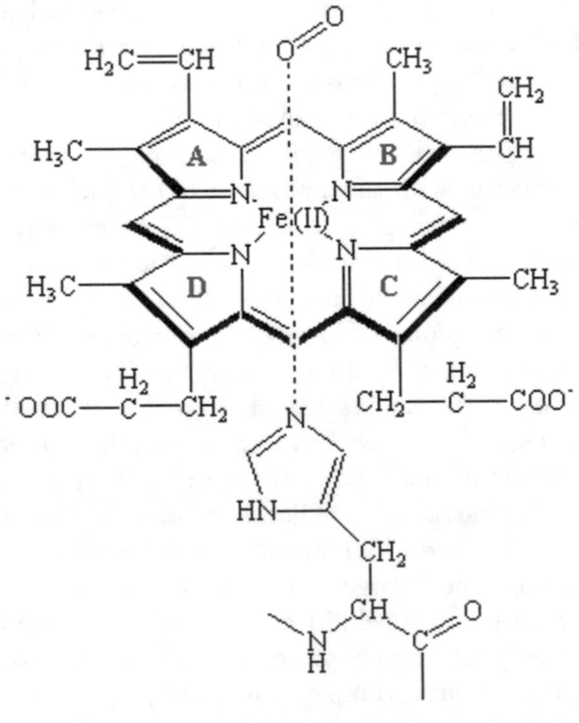

Fig. 12.2 Coordination of a porphyrin molecule to a ferrous ion to give an oxygen binding haem

The polypeptide chains of myoglobin are folded in such a way to maximise water-liking, polar residues on the exposed surface, with non-polar residues facing inwards. The charged amino acids on the outside of the spherical protein make it water soluble and able to move in the largely aqueous environment of an animal cell.

On one side of myoglobin, the tertiary structure of the globular protein forms a cleft, within which the haem group is embedded. This is situated between the 5th and 6th helices, where a hydrophobic region acts as a 'pocket' for the haem moiety. At its 5th coordination position, the ferrous ion is bonded to the nitrogen atom of the (proximal) histidine of the globin protein. The latter's second (distal) histidine group constructs the shape of a crevice so that only a small molecule—such as oxygen—can penetrate the globular protein and bond to the ferrous ion at its 6th coordination position. The oxygen is released in muscle tissue when the partial pressure of oxygen falls to a certain level.

In the case of haemoglobin, the flexing motion of the globular protein aids the entry and departure of exogenous molecules such as oxygen and carbon dioxide. Other ligands such as carbon monoxide and the cyanide ion can also bond to the ferrous ion, albeit irreversibly and fatally.

The following diagrammatic representation of myoglobin shows the globin protein as a blue ribbon, with a "ball and stick" representation of iron haem in green and oxygen in red (Fig. 12.3).

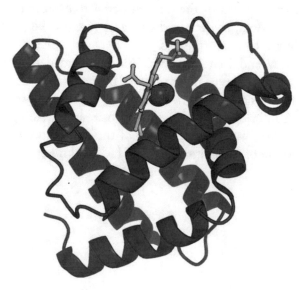

Fig. 12.3 Diagrammatic representation of myoglobin

Based on the pioneering work by Perutz, Kendrew modified myoglobin and added heavy metal atoms to five distinct sites of the myoglobin molecule. He then used the molecular position of these metal atoms as reference points to solve the phase problem and determine the 3-D structure of the protein.

Kendrew's research resulted in the production of a 3-D model of myoglobin at ever improving resolutions—6 Å in 1957 (showed the asymmetry and the orientation of the haem group), 2 Å in 1959 (showed the α-helices) and 1.4 Å in the 1960s, thereby determining virtually all the coordinates of the 2,500 atoms in the molecule. This was the first protein to have its atomic structure determined by Xray crystallography.

At 2 Å resolution, tens of thousands of data points were generated. The crystallographic calculations relied decisively on one of the first electronic digital computers (viz EDSAC) at any university, built by Maurice Wilkes of Cambridge's Mathematical Laboratory (see Sect. 15.3.5).

Whale myoglobin is a quarter of the size of haemoglobin, having only one haem group, together with 153 amino acid residues. Not counting H-atoms, myoglobin contains 1,260 atoms and has a molecular weight of 17,199 Da. In contrast, vitamin B_{12}, which was subject to X-ray analysis by Hodgkin [see Sect. 12.3], is composed of only 93 non-hydrogen atoms and has a molecular weight of 1,355 Da.

Haemoglobin is a more complex molecule than myoglobin. The former has a molecular weight of 67,000 and contains four haem groups so that each journeying haemoglobin molecule can carry four oxygen molecules. In arterial blood, about 1.5% of the oxygen being transported from the lungs is carried in the plasma, as dissolved gas. The rest (ca. 98.5%) is bound to haemoglobin in red blood corpuscles.

In 1962, Perutz and Kendrew were jointly awarded the Nobel Prize in Chemistry for their studies of the structures of these two globular proteins—haemoglobin and myoglobin. In the same year, Crick and Watson, also at the MRC Unit, in Cambridge, shared the Nobel prize in Physiology or Medicine.

In 1960, Kendrew was made a Fellow of the Royal Society and, in 1962, made a Companion of the British Empire. He was knighted in 1974.

13

Nucleosides, Nucleotides, Polynucleotides (RNA and DNA) and the Genetic Code

Almost all aspects of life are engineered at the molecular level, and without understanding molecules, we can only have a very sketchy understanding of life itself. (Francis Crick, What mad pursuit: a personal view of scientific discovery, 1988)

13.1 Précis

Scientific luminaries wanted to know more about the chemistry of bio-significant substances, as well as the chemical processes that facilitate the movement, growth, self-repair, and reproduction of organisms. Todd, Crick and Sanger were major contributors in achieving this goal.

Alexander Todd studied the chemistry of a range of natural products of biological importance. He was a colossus of twentieth century organic chemistry, having an encyclopaedic knowledge of the subject and an ability to picture complex molecules in 3-dimensions.

Although not the first to synthesize the anti-beriberi vitamin, B_1, his elegant laboratory method was adopted for commercial production. By a series of either chemical syntheses, or, degradation sequences of great delicacy and subtlety, he established the structure of nucleosides, nucleotides, and nucleotide coenzymes, for which he was awarded the Nobel Prize in Chemistry, in 1957. His results paved the way for Crick and Watson to propose a double helix structure for DNA.

© The Author(s), under exclusive license to Springer Nature
Switzerland AG 2022
F. Bailey, *Inventive Geniuses Who Changed the World*,
https://doi.org/10.1007/978-3-030-81381-9_13

Francis Crick had a broad education, leading to a degree in physics, a studentship to research cytoplasm and a doctoral degree in protein structure using X-ray crystallography. Few scientists have had such a huge impact on their adopted field of study.

James Watson, who had a background in viral and bacterial genetics joined Crick at the Medical Research Council Unit, in 1951. Initially, they drew on the chemical and X-ray results of other researchers to advance their view that DNA comprises two polynucleotide chains, wound about each other, to form a double helix. Their 1-page paper in Nature, in 1957, set the stage for major advances in molecular biology. They had found 'the secret of life'.

DNA is the 'master molecule of life', the repository of heredity information. It encodes all genetic information and is the blueprint from which biological life is created. This "operating manual" contains instructions for everything our cells do, from conception until death. It can self-duplicate, bringing about new DNA molecules that are identical to the original. Genetic information is stored and transmitted in a simple language with an alphabet of only four letters, based on nucleobases A, G, T and C.

Crick, Watson, and others postulated that messenger RNA takes instructions from the DNA, in the nucleus, to a host of ribosomes in the cytoplasm, where protein synthesis takes place. Here, only three of the four letters of the genetic alphabet are needed to determine the composition of the many proteins synthesized from the amino acids available. In 1962, Crick, Watson and Wilkins shared the Nobel Prize in Physiology or Medicine.

Frederick Sanger was raised as a Quaker and had a modest and quiet demeanour. He may not have been academically brilliant, but he was a gifted experimentalist and is one of only four individuals to be awarded two Nobel prizes, his in chemistry (in 1958 and 1980). He was responsible for two critical technical advances. Firstly, he perfected a way of unravelling the complete amino acid sequence of even the most complex of proteins. Using novel sequencing methods, he established the composition of bovine insulin, an essential step for the synthesis of human insulin, offering a major advance in the treatment for diabetes.

Second, he developed yet more ingenious ways of sequencing the nucleotides of DNA, spending ten years, carefully identifying small fragments of this giant molecule. He worked out the precise composition of the entire genome of the bacteriophage ΦX174, a relatively modest life form featuring 5,375 nucleotides.

He sequenced the first human genome in the shape of mitochondrial DNA, laying the foundation of humanity's ability to read and understand the genetic code. Genetic testing can now be used to diagnose inherited disorders. Genome sequencing may revolutionise the diagnosis of rare childhood conditions. The field of genetics has been transformed from a science of descriptive analysis, into today's powerful technology of gene therapy where defective genes are edited to cure disease. This will result in ground-breaking improvements in healthcare.

13.2 Alexander Robertus Todd, Baron Todd of Trumpington, OM, PRS, FRS, HonFRSE (1907–1997)—Organic Chemist

13.2.1 Background

Todd was born in Glasgow, the eldest son of a clerk working for the Glasgow Subway Railway Company. Having been given a chemistry set when he was seven or eight, Alex started experimenting before secondary school. When he heated iron filings and powdered sulphur together, he obtained a substance (ferrous sulphide) that was entirely different from the starting reagents. This stimulated his life-long interest in chemistry.

In 1918, he attended Allan Glen's High School of Science, where he passed the Higher Learning Certificate examination in 1924. In his final school year, he was fascinated by the beauty of the molecular structure of organic chemicals and the way their architecture influences their properties. He was determined to study chemistry at university, particularly organic chemistry. In 1928, he received a B.Sc. degree in chemistry, with first-class honours, from Glasgow University.

He then moved to Germany to be at the global centre of organic chemistry research. He graduated Dr.phil.nat. from the University of Frankfurt-am-Maine, in 1931, for a thesis on the chemistry of bile acids. He became proficient in the German language which meant he could read chemistry textbooks written in German, authored by distinguished chemists.

He returned to UK and joined the Nobel laureate, Professor R. Robinson, at Oxford University. He worked on the structure and synthesis of several flower pigments of the anthocyanin group, including pelargonin, and was awarded a second D.Phil. in 1934.

He moved to Edinburgh University and worked on vitamin B_1 (anti-beriberi factor; thiamine). Although not the first to synthesize it, his elegant and simple method became the adopted commercial route to its production by Roche. This aroused his interest in the specificity of vitamins and the reasons for their importance to living organisms.

He was accumulating ever greater theoretical knowledge and skills in the laboratory synthesis of organic substances, observing that classical synthesis and degradative studies were complementary tools in structural elucidation.

13.2.2   Nucleosides and Nucleotides

Todd was recognised as the brightest young organic chemist in the country and, in 1938, was invited to join the University of Manchester, accepting the highly prestigious Chair of Organic Chemistry. Biochemical studies elsewhere had revealed that vitamins of the B group can be converted to more complex derivatives which function as 'nucleotide coenzymes', whereupon they speed up metabolic processes. To improve his understanding of enzyme specificity, he realized that it would be necessary to better understand nucleosides and nucleotides, as a preliminary step towards the synthesis of coenzymes.

In the 1930s, no natural nucleoside had been synthesized and their structures were largely unknown. We now know that a 'nucleoside' is simply a base-sugar unit. When attached to one, two or three phosphate groups, it becomes a 'nucleotide'—a base-sugar-phosphate unit. Adenosine di- and tri-phosphate (ATP and ADP) and ribo- and deoxyribo-nucleic acids (RNAs and DNAs) are all nucleotides. ATP and ADP, however, are monomeric nucleotides (see below). On the other hand, in the case of RNA and DNA, a collection of monomeric nucleotides have combined to form a polymer—a polynucleotide chain, with phosphate as the 'chemical mortar' between units.

In the 1930s, RNA and DNA were ill-defined components of living cells, but Todd was determined to change that. Firstly, Todd clarified the structure of nucleosides using the method of unambiguous synthesis. Then he developed novel ways of adding a phosphate group to the nucleoside, by phosphorylation, to form nucleotides.

The primary structure of nucleotides was elucidated by degrading (hydrolysing) key bonds and identifying the fragments. Around 1940, Todd chemically split nucleotides apart and compared their parts with substances that had been constructed from known components, in a way that was already understood.

For his research into nucleotides, he used new experimental techniques such as paper and ion-exchange chromatography, paper electrophoresis, and concurrent distribution, peculiarly appropriate to compounds of this group.

Whilst the research had begun in Manchester, it continued when he and his group moved to Cambridge University, in 1944, when he was appointed to the Chair of Organic Chemistry, a post he held until 1971. The group formed a club known as the 'Toddlers'.

Todd and his team established that the backbone of a polynucleotide chain is a polyester chain, formed when a portion of an acid (e.g., phosphoric acid, $OP(OH)_3$) reacts with an alcohol group (from a sugar molecule in a nucleoside). Condensation occurs and water (H_2O) is ejected, its oxygen atom coming from the sugar molecule. Successive condensation reactions result in the formation of long poly-nucleotide strands.

During evolution, the sugars, ribose and deoxyribose, were, because of their 'pucker' conformation, 'selected' as ideal candidates for the physiological forms of two key families of nucleic acids. When the sugar molecule is D-ribose, then the group of nucleic acids is known as ribonucleic acids (RNAs). When the sugar is D-2-deoxyribose, the family is known as deoxyribonucleic acids (DNAs). The prefix "2-deoxy" indicates one fewer hydroxyl group (-OH), with the 2nd carbon atom carrying two H-atoms (Fig. 13.1).

Both these sugar units are based on a five-membered furanose / pentose ring. When they combine with a nitrogenous base, a nucleoside is formed. One of six heterocyclic nitrogenous bases—designated A, C, 5mC, G, U or T—becomes attached to the 1st carbon atom of each sugar, through a β-linkage. It is a β-linkage because the derivation occurs when the hydroxyl group (–OH) on carbon-1—above the plane of the ribose ring—combines with a hydrogen atom on the base, and water is eliminated.

Nitrogenous bases found in RNA and DNA are derived either from 1-ringed pyrimidine or 2-ringed purine. They are defined as 'bases' because they have one or more nitrogen atoms, each with a pair of unused electrons which will accept a proton. Pyrimidine contains nitrogen atoms at the 1

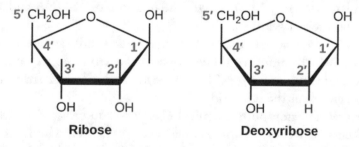

Ribose   **Deoxyribose**

Fig. 13.1 Ribose and deoxyribose as sugar components of nucleosides. *Source* http://open.bccampus.ca. Search for Concepts of Biology—1st Canadian Edition. On RHS, "Get This Book", click "readable" format (7), then "website", and "Read Book". From Contents: Chap. 9 (Introduction to Molecular Biology). Section 9.1 (The Structure of DNA). Figure 9.5. Licensed under a Creative Commons Attribution 4.0 International License.https://creativecommons.org/licenses/by/4.0/. This textbook was adapted from Concepts of Biology © 2013 by OpenStax, which is also under a CC BY 4.0 International Licence. https://opentextbc.ca/biology/. The book was adapted by Charles Molnar and Jane Gair; the figure modified by Jerome Walker and Dennis Myts

and 3 positions of a single, 6-membered heterocyclic ring (see Figs. 13.2 and 13.3). Purine comprises a pyrimidine ring fused to an imidazole ring with two more nitrogen atoms enclosed at the 7 and 9 positions. Purines are the most widely naturally occurring heterocyclic molecules containing nitrogen. When nitrogenous bases play a role in nucleic acids, they are called nucleobases.

Atom or group attachments at carbon atoms 4 and 5 of pyrimidine, determine whether the nucleobase is uracil (U), cytosine (C), 5-methylcytosine (5mC) or thymine (T). Atom or group attachments at carbon-6 of purine decide whether the nucleobase is adenine (A) or guanine (G). The various outcomes are shown below.

Nucleic acid, RNA, for example, contains nucleobases adenine (A) cytosine (C) guanine (G) and uracil (U). DNA is slightly different, containing adenine, cytosine, guanine, thymine (T) or 5-methylcytosine (5mC). The proportions of these bases and the sequence in which they follow each other along the polynucleotide chain differ from one kind of nucleic acid to the next.

A nucleoside can join to phosphate ions by condensation through the hydroxyl group on the 5th carbon atoms, giving phospho-diester derivatives, called nucleoside phosphates or nucleotides. A monoester is called a nucleoside monophosphate; a tri-ester is called nucleoside triphosphate (NTP):

Successive condensations result in polynucleotides such as DNA and RNA which are found in nearly all cells, in almost all organisms. Their phosphate backbone is said to enter the sugar molecule through carbon 5' and exit through carbon 3'. The significance of this 5' to 3' arrangement will be discussed later (see Fig. 13.2), as will the importance of DNA and RNA in heredity and protein synthesis.

In 1949, Todd synthesized the monomeric nucleotide, adenosine triphosphate (ATP) which comprises a sugar (ribose shown in Fig. 13.1) at its centre; a base (viz adenine in Fig. 13.2) at one side and a linked string of three phosphate groups at the other side.

ATP stores and transports chemical energy within cells, effectively being the 'molecular currency' of intracellular energy transfer. The hydrolysis of ATP to the diphosphate, ADP, is energetically favourable, providing free energy in processes such as motion and biosynthesis. When the organism is at rest, a phosphate group can be re-attached to ADP, utilising energy from food or sunlight, regenerating ATP. This allows ATP to store and liberate energy rather like a rechargeable battery (Fig. 13.4).

Purine

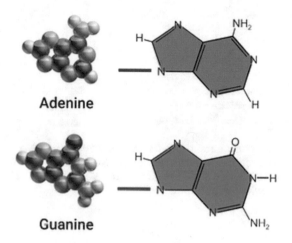

Adenine

Guanine

Pyrimidine

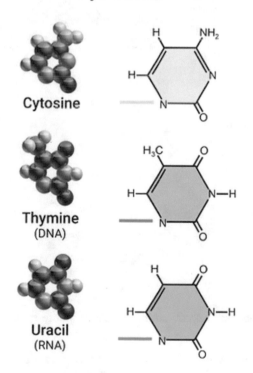

Cytosine

Thymine
(DNA)

Uracil
(RNA)

Fig. 13.2 Nucleobases of pyrimidine and purine. *Source.* By Blausen-Blausen, CC BY-SA 4.0, https://creativecommons.org/licenses/by-sa/4.0/deed.en. https://commons.wikimedia.org/w/index.php?curid=53113899.   https://commons.wikimedia.org/wiki/File:Purine_and_Pyrimidine.png#/media/

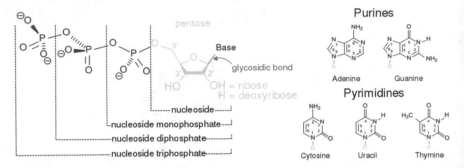

Fig. 13.3 Base-sugar-phosphate subunits of nucleotides of RNA and DNA. *Source* https://www.diffen.com/difference/Nucleoside_vs_Nucleotide. Citation: "Nucleoside versus Nucleotide." Diffen.com. Diffen LLC, n.d. Web. 6 Apr 2021. https://static.diffen.com/uploadz/b/b9/Nucleotides.png

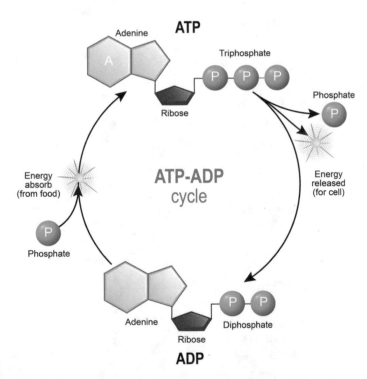

Fig. 13.4 Adenosine triphosphate-adenosine diphosphate cycle

It is noteworthy that Todd helped Dorothy Hodgkin (see Sect. 12.3) elucidate the structure of vitamin B$_{12}$. His chemical studies of the degradation products of B$_{12}$ were crucial to Hodgkin's X-ray determination of its definitive structure in 1955.

13.2.3 Summary

In summary, the main subject of Todd's research was the chemistry of natural products of biological importance. He was particularly interested in the structure and synthesis of nucleosides, nucleotides and nucleotide co-enzymes. By his systematic approach, he was able to synthesize polynucleotides, a feat now done by machine. He was also involved with the synthesis of ATP, vitamin B_1 (thiamine) and vitamin E (α-tocopherol and its analogues) as well as the structural identification of anti-anaemia vitamin B_{12}.

Todd was a superb research director with outstanding managerial skills. He was knighted, in 1954, and raised to the peerage, in 1962. In 1957, he was awarded the Nobel Prize in Chemistry, sole winner for his work on nucleotides and nucleotide co-enzymes. He was President of the Royal Society, in London, between 1975 and 1980. He was made a member of the Order of Merit in 1977 and received numerous honorary degrees and awards from many parts of the world.

13.3 Francis Harry Compton Crick OM FRS (1916–2004)—Physicist and Molecular Biologist

13.3.1 Background

Crick was born in Weston Favell, a village near Northampton, where his middle-class father managed the family shoe and boot factory. His uncle, Walter, an amateur chemist, allowed Francis to conduct experiments in his garden shed. Francis first went to Northampton Grammar School but, at the age of fourteen years, he won a scholarship to a private school in London. He studied physics at University College, London, taking a 2nd class honours degree in 1937.

In 1947, when he was thirty-one years old, he elected to change his interests from physics to biology. He joined Cambridge University, where he worked, for two years, at the Strangeways Laboratory, with a Medical Research Council studentship, on the physical properties of cytoplasm.

In 1949, he joined future, joint Nobel Prize winners, Max Perutz, and John Kendrew (see Sect. 12.4) under the general stewardship of Sir Lawrence Bragg (see Sect. 12.2). Crick pursued doctoral research into the determination of the structure of polypeptides and proteins by X-ray crystallography, gaining his Ph.D. in

1953/4. Other researchers in the Unit were trying to determine the most stable helical conformation of chains of amino acids in proteins.

The S-strain of *Streptococcus pneumoniae* is virulent and is encircled by a polysaccharide capsule which protects it from a host's immune system. The R-strain lacks a capsule and is non-virulent. In 1928, British microbiologist, F. Griffith, showed that the virulent strain could somehow convert, or transform, the non-virulent bacterial strain into an agent of disease.

In 1944, Oswald Avery and co-workers at the Rockefeller Institute reported that the transformation of these *pneumococcus* bacteria, from one type to another, occurred through the action of a 'transforming substance' that they identified as being sodium deoxyribose nucleic acid. It was postulated that DNA is the informational molecule of the cell and not protein, as was previously thought.

13.3.2   DNA Collaboration with J. Watson

In 1951, a twenty-three-year-old American student, James Watson, arrived at the MRC Unit and he and Crick began working together. They had complementary scientific backgrounds—Crick in physics, biology and X-ray crystallography, and Watson in viral and bacterial genetics.

In 1952, they had informal discussions, in Cambridge, with Erwin Chargaff of Columbia University. He told them about his two rules. Rule (1)—the relative amounts of organic bases guanine (G), cytosine (C), adenine (A) and thymine (T) vary from species to species. Rule (2)—whatever species the DNA is taken from, there is always a 1:1 ratio of C to G, and A to T. For instance, in human sperm, there is 19% cytosine and 19% guanine, 31% adenine and 31% thymine.

At a similar time, Maurice Wilkins and Rosalind Franklin, working at King's College, London were using X-ray diffraction to study DNA. They obtained high-resolution X-ray images of DNA fibres that suggested a helical, corkscrew-like shape. It is said that Wilkins, a friend of Crick, showed a key image (viz Photo 51) to the Cambridge team without Franklin's permission.

Initially, Crick and Watson did not conduct any experiments on DNA themselves. Instead, they drew on the experimental results of others. Using Linus Pauling's model-building techniques, Watson, using cardboard cutouts, showed that adenine, when 'joined' with thymine, very nearly resembled, in molecular dimensional terms, a combination of cytosine and guanine. Because the double-ring purines (A and G) are bigger than single-ring pyrimidines (C and T) the helical structure can only form when the purine bases are opposite pyrimidine bases (see Figs. 13.2 and 13.5). The

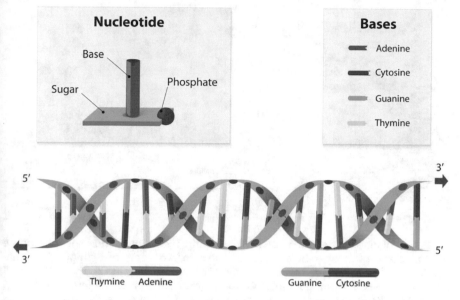

Fig. 13.5 Structure of DNA

combined length of each complimentary pair are always equal.

These two pairs of bases (T + A and G + C) can be neatly fitted at right angles between two helical, sugar-phosphate, polynucleotide backbones with hydrogen bonds between the bases forming weak linkages, at intervals, across the helices. There are two hydrogen bonds between T and A, and three hydrogen bonds between G and C. Intermolecular forces twist the 'ladder' into a double helix—imagine the rungs of a 'helical ladder'. Water-loving (hydrophilic) phosphate groups are on the outside of the double helix, whilst hydrophobic bases are on the inside (Fig. 13.6).

When a typical polynucleotide chain for DNA or RNA is shown schematically, it starts at deoxyribose or ribose carbon-5' and ends at carbon-3', with phosphate groups bridging the individual nucleotides.

Strands of RNA are shorter than strands of DNA. Unlike DNA, RNA normally appears as a single strand (ssRNA). Various specialized forms of RNA (e.g., messenger-, transport- and ribosome-RNA) are involved with the multi-step process of protein synthesis (see Sect. 13.3.5). Its chemical versatility arises because the ribose sugar of RNA has an extra hydroxyl group, whilst its molecular conformation make it more accessible to enzyme attack than DNA (Fig. 13.7).

An imaginary cross-section along the 'rung of the DNA ladder' would show that the organic base on C-3' of one helical strand is attached to the

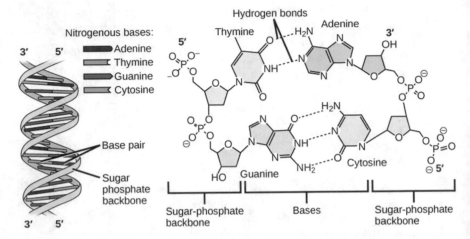

Fig. 13.6 Polynucleotide chain for DNA showing hydrogen bonds and phosphate bridges. *Source* http://open.bccampus.ca. Search for Concepts of Biology—1st Canadian Edition. On RHS, "Get This Book", click "readable" format (7), then "Website" and "Read Book". From Contents: Chap. 9 (Introduction to Molecular Biology). Section 9.1 (The Structure of DNA). Figure 9.4. Licensed under a Creative Commons Attribution 4.0 International License https://creativecommons.org/licenses/by/4.0/. This textbook was adapted from Concepts of Biology © 2013 by OpenStax, which is also under a CC BY 4.0 International Licence https://opentextbc.ca/biology/. The book was adapted by Charles Molnar and Jane Gair; the figure modified by Jerome Walker and Dennis Myts

base of C-5' on the opposing strand. Effectively, the two strands run in opposite directions—they are anti-parallel. The base pairs are inverted relative to each other (have dyad symmetry). The two helical strands are, therefore, not identical but complementary due to the pairing of specific bases. For example,

$$\text{Strand A'} = 5' - \text{GGCCAATTCCATACTAGGT} - 3'$$
$$\text{Strand B'} = 3' - \text{CCGGTTAAGGTATGATCCA} - 5'$$

The specific sequence of A, T, C and G nucleotides within an organism's DNA is unique to that individual and it is this sequence that control, not only the operations within a specific cell, but within the whole organism.

In layman's language, the DNA molecule is a right-handed spiral staircase in which each tread is of the same size, at the same distance from the next, and turns at the same rate, namely 36° between successive treads. If cytosine is at one end of a tread, then guanine is at the other, and likewise for the other base-pair. This structure implies that each half of the spiral carries the complete genetic message so that, in a sense, the other is redundant. However

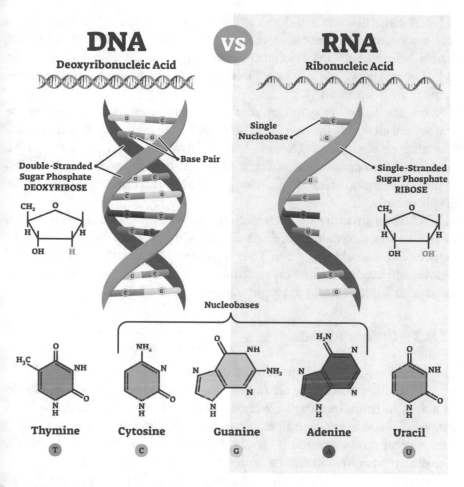

Fig. 13.7 Doubled-stranded DNA and single-stranded RNA

if one strand is damaged in any way, the other will maintain the full genetic information.

In technical terms, chemical and x-ray evidence suggest that the secondary structure of DNA comprises two polynucleotide chains, identical but heading in opposite directions—wound about each other to form a double helix with a diameter of 18 angstrom (=18 × 10⁻¹⁰ m; 1.8 nm). Both helices are 'right-handed' and have ten nucleotide residues per turn. The gap between two pairs of bases (adjacent rungs of the 'ladder') is 340 pm; the gap between 10 'rungs' is 3,400 pm.

In 1953, Crick and Watson published their observations, on DNA, in a 1-page paper, in Nature. This set the stage for a rapid advance in molecular biology.

DNA molecules can self-duplicate, bringing about new DNA molecules that are essentially identical to the original. This happens when a 'parent' double helix partially uncoils (unzips) creating a Y-shape, called a replication fork. The leading strand is oriented in the 3' to 5' direction; the lagging strand in the 5' to 3' direction.

With the aid of enzymes as catalytic agents, new polynucleotides are synthesized alongside the unfolding helix, using the 'old' strands as templates, matching up nucleobases (A with T and C with G). The replicated 'child' double helix comprises one 'old' strand linked in a complementary fashion to one 'new' strand. Thus, half the original DNA is conserved in each of the daughter molecules.

DNA has a structure sufficiently complex, yet elegantly simple enough to be the 'master molecule of life'. As Francis Crick put it, the two chains fit together like a hand and a glove. They separate and, about the hand is formed a glove, and inside the glove is formed a new hand. Thus, the DNA chemical fingerprint is preserved in the replicates of any cell division.

13.3.3   Gregor Mendel

This would explain the observations made by a farmer's son, Gregor Mendel who, in adulthood, became an Augustinian monk. He lived and worked in an abbey, in Brno, in today's Czech Republic. He became a 'kitchen-garden naturalist'. experimenting on pea plants between 1854 and 1863. Pea plants were a good model because he could emasculate them and then artificially inseminate them by transferring pollen with a small paintbrush.

Fortuitously, the pea plant displays dominant and recessive features (with no in-between characteristics) that are relatively easy to monitor from generation (progeny) to generation. Using methodical testing and statistical analysis of the pattern of occurrence, he formulated the basic principles of heredity.

He concluded that all living things pass on their key characteristics to their offspring in a predictable way. For instance, in the case of pea plants, he monitored seven characteristics, including their height (tall or short stems); the pea colour (green or yellow) and seed surface (smooth round or wrinkled) etc (Fig. 13.8).

His findings depend on the fact that genes come in pairs and are inherited as distinct units, one from each parent. Using Mendel's terminology for opposing characteristics (e.g., tall versus short plants), one trait (gene) is dominant (D), and the other recessive (R). In the case of height, for instance the offspring are either tall or short but not the average height of the two parents. The characteristics separated in an all-or-none fashion.

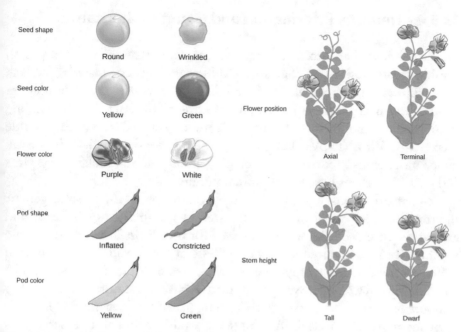

Fig. 13.8 Mendel's pea plant characteristics. *Source* http://open.bccampus.ca. Search for Concepts of Biology—1st Canadian Edition. On RHS—Get This Book—click "readable" format (7). Click "Website" and "Read Book". From Contents: Chap. 8, Introduction to Patterns of Inheritance. Section 8.1, Mendel's Experiments, Fig. 8.4. Licensed under a Creative Commons Attribution 4.0 International License. http://creativecommons.org/licenses/by/4.0/. This textbook was adapted from Concepts of Biology © 2013 by OpenStax, which is also under a CC BY 4.0 International Licence. https://opentextbc.ca/biology/. The book was adapted by Charles Molnar and Jane Gair

During fertilisation, the resulting pairings possible are—DD, DR, RD, and RR. For pairings DD, DR and RD, the dominant trait prevails. Only in the case of RR does the recessive trait manifest itself. In this progeny, therefore, the ratio of dominant trait to recessive trait is $\frac{1}{4} + \frac{1}{4} + \frac{1}{4} = \frac{3}{4}$ to $\frac{1}{4}$. Hence, 3 of the offspring will display the dominant characteristic and only one will show the recessive trait, a ratio of 3 to 1. In 1866, Mendel published his findings in the *Journal of the Brno Natural History Society* and achieved instant oblivion.

In 1909, Danish botanist, W. Johannsen, coined the word "gene" to describe Mendelian units of heredity, making the distinction between the outward appearance of an individual (phenotype) and its genetic traits (genotype).

13.3.4   Heredity Information and Genetic Inheritance

We now know that viroids are acellular particles that infect only plants, causing devastating losses. They utilise RNA as their genetic material and lack a protein coat. There are also various non-cellular viruses that rely only on RNA to transmit genetic information, infecting a host cell and commanding it to produce more virus. Human, disease-causing RNA viruses include Hepatitis C Virus (HCV), Ebola disease, SARS (see Sect. 9.4.4), influenza, polio, measles and retroviruses, including human immunodeficiency virus (HIV). RNA viruses have a high mutation rate.

Cellular organisms, however, use DNA, and not RNA, to store genetic information. The origin of DNA occurred at a critical stage in plant and animal evolution. DNA may be a derived form of RNA in which the ribose sugar of RNA was reduced to the deoxyribose of DNA, and the uracil base of RNA was methylated to thymine in DNA (see Sect. 13.2.2). Since DNA contains one fewer hydroxyl group than RNA, it is more stable—a key attribute for a molecule tasked with keeping genetic information safe.

Being more reactive, RNA adopted a multifunctional role serving as (i) a messenger (mRNA) to transcribe genetic information from DNA; (ii) a transfer vehicle (tRNA) delivering amino acids to the mRNA and (iii) a conveyor belt in ribosomes—via rRNA—translating the information brought by mRNA and tRNA to synthesise polypeptides, and proteins (see Sect. 13.3.5). RNA is the "Swiss army knife" of our cells—it can slice, dice, catalyse, build, code, replicate and transform.

The double helix of DNA is both the repository of heredity information of the organism and the vehicle for genetic inheritance. Crick advanced the hypothesis that the sequence of the bases in DNA forms a code by which genetic information can be stored and then transmitted to the next generation. The information is stored as the sequence of bases along the polynucleotide chain. It is a message 'written' in a simple language with an alphabet that has only four letters, namely A, G, T, and C (=adenine, guanine, thymine, and cytosine respectively). Crick and Watson had found 'the secret of life'.

In most cells, DNA is organised into structural units called chromosomes which vary widely in size and shape. In humans, the autosomal chromosomes are given numbers—1 to 22—based on declining size, with chromosome '1' being over three times bigger than chromosome '22'. At human conception, chromosomes of similar size line up, side-by-side. Twenty-two autosomal chromosomes from the mother and 22 from the father, together with an X chromosome from the mother, and either an X or Y from the father, merge together to give 23 pairs, or a total of 46 chromosomes, in the fertilised egg. The pea plant which Mendel studied has fourteen chromosomes.

A "genome" is an organism's complete set of DNA needed to develop and direct its activities. The haploid (egg or sperm) human genome comprises approximately 3 billion of these nucleotide pairs. So that the fertilised egg is not overwhelmed with DNA, the egg and sperm, at the point of fertilisation carry one genome each, resulting in about 6 billion base pairs in the fertilised egg. Consequently, every generation has the same number of genomes per somatic/diploid cell (equal number of chromosomes from each parent).

Scattered along a DNA molecule are particularly important sequences of nucleobases known as "genes". The latter are composed of a few thousand nucleotides. A gene carries a packet of key information that is passed to messenger RNA. The mRNA travels from the nucleus to the cytoplasm, generally with the purpose of encoding the information necessary to make a specific functional protein.

Each chromosome contains between hundreds and thousands of genes. In total, there are an estimated 21,000 genes in the human genome. Somewhat surprisingly, the genes that operate as instruction manuals for protein synthesis constitute barely 1% of the human genome. The functionality of the other 99% needs to be elucidated.

13.3.5 Protein Synthesis

Proteins are key constituents of cells and probably the most important class of chemicals in the body, playing a variety of roles. Some proteins act early in structural development of the embryo, triggering brain, heart and eyes etc. Others are responsible for physical differences (e.g., eye colour; physical height etc.). Proteins are building blocks and form structural components (e.g., muscle; skin; hair; cell walls) and are involved with cellular functions (e.g., enzymes as catalysts for biochemical reactions; hormones for signalling; haemoglobin and myoglobin as oxygen carriers).

Human proteins are derived from twenty different amino acids. In 1955, Crick proposed his 'Adaptor Hypothesis' in a privately circulated note. He suggested that each amino acid would be linked to an 'adaptor molecule', probably a nucleic acid, with twenty enzymes to perform specific linkages. Adaptor molecules are now known as 'transport RNA'.

In 1957, Crick presented a lecture on "Protein Synthesis" in which he conceptualized the link between genes and protein synthesis. Instruction for protein synthesis comes from DNA, but most of the DNA is housed in the cell nucleus, whereas protein synthesis takes place outside the nucleus, in the cytoplasm.

To facilitate transfer of instruction from nucleus to cytoplasm, Crick, Watson and others postulated that a mobile messaging system was required, with RNA as the likely messenger (mRNA). In 1961, Brenner, Jacob and Watson announced the isolation of mRNA whilst Jacob and Monod put mRNA in a theoretical context.

It was suggested that the DNA partially unzips and, alongside the appropriate gene, appears a strand of RNA. The mRNA carries instruction transcripts, from the DNA in the nucleus, to a host of 'ribosome factories' in the cytoplasm, where protein synthesis takes place. Each ribosome 'workshop' produces just one type of protein, shuffling the basic amino acids to sculpt elaborate 3D proteins. Ramakrishnan, Steitz and Moore would receive the Nobel Prize in Chemistry for their 2000 work on ribosomes—the molecular machines that make proteins.

13.3.6 Deciphering the Genetic Code. The Triplet Hypothesis

In the mid-1950s, physicist George Gamow predicted that three nucleotides, in a gene, might code for one amino acid in a protein. It would allow for sixteen ordered groups of nucleotide triplets, giving the possibility of 64 (=4^3) unique sequences of nucleotides. This would be more than sufficient for the twenty amino acids found in humans. In the 1960s, Marshall Nirenberg and Har Gobind Khorana separately presented the experimental proof. For their contribution, Nirenberg, Khorana, as well as Robert Holley were awarded the Nobel Prize in Physiology or Medicine in 1968.

In humans, the seemingly random sequence of the four bases in DNA form a code which specify the order in which some of the 20 amino acids should be combined to produce most proteins. The basis of the code is a 3-letter word selected from the 4-letter alphabet: A, C, G and T for DNA, and a slightly different 4-letter alphabet—A, C, G and U—for RNA. For instance, GAU is the code word (so called 'codon' word in messenger RNA) for aspartic acid; UUU is the code word for the amino acid, phenylalanine. Some amino acids can be enlisted by two different 'words' (e.g., both GAA and GAG will call up glutamic acid) but the first two letters are invariably common to both words (see Fig. 13.9).

To help with the translation process via mRNA, a particular codon AUG (=methionine) punctuates the start of the protein construction. Of the sixty-four possible triplet combinations, sixty-one can specify an amino acid. The

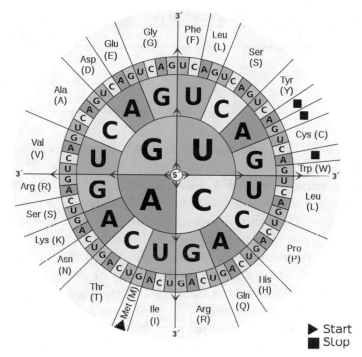

Fig. 13.9 Standard RNA genetic codon table. *Source* By Mouagip—Codons aminoacids table.png. Mouagrip has released the work into the public domain. https://en.wikipedia.org/wiki/en:public_domain. https://commons.wikimedia.org/w/index.php?curid=5986132

remaining three of the 64 trinucleotide sequences (viz UAA, UAG and UGA) are so-called "stop codons" that signal a halt to protein synthesis, triggering release of the polypeptide chain. They are also known as terminator codons (Ter) and they do not code for amino acids.

The codon 'circle' or "wheel" shown is composed of four concentric rings. Reading from ring (1) through ring (2) to ring (3) gives the letters of the codon triplet which then presents the name of the amino acid in ring (4). It also shows the starting codon (AUG = methionine) at 00:35 of the clock as well as the three terminator codons (UAA, UAG and UGA) at 00:12–00:14 of the clock.

An anti-codon is a 3-nucleotide unit of genetic code found in so-called transfer RNA. Tables of complementary codons and anti-codons are available. With a few minor exceptions, each codon has the same meaning, in any given cell, in any life form, whether the protein is being synthesized within, say, a bacterial cell or a human cell. For instance, the codon for amino acid,

Table 13.1 Illustrative codon and anticoding pairing

mRNA's codon	tRNA's anticodon
A	U
U	A
G	C
T	A

methionine, is AUG and its anticodon is UAC. Illustrative pairing is shown in Table 13.1.

Having contacted the mRNA, the ribosome travels along it reading its nucleobase sequence in chunks of 3-letter words, rather like a person reading Braille. Transport RNA (tRNA) responds by bringing the amino acid required for the growing peptide chain, to the mRNA. Beforehand, an enzyme will have recognised the shape of the specific tRNA and paired it with an appropriate amino acid. The amino acid gets attached to the head of the tRNA by forming an ester link between the last –OH group on the tRNA and the –COOH group of the amino acid.

At the base end of the tRNA will be a chunk of three nucleobases (anticodons) that are complementary (G to C; A to U etc.) with those on the mRNA's codons. The tRNA locks on to the mRNA ready to release its cargo. When two exposed amino acids are sitting side by side, a peptide bond is made between them, and the first amino acid in line breaks away from its tRNA. This tRNA can then detach itself, exit the ribosome and go hunting for another amino acid, identical to the one it carried before. With incoming tRNAs, the process of "translation" repeats until the growing polypeptide constitutes a protein, rather like a string of beads, with different colours and sizes, as is shown schematically in Fig. 13.10.

13.3.7 Summary

In summary, Crick and Watson were not the discoverers of DNA, rather they were the first scientists to accurately describe the molecule's complex, double-helical structure. In 1959, Crick was elected Fellow of the Royal Society. Rosalind Franklin died in 1958. Four years after her passing, Watson, Crick and Wilkins shared the Nobel Prize for Physiology or Medicine in 1962. Over the years, both Crick and Watson received numerous other awards and prizes for their work. Crick disapproved of royalty and, for this reason, refused a knighthood when one was offered. In 1991, Crick was appointed to the Order of Merit.

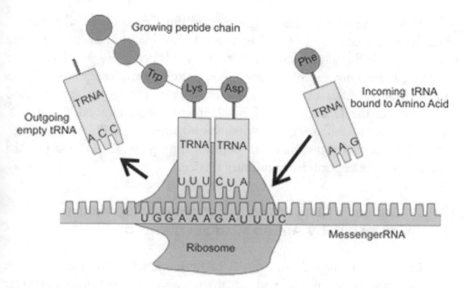

Fig. 13.10 How ribosomes direct peptide synthesis (translation). *Source* Boumphreyfr—Own work. https://commons.wikimedia.org/wiki/File:Peptide% 20syn.png. By Boumphreyfr vector conversion by Glrx—File:Peptide syn.png, CC BY-SA 3.0. https://creativecommons.org/licenses/by-sa/3.0/deed.en. https://commons. wikimedia.org/w/index.php?curid=101457889

13.4 Frederick Sanger OM, CH, CBE, FAA (1918–2013)—Biochemist, 'Father' of Genomics

13.4.1 Background

Sanger was born in Randcombe, Gloucestershire, second son of Frederick senior, a Cambridge graduate and medical practitioner. His father served as an Anglican medical missionary in China. His mother, Cicely, was a daughter of a wealthy cotton manufacturer and had a Quaker background. Soon after the birth of his two sons, Frederick senior converted to Quakerism. He encouraged his son's interest in biology.

At the age of nine, Frederick junior was sent to the Downs School, a residential preparatory school, near Malvern, run by Quakers. In 1932, he attended Bryanston School, in Dorset. Here, he particularly enjoyed scientific subjects and completed his School Certificate one year early, receiving seven credits. He spent his last school year experimenting alongside his chemistry master who had studied at Cambridge University and had been a researcher in the Cavendish Laboratory.

In 1936, Sanger began his long career at the University of Cambridge, where he took a degree in natural sciences. In his first year, his performance was average. He found physics and mathematics difficult. He was, however, a skilled experimentalist. In his second undergraduate year, he replaced physics with physiology. After obtaining a 1st class honours degree, he continued his studies, in the Department of Biochemistry, eventually working for A. Neuberger, from 1940 to 1944, when he obtained a Ph.D. As a Quaker, and pacifist, he was a conscientious objector during WW2 and was granted unconditional exemption from military service.

13.4.2   Sanger's Reagent to Determine the Structure of Insulin by Amino Acid Sequencing

Proteins play a pivotal role in cellular processes. One important protein is insulin—a peptide hormone made in the pancreas that regulates the sugar content of blood. Physical chemical evidence suggested a molecular weight of about 12,000. In 1944, Sanger started investigating the composition of bovine insulin. He used 1-fluoro-2,4-dinitrobenzene (FDNB, now called Sanger's reagent) which will react with the N-terminal amino group of a peptide or protein. This serves as a chemical tag. The reaction occurs under mild alkaline conditions, avoiding degradation of the polypeptide chain.

The dinitrophenyl (DNP) protein formed can be hydrolysed by acid to break the peptides bonds into unlabelled amino acids, whilst the yellow DNP derivative remains intact (Fig. 13.11).

Dyslexic Archer Martin and Richard Synge (Nobel prize winners in chemistry in 1952) developed silica gel column chromatography and paper chromatography to facilitate the separation and purification of complex chemical mixtures. Sanger used paper chromatography to isolate the DNP-amino acid so that its identity could be determined. FDNB has since been superseded by other reagents for terminal analysis.

In 1945, Sanger found both phenylalanine and glycine at the N-terminus of insulin, indicating that there were at least two polypeptide chains. By 1953, Sanger had used electrophoresis and chromatography to determine the amino acid sequence in the two molecular chains (A and B) that comprise insulin.

In 1955, he identified how the two chains are linked together via two disulphide bonds. These are integral to its correct geometric folding and is essential for engaging with its receptor. The amino acid, cysteine (HOOC–CH–(NH$_2$)–CH$_2$–SH) is uniquely structured to bring this about. Two disulphide

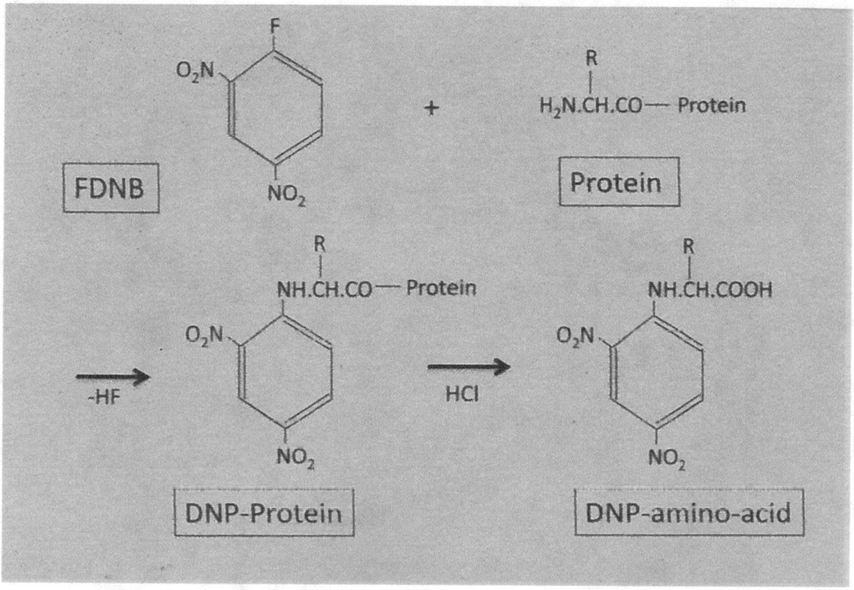

Fig. 13.11 Reaction of Sanger's reagent with peptides or proteins

bridges (A7–B7 and A20–B19) covalently connect the two chains. In addition, chain A contains an internal disulphide bridge (A6–A11). These joints are similar in all mammalian forms of insulin (Fig. 13.12).

Sanger was able to show that the chemical differences between insulin produced by different mammalian species were small. His revelation of the structure of insulin advanced our understanding of diabetes and kickstarted the development of man-made, human insulin, ending our reliance on cattle insulin.

Sanger had achieved the first sequencing of any protein. Only three years later, for his revolutionary approach to protein analysis, Sanger was awarded outright the Nobel Prize in Chemistry, in 1958. In 1969, applying X-ray crystallographic techniques, Dorothy Hodgkin and her team established the rudimentary, 3-D structure of crystalline porcine insulin (see Sect. 12.3).

13.4.3 Nucleotide Sequencing

In 1962, Sanger joined the Medical Research Council Laboratory of Molecular Biology, where Crick and Watson were working (see Sect. 13.3.2). Once they had produced an explanation of how the genetic code was

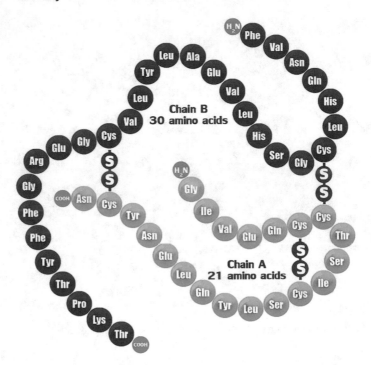

Fig. 13.12 Human insulin chains A and B liked by S–S bridges. *Source* By Zappys Technology Solutions—https://www.flickr.com/photos/102642344@N02/10083633053/, CC BY2.0 https://creativecommons.org/licenses/by/2.0/. https://commons.wikimedia.org/w/index.php?curid=91321395/, CC BY2.0 https://creativecommons.org/licenses/by-sa/2.0/legalcode

inherited through DNA, Sanger applied his flair in amino acid sequencing to decipher the detailed construction of individual genes.

DNA sequencing includes any method or technique that can be used to determine the order of the four nucleobases—adenine, guanine, cytosine, and thymine (A, G, C and T)—in strands of DNA. The sheer size of the DNA molecule called for subtler techniques than Sanger had applied to the structure of insulin.

An organism's genome is stored in the form of long rows of building blocks known as nucleotides which form the DNA molecule (see Sect. 13.2.2). An organism's genome can be mapped by establishing the order of these nucleotides. In the mid-1970s, Sanger developed an ingenious method of achieving this.

The Sanger sequencing, chain-terminating technique involves making many copies of the target DNA. The ingredients necessary to complete this cloning exercise are:

(i) A DNA polymerase enzyme to accelerate the creation of new strands.

(ii) A primer that binds to the template DNA and acts as a starter for the polymerase. The primer provides the initial 3' hydroxy group to form the 1st phospho-diester bond with dNTP nucleotides seized from the chemical mix and added to the expanding chain.

(iii) Four 'regular' deoxynucleoside triphosphates (dATP, dTTP, dCTP and dGTP).

(iv) A small proportion of four, chain-terminating, dideoooxy versions (designated ddATP, ddTTP, ddCTP and ddGTP) of those four regular nucleotides, each of which is labelled with a molecular beacon.

Di-deoxy nucleotides are related to regular deoxy nucleotides but have one key difference, namely they lack a hydroxyl group at the 3' carbon of the sugar ring. The 3' hydroxyl group normally acts as a "hook", allowing a new nucleotide to be added to an existing chain (Fig. 13.13).

Wherever ddNTP is inserted in the DNA chain, there is, then, no hydroxyl group available on which to add further nucleotides. Thus, the chain ends or terminates with the ddNTP, resulting in multiple short strands.

The sequencing method developed by Sanger involved four separate reaction tubes, each with a different, single ddNTP. In Sanger's day, a radioactive primer or ddNTP labelled with a radioactive phosphorus atom ($_{15}P^{32}$) was used for detection by autoradiography.

Today, the process of separation and detection has been simplified by using ddNTPs labelled with different coloured fluorescent dyes or fluorochromes. The fluorescent dye has a wavelength characteristic of each of the four nucleotides, namely, derivatives of adenine (ddATP); cytosine (ddCTP);

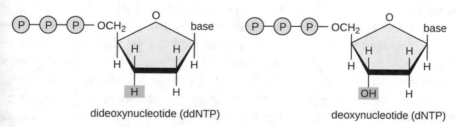

dideoxynucleotide (ddNTP)　　　　　deoxynucleotide (dNTP)

Fig. 13.13 Nucleoside triphosphates. Difference between deoxy nucleotides (dNPT) and dideoxy nucleotides (ddNPT). *Source* https://courses.lumenlearning.com/microbiology/chapter/visualizing-and-characterizing-dna-rna-and-protein/; see Fig. 9. CC LICENSED CONTENT, SHARED PREVIOUSLY. OpenStax Microbiology. Provided by: OpenStax CNX. Located at http://cnx.org/contents/e42bd376-624b-4c0f-972f-e0c579 98e765@4.2. Chapter 12 (Visualizing/characterising DNA, RNA, protein); Fig. 12.21. License: CC BY: Attribution. License Terms: Download for free. http://creativecommons.org/licenses/by/4.0/

guanine (ddGTP) and thymine (ddTTP). The sequencing reactions take place in a single tube in which all four ddNTPs are present, together with DNA sample, DNA polymerase, primer and four regular dNPTs.

A typical Sanger sequencing method consists of six steps:

1. The mixture mentioned is heated to denature the double-stranded DNA (d-sDNA) into two separate strands (s-sDNA).
2. It is cooled, whereupon the primer that corresponds to one end of the sequence can bind to the single strand.
3. The temperature is raised to encourage the DNA polymerase to synthesize new double stranded DNA by incorporating the four dNTPs as appropriate. The process is repeated over numerous cycles to allow for further replication and to maximise incorporation.
4. The elongation process continues until a termination ddNTP nucleotide is randomly incorporated, whereupon the synthetic reaction halts. Over time, as the reaction progresses, eventually there will be a ddNTP inserted at every location in the newly synthesized DNA, so that each strand synthesized differs in length by one nucleotide and is terminated by a labelled ddNTP.
5. The resulting double stranded DNA fragments are denatured into s-sDNA, having the same 5' as the original with a ddNTP at the 3' end.
6. The denatured single stranded fragments are separated, and the sequence determined.

The vessel will contain fragments of different lengths, ending at each of the nucleotide positions in the original DNA. The fragment ends will be labelled with nucleotide-specific dyes that indicate their final nucleotide.

After the reaction, the fragments are run through a long thin tube containing a gel-matrix, in a process called capillary gel electrophoresis. The negative charge of the phosphate ions facilitates movement of the fragments towards a positive electrode. Their speed of passage is inversely proportional to their molecular weight, short fragments move relatively quickly; long fragments pass relatively slowly. Each band on the gel reflects the size of the DNA strand when the ddNTP terminated the reaction.

At the end of the tube is a window subject to a laser beam. A sensor opposite detects the wavelength of the light emitted which will be characteristic of the dyed material passing the window.

The data recorded by the detector consists of a series of peaks reflecting the intensity of fluorescence, manifest as a 'chromatogram' or 'electropherogram'

Each peak can be related to a specific fragment. An abbreviated experimental protocol is shown in Fig. 13.14.

Sanger's group was the first to work out the precise composition of the entire genome of any organism, namely the bacteriophage ΦX174—a relatively modest life form with 5,386 nucleotides that link together in chains in the creation of DNA. The sequence identified many of the features responsible for producing the proteins of the nine known genes of the organism.

Fred Sanger's 1977 PNAS paper on ΦX174 (DNA sequencing with chain-terminating inhibitors; Proceedings of the National Academy of Science USA, **74** (12) 5463–5467) is one of the most cited Nobel prize winning papers across all three scientific disciplines.

Sanger then moved on to sequencing the first human genome, in the shape of DNA in human mitochondria.

Sanger's technique offers high quality sequencing for up to 900 base pairs. The method became widespread and commonplace in molecular biology. The potential value of such a technique in treating genetic disease was immediately foreseen.

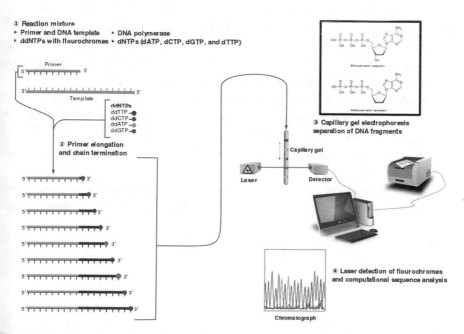

Fig. 13.14 Experimental principles of Sanger sequencing. *Source* By Estevezj—Own work, CC BY-SA 3.0, https://creativecommons.org/licenses/by-sa/3.0/deed.en https://commons.wikimedia.org/w/index.php?curid=23264166.   https://commons.wikimedia.org/wiki/File:Sanger-sequencing.svg#/media/File:Sanger-sequencing.svg

Sanger sequencing is still used for bacterial plasmids—double stranded DNA that is distinct from a cell's chromosomal DNA. However, the approach is expensive and inefficient for large scale projects. Next generation sequencing (NGS) followed involving automation, computerization, and further refinements for use in the Human Genome Project.

The Human Genome Project produced the "book of life", the operating manual for *Homo sapiens*—hominids who evolved on the African continent about 300,000 years ago. The characteristics of African DNA will give us information about the origins of humanity.

13.4.4 Accolades

In 1954, Sanger became a Fellow of the Royal Society.

Between 1901 and 2020, the Nobel Prize in Chemistry has been awarded 112 times to 186 Nobel Laureates. Sanger is the only person to have been awarded the Chemistry Prize twice, in 1958 (for his work on insulin) and again in 1980.

The Nobel Prize in Chemistry was awarded, in 1980, to three individuals, namely Paul Berg (for his fundamental study of the biochemistry of nucleic acids, regarding recombinant DNA) together with Walter Gilbert and Frederick Sanger for their contribution concerning the determination of base sequences in nucleic acids. Sanger's work transformed the field of genetics into a powerful tool of genetic manipulation and gene therapy.

He refused a knighthood (because he did not want to be called "Sir") but accepted the Order of Merit in 1986. This is a personal gift of the sovereign for those whose accomplishments may go unsung. The Welcome Trust Sanger Centre (now Institute) was opened in 1993 to continue work on DNA sequencing.

To reflect the importance of this subject area, over sixty Nobel prizes have been awarded to scientists working on DNA, genetics or genetic-related areas

14

Science of Key Building Materials—Cementitious Substances, Iron and Steel

The Rubik's cube embodies core human universals. Without words, it speaks about simplicity and complexity; frustration, perseverance, and triumphant intelligence; problem solving and creativity. This must be the reason why it could touch so many lives and inspire so much innovation. (Ernő Rubik in talks with John Wright, reported in The Sunday Telegraph, 28 July 2019)

14.1 Précis

Great Britain was blessed with iron ore to make steam engines, other machinery and tools; vast deposits of coal to power those steam engines and to keep us warm; limestone as a raw material for Portland cement, as well as clays to produce sanitaryware and pottery. Through our international networks, we could import raw cotton for fabrics and clothes, together with rubber for pneumatic tyres.

Joseph Aspdin was a pioneer in the production of Portland cement, so starting the Age of Artificial Stone. In the absence of any chemical knowledge, but with luck and perseverance, he and his sons produced a chemically complex substance with great versatility for the building industry, making modern infrastructure possible.

When mixed with sand and water, it forms mortar—an adhesive substance for bricks and stone. When combined with sand, gravel and water, concrete

Bailey, *Inventive Geniuses Who Changed the World*, https://doi.org/10.1007/978-3-030-81381-9_14

is formed which hardens to a rock-like mass, even under water. After water, concrete is the most 'consumed' product in the world.

Their patents brought the Aspdins neither fame nor excessive fortune. Now, cement is vilified on environmental grounds for the greenhouse gas emissions arising from its manufacture.

In 1709, in Shropshire, Abraham Darby succeeded in smelting iron using coke derived from low-sulphur coal, thereby creating a cost-effective process for making commercial grade iron and jump-starting the Industrial Revolution. Key factors in our Industrial Revolution were Britain's deposits of iron ore and coal to make coke.

Henry Bessemer was largely self-taught and exhibited extraordinary inventive skills. He learnt basic metallurgy at his father's foundry. At the advent of the Crimean War, in 1853, he wanted to get involved with the manufacture of more robust casings for guns and canons. He developed a large scale, inexpensive industrial process for purifying pig iron to produce steel, heralding the Age of Steel. In 1875, Britain accounted for 47% of the world's pig iron and about 40% of global steel-production.

Steel is the second most mass-produced commodity, after cement, popular because of its unique combination of workability, versatility, strength, durability, and cost. It fitted perfectly with the expanding mechanisation taking place in the latter half of the nineteenth century, facilitating great strides in transport, shipping, construction, machine tools and weapons. Bessemer brought his own projects to fruition, benefitting handsomely from their results, whilst changing the way people lived and travelled.

14.2   Joseph Aspdin (1778–1855) and William Aspdin (1815–1864)—Inventors and Manufacturers of Portland Cement. The Age of Artificial Stone

14.2.1   Family Background

Joseph was born in the Hunslet district of Leeds, in Yorkshire. He was the eldest of six children born to Thomas and Mary Aspdin. Like his father, Joseph became a bricklayer and plasterer.

Joseph married Mary Fotherby, in 1811. Shortly afterwards, he set up home and business premises in Slip Inn Yard (now Packhorse Yard) in Leeds now marked with a Blue Plaque. They had two sons—James and William—and several daughters. At the age of fourteen, William joined his father's business.

14.2.2 The Lime-Cycle

The sedimentary rocks and minerals from which lime is derived—typically limestone and chalk—are composed mainly of calcium carbonate ($CaCO_3$). The White Cliffs of Dover are composed of chalk. This is a finely grained limestone formed from the skeletal remains of microscopic sea-creatures. When calcium carbonate is heated to between 800 and 1000 °C, and the evolved carbon dioxide allowed to escape, 'calcination' occurs and calcium oxide (CaO) or quicklime results. On a large scale, this is achieved in a lime kiln. Quicklime can be preserved, in dry air, indefinitely.

When quicklime is treated ('slaked') with just enough water to hydrate it, whilst maintaining its powdery consistency, it is referred to as 'hydrated lime'. In 'wet slaking', a slight excess of water is added to hydrate the quicklime to 'lime putty'.

Pure quicklime (known as 'fat' lime) reacts vigorously with abundant water to give calcium hydroxide [$Ca(OH)_2$] or 'slaked lime' which is a non-hydraulic lime cement.

Slaked lime has an adhesive property that is exploited as a binding material in masonry works. It is used in the making of lime mortar and lime plaster. The former consists of lime, sand, and water, whilst the latter comprises lime and water, together with such things as animal hair to give extra strength.

The setting of lime mortar and lime plaster from pure slaked lime is due firstly to the release of water. This is followed by a slow conversion of the outermost layer of calcium oxide into calcium carbonate by a 'carbonation' reaction with carbon dioxide (CO_2) in the air or dissolved in rainwater. This completes the so-called lime cycle (Fig. 14.1).

Argillaceous limestone consists predominantly of calcium carbonate but includes 10–40% of clay materials. When soft, they are called "marls". On 'burning' them, a 'poor' lime results. This contains some silica and/or alumina from clay and hydrolyses (slakes) slowly to slaked lime. The latter has, in a small degree, some of the properties of Portland cement, and is known as a 'natural hydraulic' lime. When being used to make mortar, the routine processes of water loss and carbonation will occur alongside a chemical set, associated with the formation of calcium aluminosilicates (see later).

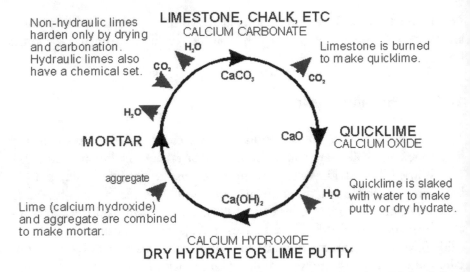

Non-hydraulic limes harden only by drying and carbonation. Hydraulic limes also have a chemical set.

LIMESTONE, CHALK, ETC
CALCIUM CARBONATE

H_2O

CO_2

$CaCO_3$

Limestone is burned to make quicklime.

CO_2

H_2O

MORTAR

CaO

QUICKLIME
CALCIUM OXIDE

aggregate

Lime (calcium hydroxide) and aggregate are combined to make mortar.

$Ca(OH)_2$

H_2O

Quicklime is slaked with water to make putty or dry hydrate.

CALCIUM HYDROXIDE
DRY HYDRATE OR LIME PUTTY

Fig. 14.1 The lime cycle. Illustration by Sara Pavia, Civil Engineering, Trinity College Dublin

14.2.3 Masonry Nomenclature; Nature of Clay

Cement is a binder. 'Hydraulic cement' not only sets by its reaction with water but forms a water-resistant product which hardens over time. Before setting and hardening, the hydrated cement will adhere to other building materials. In today's building industry, ordinary mortar for laying bricks is normally made by mixing sand with Portland cement.

Aggregate is a mix of fine and coarse material such as sand and gravel which is chemically inert. The amount of Portland cement required for a major construction project is reduced greatly by mixing it with aggregate to form concrete. Even under water, and the absence of carbon dioxide, such a mix will harden over time into a rock-like mass. The cement is the expensive part of the mix. Supplementary cementitious materials (SCM's) are by-products (e.g., fly ash; slag cement and silica flue) of other manufacturing processes. They reduce the cost of concrete and modify its properties.

The term 'clay' denotes certain earths which are highly plastic when wet. At temperatures above 900 °C, they are converted into a hard mass, unaffected by water. Clay is not a single substance but consists of fine particles from the decomposition of certain igneous rocks, bound together by the sticky substance, hydrated aluminium silicate. Common clays contain impurities such as iron oxide. Fireclay contains a good deal of silica. Clays may be found where they were originally formed, often with limestone (called marlstone) or, carried away by storm water and deposited at a distance.

14.2.4 History of Non-hydraulic Lime, Hydraulic Lime (Roman) and Portland Cement; Mortar and Concrete

One of the first 'building adhesives' was wet clay. The use of hydraulic cement/concrete probably started in the Roman Empire when cements were formed from a mixture of lime and clay. The word cement and concrete come from the Latin words, *Cemento* and *Caementicium*.

Lime mortar alone will not re-carbonate and harden under water or inside a thick masonry wall. To achieve hardening, the Romans ground together lime and volcanic ash to give an 'artificial' hydraulic Pozzolanic lime cement.

Such material was probably used to build the Pantheon, in Rome. This is topped with a gravity-defying dome, having a 43.3 m diameter. Now, nearly 2,000 years old, it is still the world's largest, non-reinforced concrete dome.

In 1774, Leeds-engineer, John Smeaton, started looking for a building material that would not be adversely affected by water. In 1793, he realized that calcination of limestone that contained clay, produced hydraulic lime, a lime that set quickly and hardened under water. This was used in the rebuilding of the Eddystone Lighthouse.

In 1812, Louise Vicat developed an artificial hydraulic lime composed of limestone and clay. The technology was used, in 1816, to build the world's first unreinforced concrete bridge in Souillac, France.

Between 1826 and 1830, C. W. Pasley, a lecturer at the Royal School of Military Engineering, near Chatham Dockyard, in Kent, conducted research into artificial hydraulic lime cement. Using chalk and Medway clay, he produced a hydraulic lime equal to the natural 'Roman' cement.

Between 1817 and 1823, Joseph Aspdin purportedly experimented, in his kitchen, with cement manufacture and arrived at an innovative building material. In 1824, he was granted British Patent 5022 for "An improvement in the method of producing an artificial stone".

Essentially, Pennine carboniferous limestone (often crushed road paving) was burned, slaked, mixed with clay, and burned again in a kiln. At the hot end of the kiln, the limestone and clay were roasted to 'cement clinker'. After cooling, the clinker was ground to a powder.

Aspdin's powder had the power of setting when mixed with water. In his view, the set-form resembled, in colour and hardness, a stone quarried in Portland, in Dorset, that was famous for its strength and quality. He gave his powder the name "Portland cement".

Aspdin formed a business with his neighbour, W. Beverley, and they founded a factory in Kirkgate, Wakefield, to make Portland cement. Portland cement is a substance that, when mixed with water, sets, and hardens independently, or can be used to bind other materials together. It can be used in the production of concrete and mortar. It bonds natural or artificial aggregates to form strong building materials that are durable in the face of extreme challenges from the environment and weather. It can be used for stuccoing both interior and exterior walls.

At the time, natural cement was widely used, and it could be manufactured more cheaply than man-made Portland cement. The growth in demand for Portland cement was slow, but it eventually developed into one of the biggest industries for the manufacture of building materials.

In 1838, after a compulsory purchase order was made by the Manchester and Leeds Railway Company, the factory was closed and Aspdin moved his equipment to a new site near Kirkgate.

Joseph Aspdin never produced Portland cement as we know it. Joseph's prototype product was made by a 'double-burning' process and was fast-setting. It is unlikely that he fired his recipe mix at a temperature above 1,250 °C and would, therefore, contain none of the key structural component, alite, present in modern Portland cement. His Portland cement was used mainly for stuccos and pre-cast moulding.

In 1841, Joseph posted a notice saying that he was no longer responsible for the debts of his son, William. The two men parted company and Joseph's other son—James—joined the business. William's career was always beset by financial irregularities and questionable business deals.

William established his own manufacturing operation, in 1841, at Rother-hithe, in London. He modified his father's manufacturing process and raw material formulation, increasing the content of limestone and burning it at a higher temperature. This gave a slower-setting concrete which was harder on solidification, albeit resulting from a more expensive method of manufacture.

Isaac C. Johnson, however, claimed that he was the creator of 'genuine' Portland cement. He also heated the ingredients to a temperature at which they partially melted.

In an advertisement, in London, William said that, "*His product was an improved version of his father's Portland cement. It is infinitely superior to any cement that has hitherto been offered to the public—. It is stronger in its cementative qualities, harder, more durable, and will take more sand than any other cement now used*".

William possibly falsely claimed that, when the roof of Brunel's ill-fated Thame's Tunnel (see Sect. 4.5.2) collapsed, tons of Aspdin's concrete were dumped in the river, to seal the breach, so that Brunel could pump the tunnel dry. Mortar made with Portland cement was then used for the relining. If correct, this was probably the first major civil engineering project using Portland cement in Great Britain. The tunnel was opened to the public in 1843.

What is more certain is that Portland cement was used to construct a 20 × 50 m concrete slab in a building, designed by I. Brunel, in 1839, to support the machinery used to fabricate the world's first screw-propelled iron ship, SS Great Britain (see Sect. 4.5.7). Chemical analysis and microscopic examination by Professor G. Allen of the Interface Analysis Centre, at the University of Bristol, confirmed the use of Portland cement. Part of Brunel's genius was his ability to identify revolutionary new materials that would build the modern world.

Portland cement was also used in the construction of the Houses of Parliament (1840–52) and the London sewer system (see Sect. 5.7.3) built by Bazalgette (1859–67). He was convinced that the use of Portland cement would make the sewers of London efficient and long-lived, by ensuring they were strong and waterproof. John Grant of the Metropolitan Board of Works set detailed specifications for the Portland cement used.

In 1846, William formed a partnership with Robins and Maude to build a cement works at Northfleet Creek in Kent, using an inexhaustible supply of local, soft chalk and alluvial clay found along the estuaries of the Rivers Medway and Thames. One of Aspdin's 'bottle/beehive-shaped kilns'—the oldest surviving cement kiln in the world—can be seen at the former Blue Circle works (National Grid Reference: TQ 6,175,274,889). It is deemed to be a Scheduled Ancient Monument. Northfleet harbour would serve to ship Portland cement around the world.

In 1860, William set up two plants in Germany to make the first 'modern' Portland cement outside Britain. William died, in Germany, in 1864.

Concrete reinforced with steel was patented by W. B. Wilkinson in 1855. The reinforced structure displays the characteristic of concrete when compression/crushing forces are applied, and the property of steel when tension/pulling forces are employed.

Successors to Aspdin and Johnson improved cement-making through systematic study, as well as the introduction of rotary kilns. In the 1880s, it was recognised that the amounts of clay and lime had to be carefully proportioned to produce high grade Portland cement. Henri le Chatelier made a notable contribution in this field of study.

In 1885, F. Ransome patented a slightly horizontal kiln that could be rotated so that the contents moved gradually from one end to the other. This facilitated more efficient mixing and better temperature control, resulting in Portland cement of more consistent quality.

14.2.5 Modern Manufacturing Process for Portland Cement. The Chemistry of Portland Cement and Concrete Derived from It. Their Carbon Footprints

In the modern manufacturing process, as the temperature approaches 1,000 °C, calcium carbonate, in limestone, and silicates in clay, break down to their respective oxides. As the temperature rises further, they combine into four main chemical compounds.

The key chemical reaction which demarcates Portland cement from other cements occurs at a temperature higher than 1,300 °C. This is when some belite (dicalcium silicate, Ca_2SiO_4; abbreviated to C_2S) combines with calcium oxide to form alite (tricalcium silicate, Ca_3SiO_5; abbrev C_3S) which is largely responsible for Portland cement's initial set and early strength, in concrete.

The molten mix, called clinker, is rapidly cooled. After the clinker is ground into a powder, gypsum ($CaSO_42H_2O$; abbrev CSH_2) is added to facilitate an extension to the workability of mortar or concrete by retarding the setting-speed, following hydration of the Portland cement.

The resulting, major components in powdered Portland cement are tricalcium silicate (approx. 50% by weight) followed by dicalcium silicate (25%), tricalcium aluminate ($Ca_3Al_2O_6$; abbrev C_3A; 10%), tetracalcium aluminoferrite ($Ca_4Al_2Fe_2O_{10}$; abbrev C_4AF; 10%) and gypsum (5%).

By controlling the proportions of these components, and by the judicious addition of certain additives, suppliers and users of Portland cement can tailor-make different types of concrete to suit specific construction requirements.

Portland cement comes to life when a controlled amount of water is added. The tricalcium aluminate is hydrolysed to calcium hydroxide and aluminium hydroxide, and these substances further react with the calcium silicates to produce calcium aluminosilicates. Crystallization begins and tiny needle-like protuberances (fibrils) extend in all directions, forming a mesh-like web, based on a framework of AlO_4 and SiO_4 tetrahedra. Within the mesh are amorphous zones where water and the inert aggregate are entrapped in concrete.

Concretes made from slaked lime become solid due to release of water followed by carbonation. The concrete from Portland cement is differen-

because it becomes solid, not because water has escaped, but because water is locked in the mesh. The process of hardening continues for years.

The manufacture of Portland cement rapidly spread to other European countries and North America. By the early twenty-first century, China and India had become the world leaders in cement production. According to the International Energy Agency (IEA), initial estimates suggest that 4.1 billion metric tons of cement were produced in 2017. China is the largest producer, accounting for almost 60% of global production, followed by India, at 7%.

On the negative side, concrete has been pilloried through its application in countless architectural eyesores. It is also vilified on environmental grounds. The cement and lime industries are unusual in that the majority (60%) of greenhouse gas emissions come, not from their fossil fuel combustion, but from the liberation of CO_2 from the calcination of limestone raw material. The mining process for ores and clay, together with the provision of thermal energy to achieve temperatures up to 1,500 °C, add further to the carbon footprint.

The rule of thumb is that, for every tonne of cement made, 1,000 kg of carbon dioxide is generated in old manufacturing plants, compared to 850 kg in modern kilns.

It has been calculated that the release of CO_2 from limestone alone contributes 5% of annual anthropogenic, global carbon dioxide. This does not include the CO_2 emissions from the fossil fuel combustion necessary for the mining of, and heating of, raw materials. The Chinese cement industry accounts for about 3 of the 5%.

The Global Cement and Concrete Association (GCCA)—founded in 2018—is focussed on sustainable development. Cement makers know that they must act because pollution regulations are tightening everywhere, whilst the cost to emit CO_2 has pressured their revenues. More costly 'green' cement, however, currently remains a low priority for down-stream developers and construction companies. The path to zero emissions requires much innovation.

14.2.6 Joseph's Death. The Aspdin Legacy

Joseph Aspdin died in 1855, aged 76 and is buried in the grounds of St. John's church, Wakefield. His patent brought him neither fame nor excessive fortune. In 1924—a century after his patent—the American Portland Cement Association visited Leeds to present the city with a bronze tablet in his memory. It hangs in a dark corridor in the Town Hall and says, "*His invention—has made the whole world his debtor*". Not far from his birthplace and manufacturing sites, the University of Leeds has become a world-leading centre for research into concrete.

Neither Joseph nor his sons had any chemical knowledge. They were alchemists but with luck and dedication, they produced a chemically complex substance with great versatility for the building industry. Joseph and William pioneered a method for industrially producing Portland cement which is one of mankind's most important manufactured materials. Currently, over half a tonne per person per year of anhydrous cement is made.

The Aspdins made modern infrastructure possible. Few schools, hospitals, or houses are now built without some concrete being incorporated. Many new, towering buildings, roads, canyon-soaring bridges, the floating bridge on Lake Washington, and colossal dams were built using Portland cement. The Sydney Opera House, the flower-like Lotus Temple in Delhi, and the 830 m tall Burj Khalifa skyscraper in Dubai, all owe their form to the material. Its derivative, concrete, is used more than any other man-made material in the world. It is said that, after water, concrete is the most 'consumed' product in the world—about 30 billion tonnes per year. Without Portland cement, the modern world would look vastly different.

14.3 Henry Bessemer FRS (1813–1898)—Engineer, Inventor, and Entrepreneur. The Steel Age

14.3.1 Background

Henry's father, Anthony, had spent several years in Holland and then France. In 1784, Anthony was made a member of the French Academy of Science for his work on optical microscopes. The family was driven to return to Great Britain, at the onset of the French Revolution, in 1789.

Henry was born on his father's small estate in Charlton, near Hitchin, in Hertfordshire. Henry received the rudiments of an education in the neighbourhood. Anthony encouraged his son's interest in things mechanical and gave him free rein in his workshop. Henry was largely self-taught and exhibited extraordinary inventive skills.

He learnt basic metallurgy at his father's foundry where typefaces were designed and manufactured. Henry's first successful business venture followed his invention of a complex, steam-driven machine for making bronze powder used in gold-coloured paint. His process reduced the price by 1/40th relative to the manual German process, dominant at the time. The profits from this business allowed him to follow other pursuits.

14.3.2   The Iron Age; Smelting and Pig Iron

The Iron Age was a stage in human history that followed the Bronze Age when local people began making tools and weapons from iron. It is the era when ferrous metallurgy became the dominant technology of metal working—superseding copper/tin. Iron is a later discovery than copper because its melting point is about 500 °C higher than that of copper, so at every stage of its processing—smelting, working and steel-making—it requires more heat. In response to added substances, steel is more sensitive than bronze. Slight variations in the composition of steel will dictate its underlying properties.

Around 1500 BC, a group of tribes known as the Hittites—living in parts of today's Turkey—developed new techniques for making iron weapons. The use of iron became more widespread when people learned how to make steel by heating impure iron with a carbon source.

Minerals of iron containing high percentages of iron are the magnetic, magnetite (ferrosoferric oxide, $FeO \cdot Fe_2O_3$), anhydrous haematite (Fe_2O_3) and hydrous iron (III) oxides ($Fe_2O_3 \cdot xH_2O$) known as limonites. These ores contain an assortment of impurities, including derivatives of sulphur, phosphorus, silicon, aluminium and manganese. Iron smelting is a process of heat application, in the presence of an agent to reduce iron oxides, liberating unwanted elements, either as a gas or slag, leaving the elemental iron behind.

One of the applied reducing agents is carbon monoxide, originally generated in an air-starved furnace, from a source of carbon, initially charcoal, made from wood. As wood became more expensive, an alternative was sought. One candidate was coke—made by heating coal in the absence of air. Most samples of coal contain appreciable levels of sulphur, giving coke that led to brittle iron. In 1709, at Coalbrookdale, in Shropshire, Abraham Darby succeeded in smelting iron using coke derived from low-sulphur coal, thereby creating a cost-effective process for making commercial grade iron.

In the process, carbon monoxide generated from coke reduces iron oxides and is itself oxidised to carbon dioxide. Purifying agents (called flux) such as limestone are now added to remove the unwanted earthy impurities (gangue) as slag. The heat generated from the burning coke decomposes the limestone to form calcium oxide (see Fig. 14.2). The resulting, basic calcium oxide reacts with acidic impurities, forming slag which comprises mainly silicates, aluminates, and phosphates of calcium. Iron oxide is converted to metallic

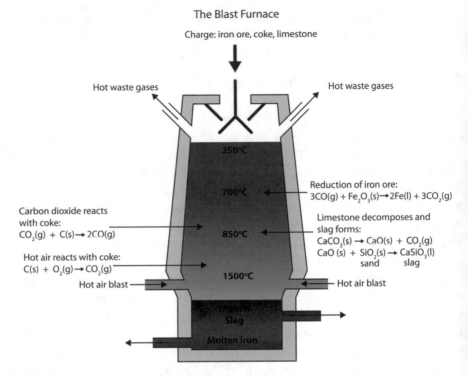

Fig. 14.2 Blast furnace

iron at furnace temperatures of about 1,250 °C, that is 280 °C below the melting point of elemental iron (1,538 °C).

When the fuel to ore ratio is raised, in a blast furnace, a higher reducing temperature can be achieved. At the base of the furnace, preheated air is introduced (blasted) through twyers (tuyères), and this reacts with carbon, in a very exothermic reaction, to produce carbon dioxide, thereby raising the temperature as high as 1,900 °C. The heat generated is carried upward by the ascending gases. Carbon dioxide reacts with additional carbon, in an endothermic reaction, to form carbon monoxide—the active reducing agent.

Descending iron oxides are converted to impure iron, formed as a liquid which trickles to the base of the blast furnace. In the hotter parts of the furnace, carbon itself acts as a reducing agent, reacting directly with any ferrous oxide. The lighter, liquid slag floats on top of the denser molten iron protecting it from oxidation. Both layers are drained periodically through a tapping point in the blast furnace wall. In this way, the blast furnace operates continuously, with fresh charge being added at the top, and molten iron and slag being tapped off at the bottom.

The impure iron can be cast into bars called "pigs"; the cast iron itself is called pig iron. The slag is used in road making (see Sect. 4.3) and can be used in cement, often mixed with Portland cement (see Sect. 14.2.3).

The inventions of Watt, Trevithick, and Stephenson (see Chaps. 3 and 4) depended on an abundant supply of a cheap, reliable metal, namely iron. Key factors in Great Britain's Industrial Revolution were its extensive deposits of iron ore, and coal to make coke. At the time, two forms of iron were the mainstay of manufacturing industry—cast iron and wrought iron.

14.3.3   Cast Iron

When pig iron is re-melted, it can be made ready for casting, that is, being poured into a mould of predetermined shape and cooled to give a 'cast iron' object. Cast iron is non-malleable; has low tensile strength and is brittle because it contains 2–4% carbon, together with other impurities.

"White cast iron" results from sudden cooling, when the carbon in solution is unable to form graphite, and iron carbide (cementite) is formed. "Grey cast iron" is made by slow cooling and consists of crystalline grains of pure iron (called ferrite) and free flakes of graphite.

When cast iron is struck hard, it tends to shatter rather than bend or dent. The family of cast irons cannot be readily forged or welded. They are used for castings that do not have to withstand great strain or violent shocks. Modern day examples include engine cylinder blocks, machine frames, stoves, drain-pipes etc. Before the advent of the steel industry, cast iron was used in a wide range of architectural applications because of its relative affordability.

In 1779, cast iron from Abraham Darby's blast furnace was used to make girders for the world's first iron bridge. The Iron Bridge was built across the River Severn, near Coalbrookdale. It became a UNESCO World Heritage Site in 1986. It can be seen at the English Heritage site in the Borough of Telford, Shropshire (TF8 7JP).

In 1878, the longest bridge in the world was built over the River Tay. Civil engineer Thomas Bouch adopted a cast iron lattice design. Unfortunately, in 1879, fierce winter gales battered the bridge; the central span buckled, and a steam-locomotive and six carriages plunged into the icy waters.

The brittle cast iron had been flexed to such a degree that it gave way. Engineers recognised that a new material was required for large scale engineering that lacked the brittle character of cast iron. It had to be more flexible and less prone to fatigue. The Tay disaster accelerated the transition from iron to steel.

14.3.4 Wrought (Worked) Iron and Its Use in the Construction of Transport Bridges Across the Navigable Menai Strait

Wrought iron (as opposed to decorative hand-made decorative steel 'iron-work') is a low carbon iron, often made by intensive hand work using small scale production methods.

The conversion of 'pig iron' into wrought iron requires the removal of impurities such as silicon, sulphur and phosphorus, as well as the lowering of the carbon content, from the 3–4% level present in pig iron, to 0.1–0.2% in wrought iron.

The chief impurities are readily oxidizable. Their removal can be achieved in the slow, operator-dependent, 'puddling process', devised by Henry Cort, in 1784, and modified by Joseph Hall, in 1816. Cast iron is melted on a bed of iron oxide in a coal fired reverberating furnace. As the cast iron melts, it is stirred, and the iron oxide oxidizes the dissolved carbon and other impurities. Some of the end-products escape as a gas; others pass into the slag. A pasty mass of semi-solid iron, mixed with slag, results. The "puddler" then forms the mass into "blooms" which are repeatedly hammered and rolled into wrought iron with a characteristic fibrous structure.

Wrought iron is nearly pure iron, containing only 0.1–0.2% carbon, interspersed with 2–3% slag. It is extremely tough; easily welded and worked under the smith's hammer. It can be forged, rolled, and shaped in various ways.

It was the first metallic material used for railway rails, hulls of ships and bridges etc. It was used for making pipes, tubes, wires, cables, chains, and other forged articles where resistance to corrosion and repeated bending stress and shock are important.

In 1826, Thomas Telford built the Menai Suspension Road Bridge linking the island of Anglesey with the mainland of Wales. Its central span was 577 ft. (176 m) long which, at the time, made it the longest suspension bridge in the world. It used a ground-breaking piece of civil engineering—the roadway deck was suspended from sixteen wrought iron chains.

Twenty years later, the Britannia railway bridge was built across the treacherous Menai Strait. A suspension bridge was considered unsuitable for a railway line since there would be too much flexing. The design team comprised Robert Stephenson (see Sect. 4.4.3), William Fairbairn and mathematician, Eaton Hodgkinson. Their revolutionary design included oblong 'box girders' in which a roof and base of multiple metallic tubes sandwiched a main tube. This gave a "railway tunnel in the sky", one hundred feet above high water to satisfy the Royal Navy's demand for a headroom sufficient to allow passage of a fully rigged man-of-war,

Thus, for a pair of railway tracks, two parallel tunnels were constructed from wrought iron plates riveted together. Both tunnels required two main spans of 140 m and two shorter spans of 70 m, giving a total tunnel length of 420 m (1,511 ft.) for each track. By creating this tubular design, the forces of compression and tension acting upon the tube as a train travelled across was dissipated across all four surfaces of the oblong tube.

When completed, in 1850, it was the longest continuous wrought iron span in the world. Sadly, a fire in 1979, irreparably damaged the rectangular box sections but not the stone-built towers.

The puddling process for wrought iron manufacture requires highly skilled labour with great physical strength and, for this reason, more easily manufactured mild steel has largely replaced wrought iron.

14.3.5 Steel: The Bessemer Refining Process

Before Bessemer's invention, steel was made by a slow process. It was a semi-precious metal, and its use was mainly restricted to cutting tools and expensive cutlery.

Following the outbreak of the Crimean War, in 1853, Bessemer wanted to get involved with weapons' manufacture. Sometimes, cannons shattered into shards when shells were fired. Guns able to withstand the firing of heavier projectiles were wanted. Guns and cannons with more robust casings were necessary, and Bessemer decided that a large-scale, inexpensive method for purifying 'pig iron' to produce steel, was the answer.

Bessemer was convinced that, if air could be brought into contact with a sufficiently extensive surface of molten pig iron, it would rapidly convert it into malleable iron.

Bessemer developed a cost-efficient, industrial process for the large-scale production of steel. In 1855, he published a paper entitled, "*The manufacture of malleable iron and steel without fuel*" and filed a British patent.

He configured his furnace to force cold air through small nozzles at the base of the converter. The chemical reactions that occur, with atmospheric oxygen, in the converter are exothermic and generate sufficient heat to keep the mix molten, so costly fuel is not required to maintain the process. In the early days of the process, violent eruptions occurred. The finished steel can be poured from the top of the tilted converter and solidifies in the form of ingots, ready for further processing.

In the original Bessemer process, impurities of silicon and sulphur were eliminated, and carbon levels were reduced, but phosphorus contamination remained, and the resulting steel was of variable quality. Iron ore sourced in Great Britain contained relatively high amounts of phosphorus.

Bessemer sold licences for his process to five British ironmasters, but users found that only relatively scarce, and expensive, phosphorus-free, Swedish iron ores could be used to produce an acceptable steel product. The rudimentary Bessemer process is difficult to control, so the quality of the steel produced can be quite variable. His credibility was damaged, and he was forced to repay the licence-fees previously collected.

Bessemer filed a patent in the USA in 1856. Unbeknown to Bessemer, American, W. Kelly had, in the early 1850s, independently arrived at a steel making process that was like Bessemer's approach. It was known as Kelly's pneumatic process—recognising the role of forced air. Kelly filed a priority claim and, in 1857, was granted US patent 17,268 superseding Bessemer's US patent, granted in 1856. In 1866, the two business groups pooled their resources, forming the Pneumatic Steel Association.

In 1856, Robert Forester Mushet found a way of improving the properties of Bessemer steel. It involved burning off, as far as possible, the impurities, and then re-introducing precise amounts of carbon and manganese. This was achieved using a carbon-containing iron-manganese alloy, called Spiegeleisen (looking-glass iron) which had been sourced from Prussia.

This improved the quality of the finished product and increased its ability to withstand rolling and forging at high temperatures (i.e., its malleability), making it more suitable for a vast array of products. In 1857, Mushet durable steel rails were delivered from Ebbw Vale steelworks to Derby Railway Station, where cast iron rails needed replacement every 3–6 months. The steel ones were still operational six years later.

In 1857, Mushet applied for a second patent based on the addition of tungsten to his crucible method, allowing steel to be hardened in air. Tungsten steel tools, self-hardened in this way, could run much faster and cut harder materials than those made from steel hardened by water quenching.

Mushet's first patent claim was filed in 1856. Since he did not pay the stamp duty that was due in 1859, his patent rights were extinguished. In that year, Bessemer built his own steel works in Sheffield to commercialise the Mushet process. He then held the rights to use each of the three critical processes for commercial steel production, namely forced air, Spiegeleisen as well as the mechanical design for the converter, including its tilting action.

Later, Bessemer reverted to low-phosphorus Swedish pig iron as raw material. It established Sheffield as a major industrial centre and the hub of the

British steel industry. "Made in Sheffield" became a hallmark of quality. Its workshops made the tools that were used to build the nation. For the first time in history, steel could be mass produced.

When the Crimean War ended in 1856, the military demand for steel declined. Bessemer set his sights on a promising new market—the railroads which required rails and bridges. In the emerging railway industry, cast or wrought iron rails limited the speed, weight and quantity of rail traffic. Steel rails, however, had much greater strength, survived over sixteen years longer and could handle heavier and faster steam locomotives and rolling stock. In 1861, the first Bessemer steel rail was laid at Crewe station; a year later at the Camden goods station, in London. In 1863, Bessemer steel was first used for ship plate in shipbuilding.

When his patent expired in 1870, after fourteen years, it was estimated that the Bessemer process had saved the world billions of dollars resulting from the reduced cost of steel manufacture. In 1875, Britain accounted for 47% of the world's production of pig iron, and about 40% of global steel production. In 1876, one third of Sheffield's steel output was exported across the Atlantic to build American railroads.

14.3.6 Further Modifications to the Bessemer Process. Enhancements for Speciality Steels

In 1876, Welshman Sydney Gilchrist Thomas discovered that, by adding a basic material, such as limestone, to the converter, the phosphorus impurities could be extracted in the form of removeable slag. With his cousin, Percy Gilchrist, Thomas perfected the dephosphorising process and applied for a patent in 1877.

Thereafter, for phosphorus-containing ores, a basic material such as dolomite (a carbonate mineral of calcium and magnesium) was used to line the base of the converter. The phosphorus is oxidised to its acidic oxide which reacts with the metal carbonates to give metal phosphates. These can be removed and used as agricultural fertiliser. This meant that iron ore from all over the world, irrespective of its phosphorus content, could be used, and a more consistent steel product resulted.

The conversion of 25 ton of pig iron by the Bessemer process takes about 15 min but this offers insufficient time to assess the chemistry of the steel and to modify its composition, if required.

Low-carbon (mild) steels contain less than 0.3% carbon. They are malleable and ductile and can be used in place of wrought iron. Mild steel is used for nails, wire and for flat sheet plates for car bodies and ship building.

Medium steels contain 0.31–0.6% carbon. They are stronger than mild steel and are used for making rails and structural elements such as beams, girders etc. High-carbon steels contain 0.61–1.5% carbon. They are difficult to cut, bend or weld and are used for making razors, surgical instruments, masonry nails, drills, and other tools.

The cost of steel production fell from about £50 per ton in the 1850's to £5/ton by the 1890s. By 1890, two thirds of all the mild steel produced was made by the Bessemer process and this made him fabulously wealthy. In 1892, there were 115 Bessemer converters in Britain alone, as well as many overseas.

Structural steel was used to build the Forth Bridge (1890) and Blackpool Tower (1894). The Forth Bridge was the world's first major river crossing made entirely of steel. About 53,000 tonnes of steel were required for its construction. It is still the second largest single cantilever bridge span (1,709 ft.; 521 m) in the world, built at a time when railways came to dominate long distance travel. It continues to serve 534 miles of uninterrupted railway track from Aberdeen to London. UNESCO described the bridge as a World Heritage Site in 2015.

In 1900, total global steel production stood at 28.3 million tons, a 200-fold increase in fifty years. Britain built 61% of the world's ships.

In 1974, the last Bessemer converter in the UK was decommissioned at the Workington Works. Today, the fundamental Bessemer principles continue to be at the heart of steel production from molten pig iron, whereby oxygen is used to remove unwanted impurities and to lower the carbon content, without the need for additional fuel.

Numerous refinements have been made to the basic process. Between 1890 and 1990, the open-hearth (Siemens-Martin) process was extremely popular because larger batches could be handled. Although the process is much slower (8–10 h) than the Bessemer process (15 min), the composition of the steel can be better controlled because chemical analyses of the mixture can be frequently undertaken. The molten steel is cast into billets.

In 1900, the electric arc furnace (EAF) process for recycling end-of-life, scrap steel appeared.

Since 1952, an increasing proportion of steel has been produced by the basic oxygen furnace (BOF/Linz-Donawitz) process, taking about thirty-five minutes to complete. Up to 30% inexpensive scrap steel can be mixed with molten pig iron and fluxes. Pure oxygen is injected, under pressure, through a cooled lance, into the molten mix. This rapidly burns away any impurities so increasing the rate of steel production. The characteristic emission spectra of the elements in the mix can be monitored and additives can be introduced in the proper proportions to give a steel with the desired characteristics.

The production of one tonne of steel is associated with 1.8 T of carbon dioxide, the majority of which comes from the oxidation of coke. In the case of aluminium, the oxide is so stable that it cannot be reduced by conventional carbon reduction in a blast furnace. Instead, molten ore electrolysis (MOE) is applied.

Attempts to make steel by a similar route are underway. In one pioneering process, the iron ore is dissolved in a mix of silicon dioxide and calcium oxide, at 1,600 °C, and an electric current passed through it. According to the principles of electrolysis (see Sect. 10.3.9), negatively charged oxygen ions migrate to the positively charged anode, and the oxygen bubbles off. Positively charged ferrous and ferric cations migrate to the negatively charged cathode, where they are reduced to elemental iron. This collects in a pool at the bottom of the electrolytic cell and is siphoned off. The reducing agent, coke is redundant, so carbon dioxide is not a by-product. Unfortunately, the process is electricity-hungry, so a source of clean electricity is essential.

To make the perfect steel, for a specific application, a range of tweaks can be adopted. For instance, (i) alloying elements can be added; (ii) the cooling/reheating temperature in the final stages of production can be manipulated, and/or (iii) cold working such as rolling, and hammering can be utilised.

Desirable characteristics such as strength (S), hardness (H), resistance to corrosion (C), stability at high temperatures (T) or wear (W) can be enhanced by the addition of certain elements. Alloying elements to make special steels include boron (for hardenability, H), chromium (for S, H, C), cobalt (for S), copper (C), manganese (H, W), molybdenum (H, S, C), tungsten (H), nickel (S, C) and titanium (T).

14.3.7 Honours and Legacy

In 1874, Bessemer was made President of the Iron and Steel Institute. He was elected fellow of the Royal Society in 1877 and was knighted in 1879. Despite being one of the pre-eminent technological innovators of all time, he received relatively little recognition from the British Establishment.

He had a prolific flair for invention, rigorous economic analysis, and clear-minded business practices. During his lifetime, he was granted about 117 patents. He brought his own projects to fruition and profited financially from their results.

He died in Bessemer House, Denmark Hill, Camberwell and was buried at the West Norwood cemetery. The Workington No. 1 Bessemer converter can be seen at the Kelham Island Industrial Museum, in Sheffield (S3 8RY).

Bessemer found a way of making steel abundantly, at a cost comparable with the softer and weaker, wrought iron, thus paving the way for mass industrialization. Steel has become the second most mass-produced commodity, after cement. It fitted perfectly with the expanding mechanisation in the latter half of the nineteenth century and facilitated great progress in transport, shipping, construction, machine tools and weaponry.

It is one of the most popular construction materials because of its unique combination of versatility, durability, workability, recyclability, and cost. Higher tensile strength steel enabled more powerful engines, as well as stronger gears and axles, to be built. Steel girders were used for bridges, railroads, and skyscrapers. Industrial steel made possible the building of giant turbines and generators, thus making the harnessing of water and steam power more widespread. The mechanised age extended across the world and changed the way we live.

14.4 The Downside of Artificial Building Materials. The Anthropocene Epoch

In the Solar System, Earth is the only planet where there is life. This is only possible because of the unique balance between some natural forces working together. Energy from the Sun powers the living world; weather systems transport fresh water around the globe. Sedimentation and powerful ocean currents circulate essential nutrients.

Since the start of the twentieth century, the anthropological mass (e.g. iron, steel, aluminium, cement. bricks, plastics and asphalt etc.) has doubled every twenty years. It is said that 2020 was the year of "cross-over", when the accumulated anthropological mass exceeded the combined biomass.

Almost every human activity is associated with the generation of greenhouse gases. Most industrial processes are energy intensive, resulting in copious amounts of CO_2 production. It is, therefore, vital to embrace developing methods for producing electricity without recourse to fossil fuels. In the case of iron, steel and cement manufacture, the intrinsic chemical reactions that take place, inherently generate CO_2, as a by-product. Overall their manufacture is associated with about 10% of global CO_2 emissions. To mitigate greenhouse gas emissions, it is essential to find innovative climate-friendly ways of producing commodity products.

The Anthropocene epoch is an unofficial unit of geological time, said to begin with the advent of the Industrial Revolution in the late 1700s. Although Homo sapiens have inhabited the Earth for a mere 200,000 years, in the last few decades, they have realized that they are the first species, knowingly to have had a major detrimental impact on the wellbeing of the planet. In this regard, they have over-fished our seas; polluted the oceans; altered the atmosphere; contributed to climate change; changed ecosystems and been responsible for an accumulation of detritus of civilization.

By deforestation and hunting, they have got closer to other, wild animals, increasing the risk of spreading disease and pandemics by contacting creatures carrying zoonotic organisms, particularly where unsanitary conditions prevail.

They have thrown the ecosystem we call home into chaos, jeopardising global species, including our own. There is a growing realisation that we all have a moral responsibility for the future of the world whilst having the wherewithal to reverse many of these deleterious changes. We are the problem and hold the solutions.

15

Communication: Telephone, Computers and WWW

The inventor looks upon the world and is not contended with things as they are. He wants to improve whatever he sees; he wants to benefit the world; he is haunted by an idea. The spirit of invention possesses him, seeking materialization. (Alexander Graham Bell. Speech to a patent congress in Washington, 1891)

15.1 Précis

In 370 BCE, Plato, in the "mouth" of Socrates, lamented the introduction of writing since he thought that it might weaken people's ability to memorise. Since then, every communication development has been vilified by some, and its positive aspects undervalued.

The first forms of long-distance communication media were the telegraph (in 1837) and, later the telephone (Alexander Bell in 1876). Both methods relied on vast networks of wires to carry information. In 1895, the controlled generation of specific electromagnetic waves, wirelessly in free space, was the foundation for wireless telegraphy (radio transmission) and the springboard for technologies that are now ubiquitous.

British scientists accelerated the speed by which verbal communication between inhabitants of different continents could occur. It was W. Thomson's application of scientific principles and technological innovations that brought about the first successful transatlantic, underwater, electric telegraph, in 1866.

© The Author(s), under exclusive license to Springer Nature
Switzerland AG 2022
L. Bailey, *Inventive Geniuses Who Changed the World*,
https://doi.org/10.1007/978-3-030-81381-9_15

This paved the way for rapid, global communication pathways, through insulated cables, leading to world-wide social and material benefits to mankind. It helped Great Britain capture a pre-eminent place in world communications, connecting Great Britain to its Empire and trading partners throughout the world. It became a key component of economic inter-connectedness.

Nowadays, over 90% of the world's data and web traffic travel close to the speed of light, through fibre optic wires, across the sea floor, in the underwater web.

Alexander Bell was groomed and educated to follow in his father's and grandfather's footsteps, studying the mechanics of speech, teaching elocution, and working with the deaf community. In 1844, Samuel Morse sent his first telegraph message. Bell wanted to transmit human speech rather than Morse code clicks. His first step was a 'harmonic telegraph' based on six steel reeds in parallel that responded to one of six specific frequencies. A corresponding reed at the end of the line vibrated in harmony. This facilitated the sending of multi-messages, simultaneously via the telegraph.

Just as the density of air varies when a sound passes through it, Bell conceived an electric current could be made to change in response to sound. Believing that human speech causes a wave-like pattern in air, he aimed to produce an electric wave following the same pattern. By 1875, Bell, aged only twenty-eight, and electrician, Thomas Watson, created a crude telephone apparatus leading to a patent application.

Bell's 1876 US patent was one of the most lucrative ever granted, making him extremely rich. The first intelligible telephone communication soon followed and, in 1877, the first 'speaking phone' was available for commercial use. The transmitter and receiver depended on the principles of Faraday's electromagnetic induction. Responding to a sound, a vibrating membrane caused pulsations in a magnet set in a coil. This induced fluctuations in an electric current corresponding to the sound. Passage of this undulating current to an electromagnetic coil, in the receiver, caused a magnet to vibrate against a membrane reproducing the original audible sound.

The telegraph and the telephone were both wire-based electrical systems but Bell's success with the telephone came as a direct result of his attempts to improve the telegraph. The reproducing of human speech, via the telephone, refashioned the way people communicate since they could converse remotely and directly, without leaving their home and without any intermediary. It became indispensable to households, businesses, and governments.

Combined with radio technology, the mobile telephone has now become a ubiquitous wire-less tool for both global communication and information exchange. It is claimed that, in 2020, 68% of the world's 7.8 billion inhabitants have a mobile device, almost half of which are "smartphones".

We have been fascinated by mathematics (Newton and Maxwell) and how its application can extend our understanding of natural happenings. In addition, to assist with the process of data collection, manipulation, and analysis, we have helped develop the world of computers.

In 1833, Babbage, and a century later, Alan Turing, conceptualised the structural architecture of a computer. At an early age, Turing developed an obsession with puzzles and codes. His code-breaking prowess would prove invaluable during WW2. His teacher described him as a genius. He went on to take a 1st class honours degree with distinction, at Cambridge University, where he was elected Fellow at the age of twenty-two.

Whilst completing his Ph.D. at Princeton University, Turing developed the notion of a universal computer machine, describing the basic principles of a computer which became known as the Universal Turing Machine. Tommy Flowers and co-workers at Bletchley Park would go on to build 'Colossus', the second functioning programmable computer, in 1944. This, together with the code-breaking expertise of Turing, meant that the Bletchley Park team was able to decipher German military messages, sent from the German encrypting machine, Enigma, and intercepted during WW2. Churchill said Bletchley was his 'secret weapon' and may have shortened the war by up to two years and saved countless lives.

The Industrial Revolution had revealed that customised machines were capable of doing what vast swathes of human beings could do, but in a more efficient manner. The work at Bletchley demonstrated that a computer machine—based on electronics—could automate the efforts of thousands of human computing assistants. This innovation spawned a technology that became inextricably woven into the industrial and social life of late-20th and 21st century life. Omnipresent computers are now so indispensable to our societies that life grinds to a halt when they stop working.

The UK emerged from the Second World War with a technological edge in electronics, computers, and programming. However, its technology did not flourish because of benign neglect of the United Kingdom's manufacturing industry. It has no Intel, Samsung, Lenovo, Hewlett Packard, Dell, Apple, Sony, Siemens or Google.

Turing himself was always ahead of his time. In 1950, he grappled with the question, "Can machines think like humans?". Six years later the term 'artificial intelligence' appeared. In 1966, the first Turing prize was awarded. This is the 'Nobel Prize' of computing. Tim Berners-Lee would be its 2016 recipient and another Briton would make a major input to the way we communicate via computers.

The Internet is a huge network of disparate computers all linked together. Enabling useful, interactive connections between networks called for common protocols. In 1990, for the benefit of his computer-using, co-workers at CERN, Tim Berners-Lee created an "internal web" of pathways for the free flow of documents. Extending his horizons, he then laid a "world-wide information web" over the pre-existing Internet. Some of the enablers were already in existence but three others were devised by Tim. They included HTML, a document publishing language of the Web; URI, a unique name and address for each resource on the Web and HTTP which allowed retrieval of linked resources.

Berners-Lee wrote key instruction codes for a computer seeking information, the Web page where the information is held, as well as codes for the computer that releases information to the client. His hypertext system quickly became a universal infrastructure for on-line communication and the foundation of many other industries. He intended his system to be powerful and immediately useful, rather than perfect. His specifications for the nuts and bolts of the World Wide Web (WWW, W3) have been refined in the interim, but remain essentially the same.

The number of websites has increased from only one in August 1991 to over 1.94 billion in January 2019. Over 4 billion (53%) of the world's population use the Internet. New industries emerged to fill in the missing capabilities for a host of commercial applications. Google emerged as the dominant provider for Internet searches. In 2017, it had indexed 135 trillion Web pages, a figure that is constantly growing.

The WWW is a communication superhighway which has fundamentally changed the way we work; shop; play and correspond with friends and family, via social networking sites, blogs, and video sharing. Whilst "cyber power" is revolutionizing the way individuals live their lives, it has started to influence the way governments protect their citizens and fight wars. The Web has become the most far-reaching anthropological study in human history. It reveals that an array of unintended and potentially harmful consequences are emerging.

Most importantly, the Web is open, non-proprietary, and free because Berners-Lee and his employer, CERN, as an altruistic gesture, elected not to patent his invention, nor to use any technology that required royalties to be paid.

Berners-Lee made it possible to communicate more effectively across the globe, unrestricted by cables. He has done more to connect the world than has ever been previously achieved. Almost as important as the invention of the wheel, digital devices and infrastructure have become to the 21st century, what new transport systems and infrastructure were in the 19th and 20th centuries.

15.2 Alexander Graham Bell (1847–1922)—Scientist, Inventor, Innovator, Phonetician and Teacher of the Deaf. 1st Telephone Patentee

15.2.1 Background

Alexander was born in Edinburgh, being the second son of Alexander and Eliza Bell. At the age of eleven, he adopted the name Graham but was known to his relatives and close friends as Aleck. His father—a phonetician—was the head teacher at a school for the deaf. His mother had a profound influence on him. When Alexander was twelve years old, Eliza started to lose her hearing, so Aleck developed a manual finger-language to tap out conversations with her.

In his early years, he was home-schooled by his parents but had one year of formal education at a private school, as well as two years at Edinburgh's Royal High School, starting when he was eleven years of age. He had an undistinguished school record, probably because he did not enjoy a compulsory curriculum. Aged fifteen, he spent a year, in London, with his paternal grandfather who made him feel ashamed of his ignorance, and from whom his love of learning was fostered.

Whilst in London, Aleck and his father visited Charles Wheatstone who had co-invented the telegraph (see Sect. 4.4.6) in 1837. Wheatstone had developed a 'speaking machine' and Aleck and his brother were encouraged to develop their own. The exercise made him familiar with the function of the vocal cords and the way in which they vibrate when the lungs push air over them.

Both his father and grandfather were prominent elocutionists and worked with the deaf community. Alexander was groomed and educated to follow a career in their footsteps. His early passion for music prepared him for a scientific study of sound and, when he was sixteen, he began researching the mechanics of speech. His musical talents, together with these background interests were ideal groundings for his journey to an electrical device that could transmit speech.

At the age of sixteen, he secured a position as pupil-teacher at the Weston House Academy in Elgin Moray, Scotland. He taught music and elocution and experimented with tuning forks and electromagnets. Between 1864 and 1865, he attended the University of Edinburgh. Further teaching spells followed, again at Weston House, as well as Somersetshire College, Bath, and Susanna Hull's School for the Deaf, in London. His family had moved to London in 1865 and Alexander was accepted for admission to University College, London, in 1868. For two years he studied anatomy and physiology but emigrated before completing his degree.

Alexander's father had suffered from a debilitating illness early in life and had been restored to health by a period of convalescence in Newfoundland. Alexander's two brothers had died from tuberculosis, one in 1867 and the other in 1870. In search of a healthier climate, Aleck's parents convinced their only surviving child to emigrate with them to Ontario, Canada.

A year later, in 1871, he moved to Boston, USA to teach at Sarah Fuller's school for deaf children (now the Horace Mann School for the Deaf). Here, he pioneered his father's system called "visible speech", using a compilation of symbols to visualise the speaking positions of the lips and tongue. In 1873, despite the lack of a university degree, Bell was made Professor of Vocal Physiology and Elocution at the Boston School of Oratory.

He also taught at the American School for the Deaf in Hartford, Connecticut, as well as the Clarke Institution for Deaf-Mutes in Northampton, Massachusetts. He married Mabel Hubbard, one of his former pupils who had become deaf when she was aged five. She was one of the first American deaf children to both lip-read and speak.

15.2.2 Morse Code and the 1st Attempts to Supersede It

Relying on the transmission of electrical pulses of specific lengths, along a single wire, Samuel Morse produced a zig zag line on a remote strip of ticker tape. His friend, Alfred Vail, developed a stylus that lifted the tape, giving dots and dashes. These could be more easily decoded, giving a primary language of telegraphy.

In 1844, Samuel Morse sent his first telegraph message from Washington D.C. to Baltimore. The coded messages were sent through wires, by means of electrical impulses, helped by relay boosters. By 1866, a trans-Atlantic telegraph line had been laid (see Sect. 7.4.2). The telegraph established a wire-based means of global communication. However, it required a trained operator and was limited to either receiving or sending only one message at a time.

The Western Telegraph Company acquired the rights to Joseph Stearn's 2-message (duplex) system, whereby one wire could be used to simultaneously send and receive a message, thus doubling its capacity. They also hired Thomas Edison to devise as many multiple transmission systems as possible, as a way of blocking entry by competitors. In 1874, Edison succeeded in developing a quadraplex system facilitating two signals in each direction.

Bell and his rival, Elisha Gray, developed systems capable of sub-dividing a telegraph line into more than four channels.

Bell's extensive knowledge of the nature of sound, as well as his understanding of music, enabled him to conjecture the possibility of transmitting multiple messages. In 1871, Bell began working on a device that would allow for the simultaneous transmission, by telegraph, of several messages set to different frequencies—the so-called harmonic telegraph, sending sounds rather than Morse code. Bell's harmonic (multiple) telegraph used six steel reeds in parallel, each responding to one of six specific frequencies. When a reed at the transmitting end of the line vibrated, a matching reed at the receiving end gave out a sound at the same pitch. Whilst working well in the laboratory, they proved unreliable in service.

15.2.3 Sound and the Transmission of Human Speech via the 'Talking Telegraph'

As well as his work on multiple transmissions, Bell wanted to transmit human speech rather than intermittent Morse code clicks in single telegraphy, or musical tunes with his own harmonic multiple telegraphy. From the latter, it was a short conceptual step for Bell to transmit the human voice and so began his second project on a 'talking telegraph'. He set to work on a mechanism which would make a current of electricity vary in its intensity, just as the air varies in density when sound passes through it.

It is noteworthy that Italian American, Antonio Meucci, began work on a 'talking telegraph' in 1849 and had, in his home, a telephone-like device he called a 'teletrofono' which he used to communicate with his paralysed wife. It is said that he shared a workshop with Bell in the 1870s. Unfortunately, Meucci did not have the financial means to pay for a definitive patent for his invention.

When an object vibrates in a medium, such as air, it produces sound. A sound is a pressure wave, advancing in the same direction as the force that produced it. Under normal atmospheric conditions, at 20 °C, sound travels at 767 mph (343 m/s), this being the speed of the disturbance as it moves through air molecules. This is about one millionth slower than the speed of light (671 million miles per hour; 299,792,458 m/s) which is why we see lightening before we hear thunder.

There is no sound in space. Sound only exists within a medium. All sound waves, irrespective of frequency, travel at the same speed in any given medium, at a specific temperature. However, when passing through different media, sound waves will travel at different speeds, depending on the nature of the medium. They pass through a solid more quickly than they do through a liquid or gas. That is why a train can be heard earlier by listening through the railway line than through the atmosphere, since sound travels seventeen times faster in steel (5,960 m/s) (Table 15.1).

When a diaphragm, in a stereo speaker, oscillates back and forth, sounds are carried by longitudinal mechanical waves. Before the diaphragm starts to oscillate, air particles are uniformly distributed in front of it. When the diaphragm moves to the right, the particles close to it are pushed to the right and collide with their neighbours. This slows them down or temporarily stops them. Such particles are bunched together (compressed) but the particles further downstream are not yet affected. Overall, the particles in the compressed zone continue to oscillate slightly to the right and left, but never far from their original position. They are like infinitesimal coupled harmonic oscillators with kinetic and elastic energy.

Next, as the diaphragm cycles to the left, through its equilibrium position, a region of low density (rarefied) air particles forms to the right of the diaphragm. The first region of compression, however, will have moved farther to the right. Once one cycle is complete, a region of compression, followed by a region of rarefaction will be propagating to the right and a new compression will start forming in front of the diaphragm.

Table 15.1 Speed of sound through different media

Material	Temperature (°C)	Speed (ms^{-1})
Air	0	331
Hydrogen	0	1286
Air	50	360
Air	100	387
Water	20	1410
Soft biological tissue	20	1540
Mild steel	20	5960

Effectively, a compression pulse ripples along the line. The added energy that the moving particles carry is passed, one to the next. This longitudinal motion of air particles creates a periodic and rhythmic pattern of high-pressure (compression) and low-pressure (rarefaction) regions, with a 'normal' (atmospheric) pressure zone between them. Since the air vibrates in the same direction as the travelling wave, they are deemed to be "longitudinal waves".

The perturbations can be depicted as a sine wave in two ways. Firstly, with time on the horizontal axis and displacement of the particles on the vertical axis. Second, with distance from source on the x-axis and pressure variations on the y-axis. Such plots have the peak points (crests) corresponding to compressions; the low points (troughs) to rarefactions and the zero points to the pressure the air would have, were no disturbance moving through it. This is what Bell would have seen had the sound input be converted to an electrical signal in a cathode ray oscilloscope (CRO) capable of analysing waveforms.

The frequency of the sound is the number of compressions (or rarefactions) passing a given point in the medium each second. It is measured in cycles per second (Hertz). The normal range of human hearing is between 20 and 20,000 cycles per second (Hz). High frequency sounds, from a rapidly vibrating source, are perceived to be high pitched (e.g., a whistle; the squeak of a mouse); low frequency sounds are low pitched (e.g., a drum; the roar of a lion). The period of the sound wave is the time taken for successive compressions to pass a given point.

The amplitude of the sound is the maximum displacement of the air particles from their rest position which is equivalent to the difference between the maximum air pressure in the sound wave, and the at-rest air pressure. When amplitude is assessed by neuronal circuits in the human brain, it is called loudness; in the electronic circuit of a microphone, it is called intensity. Sound waves with a small amplitude are perceived to be "quiet or soft". Sound waves at the same frequency but with larger amplitudes are said to be "loud". Loudness is measured by pressure variation in pascals. A healthy young person can hear sound pressures as low as 0.00002 Pa. A normal conversation produces a sound pressure of about 0.02 Pa.

Sound, like other forms of energy, can be measured in joules. Generally, we are more interested in how much sound is transmitted or received over a set time—that is the power of sound, measured in watts. Intensity of sound is defined as the energy carried by the wave per unit time across unit area. It is measured in watts per square metre (W/m^2). The sound intensity is proportional to the square of the amplitude.

Pressure fluctuations can be detected subjectively with ears, and objectively with electromechanical devices such as microphones. Our ears are constructed to sense even small wave motion. By assuming that human speech triggers wave-like patterns of air—causing the eardrum of the listener to vibrate—Bell envisaged the creation of an electric current that would follow the same wave-like pattern and 'carry sound along a wire', from the speaker to the listener. Bell wanted to find a way of producing an undulatory electric current varying in intensity, in accordance with the complex movement of the air particles constituting the original sound wave. In this entail, he worked with Thomas A. Watson, a skilled electrician.

Had Bell been able to access an oscilloscope, he could have visualized his 'undulating current' and formed representative pictures of different sound waves. For instance, a pure musical sound of a single frequency, such as a tuning fork, appears as a smooth and regular sine wave. High notes have a high sound frequency, and the waves are close together. In contrast, a human voice appears as an irregular, jagged wave because it is a combination of sounds and the simulated waves add together to give a composite wave.

In 1874, Bell and Watson worked on both Bell's pet projects. They found that tuned metallic reeds could transmit and receive, not only exact pitches, but more complex sounds. At last, they had the basis for a system that could transmit speech and its development progressed at pace.

15.2.4   Key Patents and Commercialization

In haste, Bell filed a patent on 14 February 1876, just hours before Gray filed a caveat of intent to file for a patent, within three months, "for the art of transmitting vocal sounds or conversations telegraphically through an electric circuit".

In his own patent, Bell talks of a reed attached to a stretched membrane, with a speaking cavity over it. The motion of the reed can be triggered by a human voice. The transmitter converts the vibrations of the speaker's voice into variations in the direct electric current flowing through the telephone circuit to the receiver. The latter converts the fluctuating electric current into sound waves that reproduce human speech.

On 7 March 1876, Bell was granted a US patent (174,465A) relating to "*An improvement in telegraphy*". In effect, it was the first telephone patent. At the time, Bell was only twenty-nine years of age, but the patent was one of the most lucrative ever granted since it described, not only a telephone instrument, but also the concept of a telephone system. It outlined a method of, and apparatus for, transmitting vocal and other sounds telegraphically, by causing electrical undulations similar in form to the vibrations of air accompanying vocal or other sounds. It gave him seventeen years of exclusivity of this intellectual property but, at the time, he had no working telephone device.

Luckily for Bell, the US Patent Office accepted his description of the invention and did not require a working model. On 10 March 1976, the first intelligible telephone communication was made. Bell was in his laboratory with his latest version of a telephone transmitter. Watson waited, a few rooms removed, with a reed receiver pressed against his ear and heard the words, "Mr. Watson, come here, I want to see you".

In 1876, Bell and Watson demonstrated the power of the invention at several spectacular and widely reported exhibitions, including the Centennial Exposition to celebrate the hundredth anniversary of the Declaration of Independence, in Philadelphia.

The National Museum of American History (Smithsonian Institution in Washington) has a Bell's large box telephone, built in 1876, based on a 'magneto' transmitter and receiver. Sound waves at the mouthpiece of the transmitter cause the diaphragm to move. A magnet in a coil (see Sect. 10.3.6) mirrors the movement, and this pulsating action induces a fluctuating current corresponding to the sound wave. The undulating current is conducted along a wire to another electromagnet, at the receiver end. There, the fluctuating current in the coil of the electromagnet causes the magnet to move in response, which in turn moves the membrane, producing air vibrations that can be detected, as audible speech, by the listener.

By the spring of 1877, the first 'speaking phone' was available for commercial use. Within a year, the first telephone exchange was built in Connecticut.

With financial support from his father-in-law, Gardiner Hubbard, the Bell Telephone Company was created in 1877, with Bell owning a third of the shares, making him a wealthy man. During an 18-year period, Bell's company faced over 550 court challenges to the primacy of his patent, several of which went to the Supreme Court; all of which were unsuccessful. The patentee's intellectual property rights were supported by Lord Kelvin who worked on the inter-continental cable telegraph (see Sect. 7.4.2). The Company's defence was strengthened by Bell's lifelong study of the anatomy of the ear and speech mechanics.

Thomas Edison, still working for Western Union, focussed on the weak part of Bell's system—the transmitter. He developed the 'carbon microphone'. A lozenge of lampblack carbon (known then as 'plumbago', 'black lead') was placed beneath the hard rubber diaphragm of a transmitter. As sound waves moved the diaphragm, the pressure on the carbon button electrode changed, thus varying the resistance of the electric current in the circuit. It produced a stronger telephone signal offering greater capacity for distance and clarity.

Edison's US patent was filed in April 1877, but not granted until fifteen years later, in May 1892 (patent 474,230), after a long legal wrangle.

Edison also experimented with carbon granules and was granted a US patent (222,390) in 1879. Granules of roasted anthracite coal were used instead of lampblack. Changing amounts of air pressure compress the granules to varying degrees, altering resistance.

In 1879, the Bell Company won a lawsuit against Western Union and a settlement was reached whereby the Bell Company acquired all the Western Union telephone patents, including Edison's.

Between 1877 and 1886, ownership of a telephone increased to over 150,000. After less than twenty-five years, 1 in every 50 people in America had a telephone. Bell emerged as the giant of the tele-communication business. The Bell system enjoyed a virtual monopoly in the US until 1893, when Bell's key patent expired.

Bell became a naturalised US citizen in 1882. By the mid-1880s, Bell's role in the telephone industry was marginal. His repeated appearances in court to protect his patents placed a great strain on him and this may have prompted his resignation from the Company.

In 1915, Bell was invited to make the first trans-continental telephone call, from New York, to his former associate, Thomas Watson, in San Francisco, via 2,500 miles of overhead wires.

15.2.5 Move to Nova Scotia and New Intellectual Pursuits

In 1885, Bell and his family vacated in Nova Scotia where he found a landscape and climate that was reminiscent of his ancestral Scottish Highlands. He acquired land there and built a summer house on an estate that he called Beinn Bhreagh (Scottish Gaelic for "beautiful mountain").

Bell was a specialist in the science of speech and acoustics. In later life he performed ground-breaking work on optical telecommunications (photophone), hydrofoils, metal detectors and aeronautics. He became a prominent proponent of the principle of eugenics.

15.2.6 His Passing and Legacy

Bell died at Beinn Bhreagh Hall, near Baddeck, Canada, in 1922. Nearby, is the Alexander G. Bell National Historic Site and Museum.

On the day of his funeral, the entire North American telephone system was shut down and every telephone silenced for one minute in tribute to his life.

Later, the unit of perceived sound loudness, the bel (more usually seen as the smaller unit, the decibel) was named after him. Measurement of loudness in pascals involves a broad range of small to extremely large numbers. The decibel (dB) scale is a logarithmic scale which compresses the spread of numbers into a manageable range. The higher the number, the greater is the sound level. For instance, whispering = 20 dB; background noise = 40 dB; normal speech = 60 dB; pneumatic drill = 110 dB. Exposure to 90 dB for long periods can cause hearing loss.

The telegraph and telephone were, initially, both wire-based electrical systems. Bell's success with the telephone came as a direct result of his attempts to improve the telegraph. Shortcomings with the telegraph were that it could only send coded messages, through trained operatives, at telegraph offices. Adoption of the telephone resulted in one of the greatest revolutions in communication. It became indispensable to businesses, governments, and households. People could converse directly with each other, without leaving home, at any time of day and from any distance, thereby changing human behaviour.

Replicas of his "gallows frame" telephone; liquid transmitter and "box" telephone can be seen in the 'Communicate' gallery of the National Museum in Edinburgh, Scotland (EH1 1JF). A replica of his harmonic telegraph receiver can be seen at London's Science Museum (SW7 2DD).

Since Bell's invention, there have been huge advances in electronic telephone exchanges, communication satellites, fibre optic cables, transistors and micro-circuitry enabling cellular/mobile phones to wirelessly receive and transmit the human voice. The latter combine elements of Bell's telephone system with radio technology to allow wire-less communication, wherever signal strength permits it.

By the end of the twentieth century, the telegraph, which preceded the telephone, was replaced by digital data-transmission based on computer technology. Voice-over Internet Protocol (VoIP) is the technology that converts a human voice into a digital signal, allowing a call directly from a computer or smartphone (via, for example, Skype or WhatsApp). We can now video chat across the world.

15.3 Alan Mathison Turing OBE, FRS (1912–1954)—Mathematician, Computer Scientist, Cryptanalyst, Codebreaker of Enemy Signals, Father of Modern Computing

15.3.1 Turing's Background and Education

Alan was born in Maida Vale, London. His mother, Sarah Stoney, came from a family of engineers and scientists. Turing's father worked for the Indian Civil Service and Alan's parents travelled between India and Hastings. When his parents were abroad, Alan and his brother were left with a retired army couple.

At the age of six, he attended St. Michael's day school in St. Leonards-on-Sea, a suburb of Hastings, where his headmistress recognised his intellectual gifts. At the age of ten, attempting to give him an upper-class identity, he was moved to a preparatory school—the Hazelhurst Boarding School in Frant, Sussex. An adolescence of persecution and fagging awaited him. As a dreamy, unsociable boy, he spent hours working out complex chess problems on his own. He developed an obsession with puzzles and codes.

At the age of thirteen, he moved to an independent boarding school, Sherborne, in Dorset where he mathematics' teacher described him as a genius. He paid little attention during lessons and did not attain high marks in examinations. Nonetheless, at sixteen, he wrote an outstanding précis on Einstein's theory of relativity.

Whilst at Sherborne, he formed a close relationship with another male student, Christopher Marcom, but their friendship was cut short by Marcom's premature death, in 1930, from bovine tuberculosis. Alan was open in his grief and determined to go to Cambridge University, as his friend had intended. In his 1929 school report, Alan's physics teacher said, "He must remember that Cambridge will want sound knowledge rather than vague ideas".

Alan won a scholarship to Cambridge in 1931 to study mathematics including probability and logic theory and the new field of quantum mechanics. He graduated in 1934 obtaining a 1st class honours degree with distinction. Impressed by his dissertation on 'Central Limit Theory', he was elected Fellow of King's College at the age of 22.

His PhD was obtained from Princeton University, New Jersey, in 1938 his research tutor being Alonzo Church. Whilst at Princeton, Turing read Ada Lovelace's paper on Babbage's Analytical Engine (see below).

15.3.2 Visionary Thoughts from Victorian Britons on Computer Hardware and Software

Between 1828–39, Charles Babbage was the 11th Lucasian Professor of Mathematics at the University of Cambridge. Professionals in many fields were becoming increasingly reliant on printed numerical tables, but many of these contained erroneous entries. Babbage thought it possible to engineer machines that could recalculate and print these tables infallibly.

He conceptualised two classes of decimal (0 to 9) digital (whole number) machines—the 'Finite Difference Engine' and the 'Analytical Engine'. Number values were to be represented by gear wheels. The first device was to be an automatic mechanical calculator, using only arithmetic addition to calculate and tabulate polynomial functions via a method of finite difference. Second, he envisaged the Analytical Engine to be a general-purpose programmable computing device with higher powers of analysis.

In 1834, his visionary plan for the Analytical Engine had characteristics seen in computers existing in the current digital era. For instance, it had a memory (he called 'store') separate from the central processor (he called 'mill') and had facilities for inputting and outputting data and instructions. The gear wheels were to be driven by a steam engine or hand cranked.

The program of operation, input and output data were to be stored on a punched tape—a technique derived from a punched card system used to direct fabric looms in making complex woven patterns, invented by Frenchman, J M Jacquard.

Babbage's conceptual machines were never built in their entirety due to lack of funding and, probably, because the hardware was beyond the capabilities of engineers of his time. However, based on Babbage's design drawings, modern day enthusiasts have built working prototypes of the Finite Difference Engine No "2" and one can be seen at the Science Museum, in London (SW7 2DD).

Facing lack of support in Britain, Babbage looked for funding abroad. The only public presentation that Babbage made concerning the Analytical Engine was to group of Italian scientists, in Torino. It was published, in French, by L. F. Menabrea.

Mathematician, Ada Lovelace, was the daughter of the poet Lord Byron and wife of the Earl of Lovelace. She was mentored by Babbage and she both translated and augmented the French article written by Menabrea. She realized that the Analytical Engine could go beyond numbers to handle letters and symbols. She perceived a potential transition of numbers, from arithmetic calculation alone, to computation and other areas of human endeavour, including music composition. She recognised that a device, as described by Babbage, had the potential to revolutionize the way the world works. For her involvement, she has been called the first computer programmer.

15.3.3 Other Early Notions on Computer Architecture and Operation—Postulates by Turing and von Neumann. Zuse's Z3—The 1st Fully Automatic, Programmable Computer

In 1936, before going to Princeton, Turing published his seminal paper *"On Computable Numbers, with an Application to the Entscheidungsproblem [Decision Problem]"*. He developed the notion of a universal computing machine that could, in principle, solve any mathematical problem that could be presented to it in a symbolic form. At the time, the concept was a simplistic, theoretical device that described the basic principles of a computer, becoming known as the "Universal Turing Machine" which anticipated the digital electronic computer and formalised the idea of an "algorithm".

His 'stored-program' concept was of a single processor—a single piece of hardware—that could change itself seamlessly, from a machine dedicated to a specific type of work, into a machine devoted to a completely different task. For example, it could switch from mathematical computation to chess opponent and then to word processing—its operation controlled by sequences of coded instructions (programs) stored in its memory. Turing's machine was a 'symbol manipulator'—its programs could be represented by Leibniz's base-two (binary) numbers, comprising groups of zeros and ones.

Essentially a theorist, Turing expanded his repertoire by developing practical engineering skills that would later help him communicate with engineers working at Bletchley Park. Also, during his stay in Princeton, Turing renewed contacts with John von Neumann at the Institute for Advanced Study (IAS). In 1944, von Neumann developed a logical design for computers in which a central processing unit would access both data and instructions from a single memory.

In 1941, German scientist, K. Zuse, would change the computing world for ever by unveiling the first entirely automatic computer that was controlled by programmes. The "Turing-complete" Z3 was an electro-mechanical device with 2,400 relays performing floating point binary arithmetic, with a 22-bit word length and clock frequency of 5.3 Hz. In Germany, it was used fleetingly for the statistical analysis of aerodynamic wing-flutter problems but was never used for everyday tasks. A replica can be seen at the Deutsches Museum in Munich.

15.3.4 Code Breaking Triumphs at Bletchley Park

Turing also studied cryptology—the study of ideas and cyphers which can be used to send and interpret secret messages. After war was declared in September 1939, he was invited to join the Government Codes and Cyber School (GC&CS)—a clandestine group engaged in military intelligence, based at Bletchley Park, and nicknamed Station X. His code-breaking prowess proved invaluable.

An encrypting machine called 'Enigma' was a notched 3-wheel cypher machine that could send radio messages by Morse code. Typewriter-like, every key stroke on Enigma altered the path of an electric circuit. Enigma machines were commercially available, but some had been modified by the German army for military use. During WW2, the German transmissions were monitored using extremely sensitive radio receivers at Britain's Y-stations.

M. Rajewski, a cryptologist with the Polish Cipher Bureau learnt how to decipher Enigma messages sent by the German army. He also developed a machine—the Polish Bomba kryptologiczna (cryptologic bomb)—which mimicked Enigma. In 1940, however, the Germans instituted a radical change to the Enigma system which eliminated the loophole that Rajewski had exploited.

In response, Turing re-engineered the Polish Bomba to decode intercepted radio messages more quickly. The 'Bletchley Bombe' was built by H. H. Keen at the British Tabulating Machine Company and became operational in March 1940. It was called 'Victory'. Any intelligence from Enigma decryption was called 'Ultra'. Gordon Welshman proposed an enhancement to the Bombe—incorporating a so-called diagonal board to reduce the number of steps to break the code. The new Bombe arrived six months later and was called Agnus Dei (*aka* Agnes or Aggie).

The task of the Bombe was to discover the daily key—the wheel order, wheel setting and plugboard configuration to enable the 3—5000 Enigma messages intercepted each day to be deciphered. Even the simplest military Enigma had about 158 billion billions (1.5×10^{20}) wheel settings and these were changed daily.

Together with colleagues in Hut 8, at Bletchley, Turing then turned his attention to the more complex German naval signals. A fourth wheel had been added to the basic machine to create naval Enigmas. Bletchley eventually succeeded in decrypting these naval messages, permitting the Allies to anticipate U-boat deployment, thereby contributing to the allied successes in the Battle of the Atlantic.

In July 1942, he developed a complex code-breaking technique, named Turingery which was a procedure for deducing cypher wheel patterns. It found application against the more sophisticated German cypher machine—the Lorenz Schlüsselzusatz (= encryption add-on) SZ40/42, given the British nickname, 'Tunny'.

The SZ40/42 encrypted messages were delivered more securely to, and from, German high command, across the Third Reich, not by Morse code, but by telex using a 5-bit punched tape teleprinter. The operator keyed in plain text and the receiver obtained plain text without further human intervention. There were several layers of encryption furnished via (i) the first set of five [*chi*] wheels (ii) the last set of five [*psi*] wheels (iii) the two middle [*mu*] stutter wheels and (iv) the teleprinter code based on Gilbert Vernon's enciphering technique using obscure characters. With its 12-rotor wheels, Lorenz theoretically had 16 billion billions (1.6×10^{19}) start positions. It became the communication life blood of German high command.

A brilliant 24-year-old mathematician, Bill Tutte—the son of a gardener and housekeeper—was tasked with unravelling Lorenz's obscure characters and logical structure. He had to crack a code that no one understood, created by a machine that no one at Bletchley had seen, having a range of wheel settings of the order 1.6×10^{19}. He wrote, by hand, the bit patterns for each of the five teleprinter channels.

Similarities, repetitions, and shortcuts in intercepted cyphers were helpful to cryptographers such as John Tiltman, particularly one Lorenz message that was sent twice by a sloppy operator, without a change to wheel settings but with subtle differences in content, punctuation and abbreviations. Once Tutte had identified two of Lorenz's wheel start positions, our engineers were then actioned to harness machines to emulate the Lorenz logic. For instance, in April 1943, as the Germans prepared for their third and final offensive against Soviet forces in the Kursk salient, Tunny decryption gave the order of battle. Churchill passed this information to Stalin whereupon the Germans were prevented from crossing the River Kursk.

The sophistication of code-breaking machines increased from the 'British Tunny', through 'Heath Robinson', to Tommy Flower's 'Colossus'. Flowers—the son of a brick-layer—was an ingenious electronic engineer working for the General Post Office. A key innovation of his Colossus Mark I was its 1,500 thermionic valves used instead of electro-mechanical switches. Therefore, it had no moving parts, except for beams of electrons, so the wheel patterns were generated by electronic circuits. Colossus could read 5,000 characters per second, five times as many as Robinson.

In February 1944, the world's second electronic, digital computer with programmability, albeit limited in modern terms, was ready. It started to work out the Lorenz chi wheel settings, enabling the rest of the Bletchley team to decipher messages in hours rather than weeks.

Messages decoded by Colossus showed that Hitler had swallowed the deception campaign, prior to the Allied forces D-Day landings, in June 1944, and had deployed his troops away from the proposed landing sites. It also gave the Allies key information about the defensive structure and D-Day battle plan of the German army and its air force.

The collective code-breaking operations at Bletchley have been credited with shortening the war by as much as two years, resulting in an Allied victory, and saving countless lives. However, a recent official history by J Ferris claims this is an exaggeration. Churchill described the unit as his "secret weapon". In 1945, Turing was awarded an OBE for his services to our country. Continued State secrecy, however, meant that neither Tutte nor Flowers received any decorations from the British nation.

15.3.5 Post-war Computer Hardware and Software Developed in UK

The Industrial Revolution had revealed that customised machines were capable of doing what vast swathes of human beings could do, but in a more efficient manner. The work at Bletchley demonstrated that a computer machine—based on electronics—could automate the efforts of thousands of human computing assistants. This innovation spawned a technology that became inextricably woven into the industrial and social life of late-20th and 21st century life. Omnipresent computers are now so indispensable to our societies that life grinds to a halt when they stop working.

At the end of the war, Turing joined the National Physics Laboratory (NPL) in London. Colossus was a digital computer, custom-built to solve a specific problem. Based on this theoretical work in 1936, as well as his knowledge of Colossus, Turing was convinced that he could design a general-purpose computer, and this took the form of the Automatic Computing Engine (ACE). The name incorporated the word 'Engine' in homage to Charles Babbage and his 'Analytical Engine' (see Sect. 15.3.2).

Progress with it was pitifully slow and Turing left NPL, in 1947, to take a year's sabbatical at the University of Cambridge. A pilot model ACE was eventually constructed, in 1950. With a clock speed of 1 MHz, it remained, for some time, the world's fastest computer. It can be seen in the Information Age Gallery at the Science Museum, in London (SW7 2DD).

Absence of 'memory ability' meant that each time Colossus was required to perform a new task, it had to be physically reprogrammed and rewired. Other electronic general-purpose computers soon followed the introduction of Colossus in 1944. For instance, in February 1946, the American *E*lectronic *N*umerical *I*ntegrator *a*nd *C*omputer (ENIAC) was unveiled, at the University of Pennsylvania, amidst claims that it was the first electronic computer. It had been designed and built for the US Army for military applications. It too could not store a program electronically and had to be programmed by manipulating a series of switches and reconfiguring the electronic circuitry.

Slow moving parts constrained mechanical memory. The Williams-Kilburn tube, tested in 1947, offered a solution, by enabling a higher speed, stored, electronic memory. Fresh from their war-time work on radar, F. C Williams and T. Kilburn, together with G. Tootill, at Victoria University of Manchester, utilised a cathode ray tube (CRT) to give them the memory technology required.

When an electron strikes the phosphor screen of a Williams-Kilburn tube, it creates a momentary charge that can be used to write an operation into a computer memory.

Binary information can be planted (written) as an electrostatic charge, in one of two, mutually exclusive ways, as an array of spots on the screen of the CRT. The type of charge at any spot represents a zero or a one and storage of such a single binary digit is a 'bit'. It can be sensed (read) by a pick-up plate placed in front of the screen. The spot lasts momentarily, but by re-writing (regenerating) it continuously, at electronic speeds, information can be stored indefinitely. In this way, multiple data bits can be stored and read rapidly. This was the birth of stored, computer software.

The electron beam in the CRT can be directed to any location on the screen, thus offering random access to the data, facilitating a computer with the first random-access-memory (RAM) storage device. Running a different program only involved resetting part of the program, using a simple keyboard.

In 1949, benefitting from these advances, the Victoria University of Manchester unveiled the first digital electronic computer capable of running a program stored electronically in its memory, rather than on paper tape or hard-wired in. The stored program concept meant that both data and operating instructions could be stored in a memory so that instructions could be executed with minimal human intervention.

The first elementary device was named the Small-Scale Experimental Machine (SSEM) or "Baby" but was so big that it weighed about one ton and nearly filled a room. A replica can be seen at the Museum of Science and Industry in Manchester (M3 4FP).

In 1949, Maurice Wilkes, at Cambridge University, unveiled the second stored program electronic computer, EDSAC (*E*lectronic *D*elay *S*torage *A*utomatic *C*omputer). It also marked the birth of professional programming. It was offered as a service to other departments (users). For example, it was used to help determine the structure of myoglobin (see Sect. 12.4).

The Manchester Mark I computer, which used Williams-Kilburn tubes led to the Ferranti Mark I—the world's first commercially available computer, in 1951. The tubes were used in several early computers but became unreliable with age. Turing wrote the operating manual for the Mark I.

In 1953, Kilburn demonstrated a computer that used transistors rather than vacuum valves, so facilitating an order of magnitude increase in the clock rate.

Colossus had been the starting point for British-built universal stored-program computers. Thereafter, electronic stored-program digital computers became faster and smaller, eventually fitting in a trouser pocket. Colossus is, therefore, a distant ancestor of laptops, tablets and smart phones, devices that facilitate our links to the whole wide world.

15.3.6 Turing's Thoughts on Machine Intelligence. The Turing Test

In 1948, Turing was appointed Reader in the Mathematics' Department at the Victoria University of Manchester. One year later, he was appointed Deputy Director of the Computing Machine Laboratory (there was no director). It was here that he grappled with the question, "Can machines think like humans?". Doubters believe that intelligence cannot be disembodied from a living host. Turing's 1950 paper was entitled "Computing Machinery and Intelligence".

He created the 'Turing test' to establish whether a machine exhibits intelligent behaviour equal to, or indistinguishable from, human intelligence. Can a computer imitate what a human does when a human thinks? In his imaginary 'Imitation Game", a human interrogator must determine whether a man or a woman is answering the questions posed. In the Turing Test, a computer program replaces the man. When the interrogator cannot tell the difference between human and machine, the computer is said to be 'thinking' and has artificial intelligence.

More recent CAPTCHA recognition of a distorted graphic image is a reverse Turin test'.

John McCarthy coined the term 'artificial intelligence' in 1956. Intelligence does not exist in a vacuum. Computers in isolation may be better at arithmetic but presently lack the common sense and levels of adaptability of humans. With human guidance, artificial intelligence makes it possible for machines to learn from experience, adjust to new inputs and perform human-like tasks. Able to process large amounts of data, they can be trained to perform specific tasks and react to emerging patterns in those data.

15.3.7   Turing's Final Years. His Legacy

In 1951, Turing was elected Fellow of the Royal Society.

In 1952, Turing reported a burglary to the police. It emerged that the offending petty thief was in cahoots with Arnold Murray who was in a sexual relationship with Turing. It was not until 1967 that homosexuality was decriminalized. Turing was charged with gross indecency. He avoided a prison sentence by accepting 'chemical castration' with a synthetic oestrogen, diethylstilboestrol (DES), taken orally. The humiliating trial ruined his career and his life. Because he was deemed vulnerable to blackmail, he was stripped of his security clearance, in both the UK and USA, effectively barring him from the computer field he had pioneered. He was initially embraced for his brilliance but then ostracized for being gay.

In 1954, two years after the trial, he died from cyanide poisoning. At his inquest, the coroner returned a verdict of suicide, but no suicide note was found. As attitudes to homosexuality changed, he was granted a posthumous Royal pardon in 2013. His legacy is a reminder, not only of the merits of embracing all aspects of human diversity, but also the steps society must take to become wholly inclusive.

Turing was sociable with those who shared his interests, but socially aloof with others. He was an untidy, unassuming, eccentric genius. Some say he had traits of Asperger syndrome. He, possibly, had one of the most brilliant minds of the twentieth century. He was an innovator in a whole new field of study.

His impact on computer science is commemorated by the annual Turing Award. This is the so-called 'Nobel Prize' of computing, considered to be one of the most prestigious awards in computer science. It is awarded for outstanding technical or theoretical contributions and presented by the Association for Computing Machines (ACM). The first recipient of the prize was A. J. Perlis in 1966. In 2017, T. Berners-Lee (see Sect. 15.4.7) was awarded the prize for inventing the World Wide Web.

In 2015, a new national centre of research in data science and artificial intelligence—the Alan Turing Institute—was created and headquartered in the British Library, London. Actual or replicas, some fully functioning, of key war-time devices can be seen at The National Museum of Computing (TNMOC) at Bletchley Park (MK3 6 EB). They include: Enigma, Bombe, Lorenz SZ40/42 cipher and Lorenz teleprinter, Tunny, Heath Robinson, Colossus II (with 2,500 valves) and EDSAC.

A new Bank of England £50 note featuring Turing appeared in circulation on 23 June 2021—the day he was born in 1912. It will include one of his 1949 quotes: "*This is only a foretaste of what is to come, and only the shadow of what is going to be*". Alan was chosen for this Bank of England accolade from a list of 989 British figures considered to have made significant contributions to science.

15.4 Timothy John Berners-Lee OM, KBE, FRS, FREng, FRSA, DFBCS (1955–)—Computer Scientist and Inventor of the World Wide Web (WWW)

15.4.1 Tim's Early Life and Career Pre-CERN

Tim was born and grew up in the leafy Borough of Richmond-upon-Thames, London. He shares his birth year, 1955, with both Steve Jobs and Bill Gates. Both Tim's parents were mathematicians who had worked on the first commercially available computer, the Ferranti Mark I (see Sect. 15.3.5) and would shape who Tim was to become.

He attended Sheen Mount Primary School in East Sheen, London. At the age of eleven, he went to Emanuel School in Battersea, then a direct grant grammar school. It was situated near Clapham Junction railway station which may have triggered his adolescent hobby of trainspotting. Hopeless at sports, he learnt about electronics by tinkering with a model railway, coupling the movement of trains with whistles made from transistors. He made gadgets to control the movement of his trains.

On a bookshelf was a musty old book entitled "Enquire Within Upon Everything"—a comprehensive guide to the necessities of domestic life, published in 1856. Tim found the title suggestive of magic, the book serving as a portal to a world of information. The name 'Enquire' was his precursor program to the World Wide Web.

Between 1973 and 1976, Berners-Lee studied physics at Oxford University, graduating BA (Hons) with a 1st class degree. Whilst at university, he made a rudimentary computer from a Motorola M6800 processor, transistor-transistor logic (TTL) gates and spare parts from an old TV set, bought from a repair shop. After graduation, he continued to work with the M6800, learning to program it more effectively.

He joined Plessey Telecommunications Ltd, in Poole, as an electronic engineer. Plessey was a manufacturer of telecom equipment. Tim worked on distributed transaction systems, message relays and bar code technologies such as a bar-code reader for shelf-stocking at Sainsbury's supermarkets, and a bar-code scanner for library books.

In 1978, he moved to D. G. Nash, a start-up company in Ferndown, Dorset owned by his two friends, Dennis Nash, and John Poole. One of Tim's tasks was to programme the microprocessor used to 'drive' the head of a dot needle printer. He also worked on a multi-tasking operating system.

After a six-month period in Geneva, he returned to Bournemouth, Dorset in 1981. John Poole had formed a new company—Image Computer Systems Ltd. and Tim became one of its directors, having responsibility for technical design. Between 1981 and 1983, he worked on a project entitled "Real-time remote procedure call", thereby gaining experience in real time control firmware, graphics and communications' software. He was seen to be a rather eccentric, fast speaking, fast reasoning, blue sky thinker.

15.4.2 Computer Science, History, and Terminology Relevant to the Internet and World Wide Web

The concept of associative trails between documents was aired by Vannevar Bush, in 1945. He envisaged a machine that could help humanity achieve a 'collective memory'. In 1963, Ted Nelson conceived the idea of 'hypertext'—a document management system to index and organise his collection of notes He had a vision of a 'docuverse' where all data was stored once. Navigation through the linked content would be non-linear. This was more than text, it was 'hypertext'.

In the early 1960's, Paul Baran, working for the RAND Corporation described a method for dividing information into "message blocks". The concept found little resonance until the independent work of Donald Davies at the UK National Physics Laboratory (NPL).

In 1965, Davies was a member of the ACE computer project (see Sect. 15.3.5). In view of the limited time slots to transfer data between computers at different sites, he conceived the idea of a network of inter-connected data terminals for which the data travelling between them was broken up into small chunks (packets). These were more easily trans-ferrable and could be readily reconstructed at their destination. Adopting his principles, the NPL Data Communications Network entered service in 1969.

In the same year, the idea about packet switching was adopted by the US Department of Defence for their ARPANet. In 1978, the first *I*nternational *P*acket *S*witched *S*ervice (IPSS) was developed and then adopted worldwide in the 1990s.

The US *A*dvanced *R*esearch *P*rojects *A*gency (ARPA) was created in 1958 to facilitate research in technology with potential for military applications. In 1969, when America was concerned about nuclear attacks from the USSR, the *A*dvanced *R*esearch *P*rojects *A*gency *Net*work (ARPAnet) appeared with the aim of distributing digital data between geographically dispersed super-computers. Sponsored by the US Defence Department, it connected military and academic networks, and provided the technical foundation for the internet.

The ARPAnet used the revolutionary idea of 'packet switching', whereby chunks of data were broken into smaller packets that could take any available route to reach its destination. Computer centres, or 'nodes' were connected, thereby making this interconnected network more resilient to nuclear attack. In addition, it stopped any interceptor collecting the whole data set. However, it was a 'closed system' and could not link to other networks because of the absence of a universal computer language. A solution to this shortcoming would be found by Cerf and Kahn.

In 1972, the first network e-mail was sent by Ray Tomlinson via ARPAnet. He chose the '@' sign to separate the user from the host.

In 1974, Vinton Cerf and Robert Kahn co-designed a suite of communica-tion protocols (*T*ransmission *C*ontrol *P*rotocol/*I*nternet *P*rotocol, TCP/IP). This facilitated a sharing of resources by allowing different types of computers, all over the world, to recognise each other, using packet-switching between the nodes. They published "A Protocol for Packet Network Inter-connection". It resulted in a "network of networks", forming the basis of inter-networking (the Internet). The ARPAnet adopted TCP/IP on 1 January 1983.

Cerf and Kahn elected not to patent the technology. IP addresses are assigned to all networked computers. The system is user-driven; it is anony-mous; there is an absence of central ownership or control, but a lack of built-in security.

Having left ARPA, Vinton Cerf joined MCI Digital Information Services and created the first commercial internet e-mail system (MCI Mail) in 1983. It allowed electronic, text-based messages to be sent to other MCI Mail users.

The *Domain Name System* (DNS) was invented by P Mockapetris and J Postel in 1983. They instituted host names such as .com, .gov, .edu etc. It is a system that converts domain names to machine-readable *Internet Protocol* (IP) addresses that are assigned to each device in a computer network. For instance, easily remembered www.cern.ch is transformed to the unmemorable IP address 192.65.187.5 for CERN ('www' signifies the web server at the CERN Laboratory in Switzerland [ch]).

By 1985, the Internet was well established as a computer technology that supported a broad base of researchers and was beginning to be used by other groups. For home or business use, the Internet now consists of millions of computers, all over the world, connected via modems, to each other by copper cable, fibre-optic cable or wire-less links, provided by Internet Service Providers (ISP's) such as AT&T and Comcast Xfinity in USA; BT and Sky Broadband in UK. The ISPs have specialised computers, called routers, that move the packets of data from their source to their destination. ISPs connect with other ISPs at *Internet Exchange Points* (IEPs).

The association embraces a multitude of stand-alone networks. Unfortunately, in the mid-1980s, documents held electronically in one network could not be readily transferred to, and read by, another.

15.4.3   Berners-Lee's 'enablers' Between Computer Networks

Tim wanted to go beyond an assortment of uncommunicative closed systems to one that was automatically available to everyone, publicly visible, irrespective of network. Enabling connections between networks called for common protocols such as shared communication standards, procedures, and formats. Some of these 'enablers' were already in existence (e.g., TCP/IP; DNS) via the infrastructure of the Internet but others were devised by Berners-Lee and include HTML, URI/N/L and HTTP. These three enablers (shown as 1 to 3) are elaborated upon below:

1. **Hyper-Text Mark-up Language (HTML)** is the publishing format for creating Web pages and Web applications on the World Wide Web. Text can be written as normal after which it is 'marked-up' using HTML tag and uploaded to a host computer running Web server software. A Web **server** is a computer that hosts a Web site. The latter can be a single page

or collection of Web pages which are documents that can be displayed on screen.

Static Web pages of text alone were soon overtaken by more sophisticated pages. Web developers started to use more versatile tools including scripting languages such as **Java Script** for making it interactive, and software for styling the page, streaming, and viewing video, audio, and multimedia. For instance, **Cascading Style Sheets (CSS)** control the layout, defining how elements are to be displayed on screen, paper or other medium.

Because of these refinements, Web pages can now contain pictures, animated images, movies, song, and other links.

2. A **Uniform Resource Identifier (URI)** is a string of characters that unambiguously identifies a specific resource (e.g., Web page or image file) by name (N), location (L) or both. This relies on two subordinate sets of the URI, namely the **uniform resource name (URN)** and the **uniform resource locator (URL)**. The URN defines an item's identity, whilst the URL provides a method for finding it, at a unique address, on a certain computer network, and a mechanism for retrieving it.

3. **Hyper Text Transfer Protocol (HTTP)** functions as a request-response protocol in the client–server computing model. For example, a Web **browser** is the client, whilst an application running on a computer hosting a Web site is the server. The client and host communicate using an agreed language called HTTP which is why Web page addresses begin with http://.

'Hypertext' implies that texts are embedded in other texts to provide fuller explanations and/or background material. One can move anywhere on the Web by clicking embedded 'hyper-links'. Following a trail of links is known as 'surfing the Web'.

HTTP allows HTML documents to be requested and transmitted between browser and Web servers, via the Internet. The client browser submits a HTTP request message to the server. The request is sent as a 'packet', a virtual parcel carrying lots of information, including the IP addresses of both the sending and receiving computers. The server parcels up the information requested and returns new 'packets', not necessarily together, nor via the same route, but including precise instructions about how to reassemble them.

HTTP has become the request-response protocol of choice, not just for browsers, but for virtually every Internet-connected software and hardware application. TCP/IP provides the infrastructure needed to transfer files across networks, from one program to another.

15.4.4   Browsers and Search Engines

A browser helps to access, retrieve, and display Web sites. We use a browser from a device such as a desktop, laptop, tablet, smartphone, or gaming console. Until 1995, when Microsoft's Internet Explorer appeared (bundled with Windows), the Mosaic-Netscape browser was one of the most popular.

Nowadays, we know browsers as 'apps' for exploring the Internet. These include proprietary products such as Microsoft's Internet Explorer (first launched 1995; current dominance = 3%), Norwegian Opera (1996; current share = 3%) Apple's Safari (2003; 15%), Mozilla's Firefox (2004; 5%), Google's Chrome (2008; about 62% dominance), China's UC Browser and Microsoft's Edge (2015; 2.7%). Google's monopoly will increase further from 2020, when Google's open-sourced software, Chromium, becomes the software of choice for Chrome, MS Edge, Opera, Brave, Vivaldi and Avast secure browser.

One uses a browser to get to a **search engine** which helps one find Web pages by using its software to comb Web sites, searching for matches to key words or concepts. If the server is unable to find the requested page, it sends an error message to the client, namely "Error 404—page not found". Search engines include Google (with 70–80% of search market share in 2019), Microsoft's Bing (ca. 10%) and Yahoo.

15.4.5   Berners-Lee's Short- and Longer-Term Careers at CERN

In the latter half of 1980, Tim worked as an independent contractor at CERN (*C*onseil *E*uropéen pour la *R*echerche *N*ucléaire; European Organization for Nuclear Research), in the proton cyclotron division, programming minicomputers controlling the cyclotron.

During this 6-month contract, Tim proposed a project using a simple hypertext system to improve the sharing and updating of information generated by nuclear scientists. To demonstrate its feasibility, he built a prototype called ENQUIRE—the first program for storing information using random associations. This was the predecessor of the World Wide Web.

When the 6-month contract ended, Tim returned to Bournemouth to work for John Poole's new company. Then, in 1984, he was offered a fellowship at CERN, in the computer services section, to work on distributed real-time systems for scientific data acquisition and system control. Among other things, he worked on FASTBUS system software and designed heterogeneous remote procedure call system.

At the time, the Internet already existed as a series of computers linked by cables. By 1989, CERN was already the largest Internet node in Europe. It had a permanent workforce of about 2,500 people with 10,000 or more physicists from many countries, using CERN's facilities, and trawling its data. They brought with them different hardware, software, and individual requirements. Any attempt to transfer information between them in a frictionless manner appeared impracticable.

Tim's personal mission was to organize and pool together data from physicists experimenting with particle accelerators. He wanted a mechanism for information dissemination and a medium for the collaboration and interaction between individuals and their computers, regardless of geographic location and computer facilities. Authors would be able to create links between documents, stored electronically, whilst readers could follow the links from one document to another.

Rather than downloading individual files, Tim suggested that it might be possible to form computer links to text-documents and diagrams, so that single, or chains of, documents could be inspected at source, by any reader, at his/her workstation. Tim decided that CERN was an ideal site for the development of a "web" of hyper-texted documents.

In March 1989, he made a proposal for an information management system. His concept was to allow departmental computers to communicate through a new layer, on top of the Internet, using a document publishing language of his devising, namely (HTML). Wherever possible, he would build the new whole system by re-combining existing tools. The pencil-written response to his proposal was, "Vague but exciting".

In May 1990, Tim was authorised to test his ideas by writing an application to run on a NeXT computer. This was a high-end personal computer utilising a novel Unix-based operating system. The computer company NeXT Inc. had been founded by Steve Jobs, in 1985.

Tim used the NeXT computer system and its development tools to write the first Web server software. The world's first Web server was nxoc01.cern.ch, later called info.cern.ch. The NeXT Cube computer, screen, keyboard, and mouse used by Tim is currently on loan, from CERN, and on display (object no. L2014-4158) in the Information Gallery, at London's Science Museum SW7 2DD). Tim's workstation carries a sticker that states, "This machine is a server, do not power down".

Tim also used the NeXT computer to develop the first Web browser/editor, allowing the first Web browser to couple with the first Web server. The work was started in October 1990, and the program, the Worldwide Web (WWW;

W3), was made available within CERN, via the local network in December 1990. The name suggested that they had created something revolutionary.

The first uniform resource locator (URL) for the very first Web page was: http://nxoc01.cern.ch/hypertext/WWW/TheProject.html. For years, it has been dormant, simply redirecting followers to the web host root of http://info.cern.ch. It gave general information about the WWW program, outlining protocols and explaining how to create web pages. It asked new subscribers to add a URL to new pages, offering a universal read/write situation in a 'permission-less space'.

Tim's 1990 browser ran on a rare NeXT computer and CERN refused to fund other versions. Therefore, in August 1991, Berners-Lee posted a short summary of the WWW (W3) on the alt.hypertext newsgroup—a site for hypertext enthusiasts. This marked the debut of the web as a publicly available service and invited collaborators, via the Internet at large, to join his project. It was particularly directed at academics, computer nerds and geeks.

15.4.6   Expansion of the WWW

Tim gave a copy of his server software to Paul Kunz who launched the first web server outside of Europe, at the *S*tanford *L*inear *A*ccelerator *C*entre (SLAC), in December 1991.

Tim and his assistant, Jean-François Groff, rewrote components of the original WWW browser in portable C code. They released a code library, libWWW—a tool kit allowing participants to create their own Web browsers, utilising computers other than NeXT. For instance, Nicola Pellow and Robert Cailliau at CERN created MacWWW browser, the first browser for the classic Mac OS. Others responded, creating UNIX and PC browsers. Unix ViolaWWW and MidasWWW were initially the most popular browsers, eclipsed later by Mosaic which was first released in January 1993.

NCSA Mosaic was a browser supported by a major institution—the *N*ational *C*entre for *S*upercomputer *A*pplications (NCSA) at the University of Illinois. Free to download, it was relatively easy to install. Because it was user-friendly, it started the Web on the road from an academic setting to a mass medium and blockbuster success.

Written by Marc Andreesen and UNIX expert, Eric Bina, at NCSA Mosaic soon added graphics incorporated with text, rather than on a separate page. The IMG tag was announced in February 1993, allowing users to seamlessly navigate between locations, mixing text, graphics, and sound. The IMG tag creates a holding space for the referenced image in a HTML page.

In December 1994, Mosaic was replaced by Netscape Navigator, a proprietary, cross-platform browser, facilitating simultaneous downloading of multiple images, including JPEG image formats.

On 30 April 1993, Tim's browser was placed in the public domain. The technologies that Tim had applied to make WWW became 'open source', with CERN allowing anyone to use the Web protocols and code for free. People started to create their own websites for personal and on-line businesses. The number of Web servers grew from 50 in January 1993, to 500 in October. In April 1994, CERN catalogued 829 servers in its 'Geographical Registry'.

In 1994, UK telecommunication companies started to provide access to the Internet. Consequently, the WWW became accessible from people's homes.

By the end of 1994, the Web had 10,000 servers, 2,000 of which were commercial. It had 10 million users and the traffic handled was equivalent to shipping the collected works of Shakespeare every second.

Coinciding with the growth of personal computers, Tim's invention quickly spread across the planet. The number of 'hits' increased exponentially, by a factor of ten, every year. With its fathomless availability of information, the WWW has revolutionized the way we work, shop and play. It is the most significant invention for written communication since the printing press. It has become an international initiative for global information sharing, being a combination of all resources and users on the internet that are using the hypertext transfer protocol (HTTP).

15.4.7 Berners-Lee's Departure from CERN. Collective Efforts to Improve the Web

In October 1994, Tim left CERN and formed the World Wide Web Consortium (W3C) at the *Laboratory for Computer Science* (LCS) in Massachusetts. In 2003, the LCS merged with the Artificial Intelligence Laboratory to become the *Computer Science and Artificial Intelligence Laboratory* (CSAIL) at Massachusetts Institute of Technology (MIT).

The difficult part of the Web's adoption was getting users to accept common principles and standards (e.g., HTTP; global addresses [URL's] and HTML). To this end, the W3 Consortium is an open forum of companies and organisations that joined together to set standards and protocols for Web usage and to make suggestions as to how inter-operable technologies can be developed to improve the Web. Ideally, any inherent bias to one product or one manufacturer should be avoided but currently, there are many more "Googlers" registered than the next largest group—Microsoft.

In 1996, Tim became a *Distinguished Fellow* of the *British Computer Society* (DFBCS).

Whilst living in Boston with his young family, he was influenced by the Unitarian Universalist philosophy (see Priestley, Sect. 7.3). He saw parallels between this philosophy and the Internet/Web. They are both decentralised, with no hierarchy, tolerant and liberally minded, aiming for harmonious inter-working.

In 2001, Tim was elected Fellow of the Royal Society and, in 2004, he was knighted (Knight Commander of the Most Excellent Order of the British Empire, KBE).

In 2007, he was awarded the Order of Merit, an honour gifted by the Queen for which the number of living members never exceeds twenty-four.

In 2009, Sir Tim helped create data.gov.uk, designed to make official information more accessible to the UK public.

In 2009, he was the founding director of the Web Science Trust (WST) to promote research and education in Web science.

Also, in 2009, the non-profit making WWW Foundation was formed. Its purpose is to advance the Web, to allow all peoples to use the Web to communicate, collaborate and innovate freely. The Foundation works to fund and coordinate efforts to defend the open Web and further its potential to benefit humanity.

In 2102, Tim's face was seen by millions across the world when he took part in the opening ceremony of the London Olympics.

Also, in 2012, the *Open Data Institute* (ODI), headquartered in London, was co-founded by Sir Tim and Sir Nigel Shadbolt, an artificial intelligence expert. It aims to show the value of open data, and to advocate for the innovative use of open data, to effect positive change across the globe.

In 2013, Sir Tim led the Alliance for Affordable Internet (A4AI). It includes Google, Facebook, Intel, and Microsoft and was created to make the internet more accessible to the Third World.

In 2016, Sir Tim was appointed a Professor in the Computer Science Department at the University of Oxford.

In 2017, he was awarded the *Association for Computing Machinery* (ACM) Alan M. Turing Prize for inventing the World Wide Web, the first web browser and the fundamental protocols and algorithms allowing the web to scale. The Turing Prize—the "Nobel Prize of Computing"—is regarded as one of the most prestigious awards in computer science.

Had CERN or Berners-Lee maintained ownership of the WWW, and rented access, domain names and the means to traffic information and stream advertising through it, an astronomical income would have followed. Instead of collecting royalties, the innovation of HTTP and WWW was given to society for free. All societies have benefitted from their altruistic decision. Tim has continued to support both Web neutrality and Web privacy.

15.4.8 WWW Warts and Thorns and How to Mitigate Them

The WWW is a virtual world of inter-connecting communities. It can spread the glories and marvels of our world. Free to humanity, with no approval system, its content and usage mirrors its users—the majority seeking access with only good intentions. As a utopian dream, it was intended for a decentralised, open, caring, society. Indeed, the 'surface/visible web' mainly encapsulates the characteristics of the vast body of humanity it serves. Based on a liberal approach to access, most of its activity is constructive, being a vehicle for good, allowing society to express its best hopes and aspirations.

In the main, it has been a vehicle for positive change across the world. For example, it has connected people who were once socially isolated; it has helped small businesses reach new customers; it has assisted fundraising efforts for charities and enabled social justice campaigns. During the COVID lockdown, it has facilitated on-line shopping, socializing, video conferencing between relatives and businesses, and virtual schooling.

It has become the most far-reaching anthropological study in human history. This shows that the WWW is witnessing a growing number of unforeseen consequences and undesirable features (what Tim calls 'thorns'). Some unintended harms are now arising. An unbridled freedom of speech encourages people, anonymously to reveal their true selves.

A growing number of its inputs and outputs serve a minority with malevolent intent. Examples are cyber-criminals; swindlers preying on the gullible; phishing attacks; online 'romance fraudsters', implanters of computer viruses; ransomware by criminal hackers; those with perverted minds—cyber-stalkers, child groomers, paedophiles and pornographers exploiting their vile interests; fake news; disinformation and conspiracy theories; political and religious extremists; gamblers squandering their savings; internet trolls; racist abusers, users who inappropriately abuse our personal data.

Perhaps, the time has come to unmask such evil users by obliging them to register, in a verifiable manner, so they can be easily traced if they break the law. Online perpetrators of hate must be shut down.

15.4.8.1   Criminal Activities. The Deep and Dark Webs

Cyber-hacks have eroded people's trust in online security. Before its infiltration by law enforcement agents, EncroChat offered a secure, encrypted, messaging service for international crooks. Ransomware is inserted into business and private operating systems as a way of extorting money.

The 'deep web' has about six times as many documents as the 'surface web' but it cannot be accessed by conventional search engines because the pages are not indexed. It consists of everything from unpublished blog posts to public websites, password-protected content, subscription-based websites and places that are surfed via browsers such as "The Onion Router (TOR)" which render a personal computer anonymous.

The 'dark web' exists within the 'deep web' and is sometimes used for illegitimate activities. For example, fraudulent COVID vaccination certificates are being offered for sale. Image-posting sites such as 8chan are among the most offensive, violent, and bigoted on the dark web.

15.4.8.2   Disinformation/Fake News. Political Ramifications. Threats to Democracy

For centuries, we have fought against tribalism, replacing it with free thought and expressive individualism. But now, some of us are addicted to smartphones and 'thought-providers' which act as ideological 'echo chambers' and political 'potholes' for like-minded citizens. It is important to maintain open minds—consider opposing views and show greater tolerance to those who legitimately stray outside the mainstream or our view of the world.

Some users are trying to undermine our democratic processes by peddling fake news and falsehoods, polluting the public sphere, creating divisive rhetoric, coarsening public discourse, segregating us from contrary views and polarising opinion. Enemies of the West promote disinformation designed to divide, distract, and demoralise our citizens. They are attempting to subjugate democracy.

Curating the voices of democratic leaders has its own dangers. We must place more reliance on reliable, truthful, and accurate reporting; be more circumspect with social media and maintain our objectivity if democracy is to survive. Well-researched facts are preferable to unsubstantiated speculation.

Political campaigners collect user data with a view to influencing voting behaviour, subverting democracy. There are computational propaganda tools to gather fake followers or spread manipulated media reports to garner vote support. There is a danger that we are heading to an Orwellian future where a handful of corporations and government departments monitor and control our lives.

Whilst artificial intelligence is becoming 'more human', human intelligence is becoming more artificial, more conformist and rule bound. Social media accelerate the dissemination of information, assisted by hashtags and retweet buttons. They sometimes make it too easy to transfer provocative posts and take harmful content to the vulnerable. Micro-targeting allows disinformation and discord to be spread unchallenged by counter speech. Excessive screen time can make us more solitary, not more social.

15.4.8.3 Trolls, Cyber-Stalkers and Concerns for children's Safety

The milk of human kindness and mutual respect has partially evaporated from public discourse. There is a vogue for public shaming and ostracism. Those subject to internet trolls can suffer from anxiety and loneliness and their mental health may be affected. Stalkers are using social media to monitor their victim's movements.

At the heart of Web technology is a democratising mechanism, where all users are dealt with equally, and there are no gatekeepers. Thus, children are treated like adults. However, the generation of children addicted to smartphones and computer-gaming appear to have fewer human interactions involving face-to-face contacts. A lack of 'like responses' on social media can lead to their feeling of inadequacy.

"Likes" on Facebook and "streaks" on Snapchat generate a digital footprint. A growing, morphing footprint results in more personalisation of follow up content and more delivery of targeted advertising. Messages may be increasingly extreme, explicit, and harmful. Whilst 'tech' is not the cause, it can be the accelerant to eating disorders, as well as self-harming, even suicidal, tendencies, of young people.

Juveniles particularly may unintentionally lock out the opinions of those with opposing views, resulting in a blinkered outlook on the world. This results in a narrowing of what should be a critical mindset, as an adolescent approaches adulthood.

15.4.8.4   Big Brother is Watching. The Power of Tech Giants. Pros and Cons of Censorship

We should caution that such rapid and widespread communication channels may be allowing our 'data footprint' to follow where search engines, smartphone operating systems and social media platforms want to lead us. Email (spy) pixels allow marketeers to track the behaviour of a user after he or she opens an email. Monopolistic control of the digital world is in the hands of a few technology behemoths. Evolving algorithms learn to match our recorded 'likes' to customise our newsfeed and "mine" our personal data. Having captured our attention, the data harvested can then be monetized to maximise revenue income streams.

Social media companies wield the power to censor views with which they disagree, whilst promoting opinions they support. For contentious items published on their platforms, they enjoy liability exemptions unavailable to traditional, regulated publishers.

Technology companies may have to be constitutionalized, providing more checks and balances for the public good. The internet must be made to work for us, rather than enslave us. More citizens and governments are calling for codes of practice and regulatory scrutiny to provide privacy protection, 'digital well-being' and the automatic rapid removal of offensive posts.

To protect the public from their failings, professions such as medicine and law saw the need to develop ethical standards. Should the digital world do the same, accepting the need to protect the citizen from its deficiencies?

Even some Silicon Valley luminaries are calling for better governance of the WWW. However, others ask, "Will restriction of the franchise lead to a constraint on our freedoms by platforms censoring every controversial post?". Is it better to have free speech on the internet, with all its attendant disagreeableness and offensiveness, than censorship which, historically, is linked to a nastiness far more profound?

Should we rely on governments to be proactive in this sphere? Can all governments be sufficiently impartial to help decide what should, and should not, appear on the Web? For instance, there are moves across the world to restrict free access to the WWW, even to censor parts of it.

Some authoritarian states have started to monitor their citizens through online surveillance tools that are both invasive and discriminatory. Their citizens are watched, listened to, recorded, and tracked more intrusively and pervasively than ever before. This is anathema to many in the West.

China is developing a "social credit" system based on credit worthiness, financial probity, and other social norms, as defined by the Chinese Communist Party.

Watching the movement of their citizens is Orwellian but has proved helpful in the battle against the spread of COVID-19. Even some liberal democracies have sacrificed some privacy and tolerated a degree of real-time surveillance, by the state, to monitor the movement of, and to alert the contacts of, infected citizens. If responsibly managed, this is far removed from a pervasive panopticon system and should help avoid any future viral or ideological contagion.

The EU has amended its copyright rules to meet the digital age. News aggregators will have to pay for snippets of content and links to news articles. Web platforms will become liable for hosting content that breaks copyright. Some in the tech lobby claim that any regulation amounts to censorship which is anathema to the core ethos of Silicon Valley. They further contend that the Web might collapse under an accumulation of litigation.

Governments are showing more muscularity in their approach to harms and competition. For instance, the UK Government is working on an "online harms bill" to compel tech firms to remove hate speech, and to impose on them, a statutory duty of care to protect children from the social and psychological impacts of social media. They will be required to purge on-line images and stories that glorify self-inflicted injuries. Age-appropriate standards will apply to all users. Legal action may be taken against errant social media firms who do not expeditiously remove child abuse content. In response, social media are developing artificial intelligence to recognise and remove offensive material before it is even posted.

We may arrive at three forms of the Web—effectively a "splinternet"—based on geography. The first format will be totally open, unfiltered but more integrated (in USA); the second, constitutionalised (Europe) and the third, censored by state control (undemocratic/authoritarian regimes) and insular.

15.4.8.5 Accountability of Tech Behemoths. Centralisation of Data

The core of the problem with the abuse of personal data is the acute centralisation of our interaction with the Internet. Whilst free to users, its history has been dominated by changing monopolies. First it was the Web browser (viz Netscape Navigator), then the desktop computer and its operating system (viz Microsoft), followed by the browser (Google) and social platform (Facebook).

Currently, Google is the dominant provider of internet searches, whilst Facebook, together with its Instagram and end-to-end encrypted WhatsApp, is the dominant social media company. It is said that Google directly influences more than 70% of internet traffic; 88% of the worldwide search market; 75% of smartphone operating systems and 37% of digital advertising. Google is gradually integrating its browser with its array of online apps and services. Facebook has 2.32 billion registered users—almost a third of the world's population.

It was Berners-Lee who invented the World Wide Web and charged no royalties for its use, but it is the FAANG's (namely, Facebook, Amazon, Apple, Netflix, Google [Alphabet]) and others who 'share the spoils'. At the end of 2020, the combined market capitalisation of Apple, Microsoft, Amazon, Alphabet (Google) and Facebook was $7.6trn. Their combined revenues amounted to about $900bn. We should guard against their omnipotence, ensuring that they do not become dangerous oligarchies.

Because our search and browsing history is continuously monitored, data has become the 21st century's 'gold'. The chief executives of the four big tech companies have become 'cyber barons'—'emperors' of the on-line economy, wielding huge, unaccountable power.

The more information a tech company can accumulate on our personal data, tastes, and habits, the more valuable the profile becomes. Increasingly sophisticated algorithms used by Facebook, for example, help determine the photographs we see, the issues we are made aware of, and the news articles that we are invited to read. Received messages, sometimes subliminal, modify our spending habits; our behaviour; even the way we vote.

15.4.8.6   Berners-Lee's Action Plan. Constitutionalizing Big Tech. A Web Magna Carta

Tim's powerful voice is currently working to overcome some of the WWW's shortcomings. Whilst encouraging the good things, we must mitigate the negative aspects. The root of some of the failings is that users' data is amassed by internet service providers or the owner of the website or app, whereupon data breaches and violations of their users' privacy can then occur.

Fearing that harvested data and app usage patterns are being sold to advertisers and lobby groups, Sir Tim has expressed the view that ISPs should neither control nor monitor customers' browsing activities without the users' express consent. He has been critical of the increasing dominance of proprietary social media platforms and bemoaned the appearance of fake news online and the lack of editorial accountability.

March 2019 marked the thirtieth anniversary of the WWW. Sir Tim had become increasingly disillusioned with aspects of the Web, stating that global action, via new laws and systems, was required to tackle the Web's downward trend to a dysfunctional future. He argued that any mid-course correction to the Web should be proportionate and subject to the rule of law.

He jointly chairs the *D*ecentralised *I*nformation *G*roup (DIG) which is investigating decentralised architectures that facilitate genuine data ownership via the so-called SOLID project so that information can be shared in a way that is compliant with regulation and user preference.

In the 'SOLID' approach (derived from *So*cial *Li*nked *D*ata), the user can choose to place his/her data in a personal cloud storage area called a Solid 'POD' (*P*ersonal *O*nline *D*ata store), thereby giving control to the user as to where his/her data resides and how his/her data is deployed. Rather than sitting in a 'silo' controlled by digital behemoths, our personal data will be held, in isolation, in our own Pod. Thereby, our computer becomes our own personal server, making it possible to permit access to trusted contacts, whilst revoking access to undesirable third parties. Each user will be able to see how his/her data is being used and with which 'apps'.

A new company, 'Inrupt' has been formed to develop a new browser and the software backbone of a new web to provide products and services for businesses and individual users who wish to implement Solid, thereby taking power from tech corporations and restoring it to the individual.

In addition, he has called for a "Contract for the Web", its 'Magna Carta' that protects the open web as a public good and basic right for everyone, where users' fundamental rights to privacy are protected. Its launch date was 25 November 2019 (see www.contractfortheweb.org).

15.4.9 Summary

Sir Tim is not just British by nationality, but British by temperament. He is reserved and modest and avoids the limelight. He thinks and speaks in a staccato manner. He is a computer programmer with an array of practical experiences that happened to converge with his ambitions for the Web. He allied proven technologies with new ones to form the framework of a system intended to be powerful and immediately useful, rather than perfect.

In 2019, half the planet was on-line, with many of its citizens making regular use of the WWW. Berners-Lee has been described as one of the 20th century's most important figures He has been widely honoured for his work, receiving a remarkable array of international awards and prizes (see https://www.w3.org/People/Berners-Lee/Longer.html).

16

Solving Crime Via Forensic Science

It's good to know where you come from. It makes you what you are today. It's DNA, it's in your blood. (Alexander McQueen)

16.1 Précis

We are a law-abiding nation and believe in justice for all. There are times when the available evidence such as physical appearance, dermatoglyphic fingerprints (ridge patterns) and dental charts are insufficient for criminal investigators to proceed with a case. New and novel techniques were required to solve some crimes. To help both safeguard our liberties and fight crime, we have developed DNA fingerprinting and profiling (Jeffreys) and quantitative forensic soil science (Dawson).

Alec Jeffreys was an obsessive, precocious child. His interest in science started as a schoolboy when he conducted experiments in the sitting room and dissections on the kitchen table. After seven years at Oxford University, he moved to the University of Amsterdam where he and a co-worker developed a method for producing a physical map of the β-globulin gene in a rabbit's genome. A gene is a continuous or discontinuous section of DNA. Each person has a unique sequence of DNA, a signature trademark. It is noteworthy that the latest research suggests that environmental influences may cause monozygotic twins to have slightly different genomes.

Bailey, *Inventive Geniuses Who Changed the World*,
https://doi.org/10.1007/978-3-030-81381-9_16

After moving to Leicester University, he focussed on the inherited variation in the human gene, seeking to trace genes through family lineages. Although 99.9% of human sequences are the same for every person, Jeffreys realized that the remaining 0.1% is enough to distinguish individuals, one from another. Using sophisticated chemical techniques, he demonstrated that biological identification or DNA fingerprinting can be used to resolve issues of identification and kinship.

He then concentrated on the forensic field and refined DNA fingerprinting to give DNA profiling for which only a single repeated strand of DNA is counted. An individual can thereby be identified from the tiniest trace of their saliva, sweat, blood or semen, leading to a big saving in police investigative time.

The subsequent impact of these two techniques on solving paternity and immigration cases, catching criminals, whilst freeing innocents, has been extraordinary, and has impacted on the lives of millions of people worldwide. This application of molecular biology is invaluable to our justice system and helped private citizens find truth and resolution. It is also being used in 'non-human' crime (e.g., trade in rhinoceros' horns). As well as transforming forensic science, it has revolutionized the fields of biology, biodiversity, ecology, archaeology, genetic disorders and predispositions, livestock breeding and pedigree authentication. Jeffreys is very deserving of a Nobel Prize.

Growing up on her father's potato farm, Lorna Dawson saw how different soils affected crop yields and how different crops required different soil types for optimal growth. Her undergraduate studies in geography exposed her to other disciplines, including biology, chemistry and statistics which would prove crucial to her future career.

Every contact leaves a mark. The landscape leaves its mark on us. Lorna recognised that soil embodies a signature matrix which she was able to link to location. Trace samples of soil, taken at a crime scene, or, from an item involved with the crime, can be subject to accredited methods of analysis, interpretation, presentation, and explanation. She has used quantitative forensic evidence, not only to snare suspects, but to clear the innocent. It has helped with crime reconstruction and has been admissible in court.

16.2   Alec John Jeffreys CH KBE FRS MAE (1950)—Geneticist and Forensic Scientist; Father of Human and Animal DNA Fingerprinting and Profiling

16.2.1   Background

Alec was born to a middle-class family and has one brother and one sister. His mother was a housewife whilst his father worked as a designer in the automotive industry. He spent the early years of his life in suburban Oxford and then moved to Luton when he was six. At the age of eight, his father gave him a brass microscope, and a chemistry set put together by a chemistry graduate. These gifts stimulated his interest in both biology and chemistry.

Alec expanded his chemistry set to include hydrochloric, sulphuric, and nitric acids and metallic sodium. He soon discovered that sodium reacts violently with water and he wears a beard to hide a sulphuric acid scar resulting from an experiment to produce hydrogen cyanide gas.

He describes himself as an obsessive and precocious child, fascinated with explosives. He conducted experiments in the sitting room and dissections on the dining room table. By the age of twelve, he was carrying out syntheses in his kitchen, the standard of which was possibly at undergraduate level. He tried to build a model of myoglobin (see Sect. 12.4) from balsa wood.

Jeffreys was a pupil at Luton Grammar School and then Luton Sixth Form College, leaving there with a scholarship to read biochemistry at Oxford University. On completion of his first degree, he realized that genetics was the subject he wanted to pursue, and this became the primary focus of his DPhil. He studied the mitochondria of cultured mammalian cells in the Genetics Laboratory at Oxford University.

After seven years in Oxford, a post-doctoral fellowship beckoned at the University of Amsterdam. He, and Richard Flavell, applied molecular biology, in an attempt to detect and clone a single copy of a mammalian gene, specifically the β-globin gene in rabbits. They developed a method for detecting the β-globin gene but were unable to clone it.

16.2.2   DNA Genetic Fingerprinting

DNA molecules (see Sect. 13.3.2) carry the genetic code that determine the characteristics of living things. Apart from identical twins, each person has a unique sequence of DNA, a signature trademark of every person, making us different from everyone else. On average, there are between four and five million genetic differences between two people. A gene is a continuous or discontinuous section of DNA. When a gene is activated, its information is copied into a single-stranded (messenger) RNA molecule (see Sect. 13.3.5) which translates the information into a specific protein.

Chromosomes are strands of DNA. Human cells have 23 pairs (= 46) chromosomes, and the human genome carries an estimated 30,000 genes. Meiosis is a process of cell division, whereby a single cell divides twice, resulting in four daughter cells with genetic information half that of the parent cell. These cells are our reproductive cells—female egg cells and male sperm carry only one chromosome, at random, from each of the 23 pairs. Essentially, they are 'half-cells'.

When these merge, on fertilisation, the outcome is a 'complete cell' with the full complement of forty-six chromosomes, carrying a hybrid 'instruction package' from both parents. The coupling of genes from both parents stirs the pool of diversity. The child inherits genetic gifts and shortcomings from both parents, and a chance that these genetic features are combined in new and original ways. That is why children in the same family look a little like each other, and a little like each parent but are not identical to them.

In their experimentation, Richard and Alec broke down cells and extracted their DNA. They then used restriction endonuclease enzymes to cut double-stranded DNA into fragments, wherever a specific short sequence of DNA bases occurred. The DNA fragments were separated by gel electrophoresis, and then transferred to nitrocellulose filter paper.

Detection of the fragments was achieved by hybridizing them with a complementary DNA strand labelled with radioactive phosphate. X-ray film was used to detect the bound radioactivity, revealing DNA fragments containing the gene. By comparing these fragments produced by different restriction endonucleases, they could produce a physical map of the β-globin gene in the rabbit genome. The findings of their work were published in 1977.

In 1977, Phillip A. Sharp and Richard J. Roberts, working on the cold causing virus, adenovirus, independently discovered that some genes are discontinuous, being present in DNA as several, well-separated segments. Such genes are called 'split-genes'. For their pioneering work, using electron microscopy, Sharp and Roberts were jointly awarded the Nobel Prize in Physiology or Medicine in 1993.

In higher organisms, therefore, a gene may consist of several segments (called 'exons') separated by intervening DNA, called "introns". With ou present knowledge, the in-between segments appear not to code anything and are regarded as 'junk DNA'. Where the gene is split, the messenger RNA must be processed by 'splicing' before it can be translated into a protein. The 'exome' is that part of the genome, composed of exons, that remains within the mature RNA, after introns have been eliminated by RNA-splicing.

Whilst minor alteration (mutation) in the genetic make-up results in gradual change to the species, rearrangement (or 'shuffling') of gene segments to give new functional units could accelerate evolution.

In the same year (viz 1977) that Sharp and Roberts made their discovery, Jeffreys and Flavell, using their novel experimental approach, added to our understanding of split-genes by independently discovering an intervening sequence of about 600 organic bases within part of the rabbit globin gene.

Also, in 1977, Alec, aged twenty-seven, was offered a temporary lecture-ship at the University of Leicester. He was to remain there for the rest of his illustrious 35-year career, rising to the post of Professor of Genetics, in 1987, and Royal Society Wolfson Research Professor, in 1991. He retired in 2012.

He initially focussed on the inherited variation of human genes, seeking to trace genes through family lineages. Although 99.9% of human sequences are the same for every person, Jeffreys realized that the remaining 0.1% is enough to distinguish individuals from one another. Using a combination of molecular biology and genetics, he accidently came across individual varia-tions that make every person's DNA (other than identical twins) one of a kind. The number of times that various DNA segments repeat themselves vary substantially between individuals. This ensures that the chances of two people having the same variables are infinitesimally small.

When maternal chromosomes (from the egg cell) and paternal chromo-somes (from the sperm) line up, the respective DNAs are "shuffled", and that shuffling can go wrong and cause variation. Jeffreys devised an experiment to see whether he could monitor those repeat variations in different individuals and their relatives.

In 1984, Jeffreys experimented with DNA samples taken from his female laboratory assistant, as well as those of her parents. DNA fragments with varying numbers of repeating sequences were placed into wells of a gel elec-trophoresis container. When an electric current was applied across the gel, DNA fragments moved at different speeds depending on their molecular weights. When reacted with a radioactive probe capable of detecting these repeats, a "bar code" of blurry marks, at various places along the length of an X-ray image, could be seen.

He noted that the bands of DNA fragments, arising from his assis-tant, were a composite of her parents. He had discovered highly variable, repeated segments of inherited DNA which would serve as valuable informative genetic markers. The differing bands can be traced back through the ancestry of the individual. He had demonstrated that biological identification, or "DNA fingerprinting", can be used to resolve issues of identity and kinship.

His technique was the accidental product of blue skies research and was patented (e.g., GB2166445 and US5175082A). He, his technician, and his wife then brainstormed how this technology could be applied. Ideas included paternity testing, immigration disputes, forensics, sexual assaults, ecology, biodiversity, and non-human crimes (e.g., the trade in rhinoceros' horns).

For instance, in 1985, a group of lawyers were challenging the deportation of a boy who, the Home Office said, was not the son of a British woman. However, DNA samples from the boy showed every genetic character in the boy was present, either in the woman, or as paternal characters in his three undisputed siblings.

Between 1985–7 Jeffreys' laboratory was the only one in the world capable of conducting DNA paternity and maternity tests. Then, in 1987, ICI was awarded a licence to establish Cellmark—the world's first commercial DNA testing company.

All rights to Jeffreys' patents on DNA genetic fingerprinting are vested in the Lister Institute of Preventive Medicine, where Alec was a Research Fellow between 1982 and 1991. Through an agreement with the Lister Institute, ICI obtained the exclusive rights to the DNA fingerprinting technology discovered by Alec. His testing method was developed with ICI, at their laboratories in Abingdon, UK and Maryland, USA. A share of the significant patent royalties flowed to the Lister Institute to help fund biomedical research.

16.2.3   DNA Profiling

Next, Jeffreys moved to the forensic field of "DNA profiling", and to a refinement of the "DNA fingerprinting" technique. This is a laboratory technique used to establish a link between biological evidence at a crime scene and a suspect in a criminal investigation. Only a single repeated region of DNA (mini satellite) is counted, instead of the hundreds detected in DNA fingerprinting.

A mini satellite is a repetitive variant tract of DNA that is repeated 5-100 times. The tract ranges in length from 10 to 60 organic base pairs. They are notable for their high diversity in the population, being loci where most variations between people occur. This technique is quicker to use and requires smaller samples. It can be used to pinpoint an individual from the tiniest trace of their saliva, sweat, blood or semen. It led to a big saving in police investigative time.

In 1986, it was first applied to prove the innocence of Richard Buckland who had been accused of the rape and murder of two Leicestershire girls. It led to the world's first 'DNA manhunt'. Eventually, this led to the arrest and conviction of Colin Pitchfork.

The techniques used in the mid-1980s have been superseded by faster and simpler approaches. The development of DNA amplification by the polymerase chain reaction (PCR; see Sect. 13.4.3) opened new approaches to forensic DNA testing. PCR increases the amount of DNA for typing. It facilitated automation, greatly increased sensitivity, and a move to alternative marker systems, by looking at micro-satellites or short tandem repeats (STR's). Micro-satellites are repetitive tracts of DNA in which 2–6 base pairs show a high level of polymorphism (variability) and are repeated 5–50 times.

DNA profiling can be performed in about twenty minutes. If real-time applications could be achieved in, say twenty seconds, then this could be used at Immigration Control at airports and seaports.

STR profiling was further refined by Peter Gill at the Forensic Science Service. This led to the UK National Criminal Intelligence DNA Database in 1995. Since 2014, the DNA-17-test has been used. Sixteen micro-satellites, together with a sex marker, are used giving a discrimination power of one in over a billion. The database held the genetic profiles of both convicted criminals and suspects.

By 2009, about 10% of the population was on the DNA database, making it the world's largest collection of DNA information. Some groups, however, saw the storage of genetic profiles of suspects as an infringement of civil rights and the Protection of Freedoms Act 2012 now means that only convicted people will have their DNA profiles retained indefinitely. Alec believes in the right to genetic privacy.

Since DNA can be obtained and analysed, even centuries after a person's death, the latest technology has helped to solve war crimes and historic puzzles (e.g., the death, in Brazil, of Josef Mengele, the German SS physician at Auschwitz; the authenticity of the bones of the last Tsar of Russia, Nicholas II and his family; verifying the authenticity of Dolly, the cloned sheep).

16.2.4   Decorations and Awards

Jeffreys interest in science started as a school-boy hobby. He sees himself as a bench scientist whose applied discipline was underpinned by fundamental, curiosity-driven, background science. He has been described as "An ordinary guy who did something extraordinary".

In 1986, Jeffreys was elected a Fellow of the Royal Society. In 1989, he became a Member of the Academia Europaea (MAE)—having demonstrated sustained academic excellence. He was conferred Honorary Freeman of the City of Leicester, in 1992, and given a knighthood for services to genetics in 1994. In 2014, he was awarded the Royal Society Copley medal—said to be the world's oldest scientific prize. This was bestowed for his pioneering work on variation and mutation in the human genome. He was conferred Order of the Companions of Honour in 2016. Only sixty-five people in the world can hold this distinction at any one time. A long-overdue Nobel Prize continues to elude him.

16.3   Lorna Anne Dawson CBE, FRSE, FISS, FRSA, CSci (1958–)—Forensic Geologist; Soil Sleuth

16.3.1   Background

Lorna grew up on her father's potato farm, near Forfar, in the county of Angus, south of Aberdeen. She used to love working outside in the fields and earnt pocket money by harvesting tatties (potatoes) in the autumn tattie holidays, and by picking strawberries and raspberries in the summer holidays. She saw how different soils affected the yields of diverse potato varieties and how different crops required different soil types for optimal growth.

Lorna gained a geography degree from Edinburgh University, in 1980. Her undergraduate studies exposed her to other disciplines, including biology, chemistry and statistics which would prove crucial to her future career. She moved to Aberdeen University to undertake a PhD in soil science. She recognised that soil embodies a signature matrix of biological, chemical, and physical features. With her geography speciality, she was able to link soil science (pedology) with location.

Later, she realised that to be an expert witness in legal proceedings, she should obtain relevant qualifications. Therefore, she obtained a post-graduate diploma in civil and criminal law from Cardiff University and trained as an Expert Witness.

16.3.2   James Hutton Research Institute

She started working at the Macaulay Soil Research Institute which became the Macaulay Land Use Research Institute, an international centre for research and consultancy on the environmental and social consequences of rural land uses. She investigated soil-root interactions and endeavoured to help farmers improve the nutritional quality of agricultural products, whilst minimising the impact of fertilisers on the natural environment. The Institute holds soil samples from across Scotland which are referenced geographically.

In 2011, the Macaulay Land Use Research Institute (MLURI) combined with the Scottish Crops Research Institute (SCRI) to create the James Hutton Research Institute, in Aberdeen and Dundee. Her subject matter expanded to include quantitative forensic soil science, advancing and testing methodologies and working to disseminate her techniques to others. Lorna is Head of the Soil Forensics Group at the Institute.

David Barclay was Head of Physical Evidence at the National Crime and Operations Faculty (NCOF), a national crime agency that helped UK police departments with the most serious crimes. It is now the National Crime Agency (NCA). In the context of criminal law, the application of forensic science is governed by legal standards of admissible evidence and criminal procedure. Barclay wanted to ensure that UK investigators drew on the most robust science whether, or not, it supported the prosecution's case. Dawson has been able to use her forensic techniques, not only to snare suspects, but to clear the innocent.

Barclay and Dawson assembled a team of scientists, investigators and lawyers which obtained a grant from the UK Engineering and Physical Sciences Research Council (UKEPSRC) to develop accredited standards for using soil-science techniques in forensic investigations. The project was called "Soilfit". Lorna now runs one of only a few laboratories in the world dedicated to forensic soil science and ecology.

She is part of a multi-disciplinary, multi-faceted team. Disciplines covered at the Institute include pedology (study of soils in their natural environment), geology (study of rocks and similar substances that make up the earth's surface), taphonomy (study of the processes that affect fossilisation), paly-nology (study of pollen grains and other spores); botany (study of plants) and entomology (study of insects).

Soil is a complex mix of mineral particles, organic materials, air, water and living organisms. Every area of soil has its own fingerprint, composed of various characteristics, derived from the rocks on which it lies, the vegetation that has grown in it, and the micro-organisms that live in it. Each type of rock has its own mineral composition that is characteristic of a specific geograph-ical location. However, soils are continuous and are not categoric. They vary is many aspects, horizontally across a landscape and vertically down a soil profile.

Experts can test for a potential link between a suspect and a crime scene. They do this by examining any soil found on, for example, clothing, footwear, or treads of vehicle tyres. Forensic examination of soil evidence considers various soil attributes, such as colour, particle and trace-particle size, mineralogy, elemental composition, organic profile composition and palynology.

Soil colour is a useful preliminary screen but is indefinite. Mineralogical techniques define the general geological region of interest, whilst palynology provides site specific information relating to likely vegetative history. All these techniques, however, require specialised equipment and expert knowledge, are time-consuming and costly.

Dawson's work ranges from wildlife and environmental crime to civil and criminal law. She developed the application of soil organo-mineral markers in forensic investigations. In the initial phase, soil is examined under the microscope and vegetation, hairs, and anthropogenic materials (e.g., brick, concrete, paint, metal, glass and/or plastic) are identified and picked out. Unknown fragments of vegetation can be identified via plant-DNA analysis. Each set of fingerprints is compared to a growing reference data base which then provides a ranking of potential areas of interest.

The Institute can offer advanced methods of analysis for inorganic substances. These include scanning electron microscopy (SEM EDS) to study both morphology and material composition; X-ray powder diffraction (XRPD) to identify and characterise mineral crystals and profiles; X-ray fluo-rescence (XRF); inductively coupled plasma mass spectrometry (ICP-MS), as well as Fourier transform infra-red (FTIR) spectroscopy—offering a chemical fingerprint for both organic and inorganic substances.

Organic analysis can also be undertaken via gas chromatography mass spectrometry (GC–MS) requiring a sample size of only 20 mg. Isotope analysis, including carbon-13 and nitrogen-15 is optional but can be useful in distinguishing the origins of foodstuffs.

Elemental analysis may also be useful to identify the provenance of foodstuffs and drinks. Bacterial analysis and diatom plankton analysis may also be called upon where environments associated with aquatic habitats are involved.

Dawson saw opportunities to develop more focussed techniques by examining the organic matter in soil, that which is composed of dead and decomposing plants and organisms. Whereas inorganic components may be broadly the same over kilometres, organic characteristics vary on a scale of centimetres or metres, offering a finer and more appropriate spatial scale of resolution for a contact point location. When combined with soil-survey databases, that extra level of resolution assists investigators to use soil attached to a suspects' shoe or tyre to locate a crime scene or burial site and to provide evidence in court of likely contact with a location.

Lorna is now part of an international collaboration that is developing and testing methodology to profile microbial communities using DNA. Forensic microbiology is emerging as an independent discipline, ranging from specific culture methods to whole genome sequencing. Exploring the application of modern molecular biology as a potential tool in the arsenal of forensic analysis may prove to be an instrument, as powerful for forensic investigation, as it is in human DNA profiling (see Sect. 16.2.3).

Perfecting the microbial genetic profiling technique to guarantee robust and reliable data is now the challenge. Soil fungi may represent a robust target. However, the spatial scale of resolution requires further testing before it is considered for the evaluative stage of an investigation. Meanwhile, it is used in intelligence questions such as where a potential burial site may be located.

16.3.3 Illustrative Criminal Investigations

In 1977, when Lorna was studying in her dormitory, in Pollock Halls, at Edinburgh University, two local, 17-year-old girls—Helen Scott and Christine Eadie—disappeared. They were last seen alive leaving the World's End pub in Edinburgh's Old Town. They were brutally murdered, but the crime remained unsolved for about thirty years. In 2007, Angus Sinclair was acquitted of the murders, due to the lack of sufficient evidence to convict him.

In 2014, in the first double jeopardy case in Scotland, re-analysed human DNA from inside the knots, binding the girls' limbs, showed that Sinclair had helped tie them. Semen, taken from Helen Scott's coat, where she was lying, indicated a recent sexual event—countering Sinclair's alibi of being on a fishing trip.

Moreover, small aggregates of soil, on the soles of Helen's feet, contained traces of plant wax that both matched the field where the girl's body was found and the grass verge where the transport vehicle likely stood. Additionally, husks and grains recovered from the dirt on her feet suggested that she had been walked into the field, to her death. The case was a triumph for forensic science and Dawson had made a small but crucial contribution to the conviction of a serial killer.

With her analytical tools for soil characterisation, Dawson was also involved with the Christopher Halliwell case. Each sample of soil has its own unique signature, particle size, colour (potentially reflecting the amount of iron oxide) and quantity of organic matter.

In 2011, Halliwell was charged with the murder of Sian O'Callaghan, as well as the murder of Becky Godden who disappeared in 2003. However, in the original trial only the murder of Sian was considered because the investigating officer breached police procedure and protocol by denying Halliwell his right to a solicitor. In later years, soil, recovered from the field where Becky's remains were discovered, were highly comparable to soil samples taken from tools found in Halliwell's garden shed. At his second trial in 2016, Halliwell was found guilty of Becky's murder.

Over the years, Dawson and her team have unearthed vital clues to help convict some of the UK's most evil killers. Her field and bench work have built-in systems of accreditation. She is a registered expert with the National Crime Agency. She collaborates with police forces, lawyers, and forensic service providers, sometimes presenting key evidence in court.

She also trains fellow scientists overseas in the skills of a forensic soil scientist. Countries in which she has worked include Brazil, Argentina, Denmark, Russia, and Switzerland.

She has been called upon to present her findings to juries of lay members of the public with diverse educational backgrounds. This requires a style of presentation that the jurors will readily comprehend.

16.3.4   Other Interests and Awards

Lorna also works with the entertainment industry, aiming to achieve realistic storylines in crime thrillers such as BBC's Silent Witness and STV

Vera. She has collaborated with crime authors, including Ann Cleeves, Val McDermid, Stuart MacBride, and Mark Billingham. She is a Fellow of the Royal Society for the Encouragement of Arts, Manufacture and Commerce (FRSA). Fellowship is granted by RSA judges to those individuals who have made outstanding achievements to social progress and development.

She is also a Fellow of the Institute of Professional Soil Scientists (FISS), the latter having merged with the British Society of Soil Science (BSSS). Lorna is a Chartered Scientist (CSci). She was presented with a Pride of Britain Recognition Award in 2017 and awarded the Honour of Commander of the British Empire (CBE) in 2018. In 2019 she was elected Fellow of the Royal Society of Edinburgh (FRSE).

17

Looking Forward to Challenges and Opportunities

Only in Britain could it be thought a defect to be "too clever by half". (Quote by Prime Minister John Major in the Observer, 7 July 1991)
Your mind is like a parachute: if it isn't open, it doesn't work. (Buzz Aldrin, quoted in Business Insider, by Ivan De Luce, on 15 July 2019, 50-years after Apollo 11's blast off)

17.1 Will Our Present Educational System Engender Future Innovative Geniuses?

The scientists recorded herein were often child prodigies, some of whom may have matured into eccentric adults, totally absorbed in their subject, occasionally to the exclusion of normal human interaction and activity. By our current 'tick-box' attitude to education and a socialist approach to an egalitarian society, we may be side-lining and neglecting such children, at the expense of future scientific and technological breakthroughs.

Approaching 50% of those leaving school with A-level qualifications now go into higher education. In 2020, about one third are likely to go to university, some of them studying subjects with little conspicuous purpose. As degrees become ever more commonplace, the graduate premium shrinks. Many graduates emerge with crippling debt, some with unmarketable academic degrees. Nonetheless, those with academic qualifications have a higher employment potential than those with innate intelligence but

© The Author(s), under exclusive license to Springer Nature
Switzerland AG 2022
Bailey, *Inventive Geniuses Who Changed the World*,
https://doi.org/10.1007/978-3-030-81381-9_17

no certificates. Rating people by how adept they are at studying courses and excelling at examinations is a blinkered way for society to proceed.

It is questionable whether today's educational system is sufficiently adept at recognising, catering for and nurturing our intellectually gifted and/or vocationally talented children—those who, one day, could be extending the frontiers of our scientific and engineering knowledge. The belief that a scientific qualification is superior to a technical one is flawed. We forget at our peril that other qualities such as manual dexterity, creativity and practical knowledge are qualities worth admiring and fully exploiting.

We may have a rich spectrum of latent talent waiting to be unlocked, but we need to address the imbalance between academic and technical qualifications, thereby promoting a fairer society. For many students, a work-based technical apprenticeship or professional qualification may offer greater returns than a typical 3-year bachelor's degree. The current funding system fails to sufficiently promote diplomas and certificates for practical skills, as alternatives to academic degrees.

We ought to improve our ability to identify and capture high intelligence in children from modest or humble backgrounds to reveal their potential for great accomplishments as adults. Whilst we have 'elite' schools for the offspring of the rich, we may benefit from customised 'elite' schools for aspiring scientists and engineers from underprivileged homes, where teachers and mentors can unleash their pupils' innate potential, without quelling their genius. Young minds need to be opened and fired up so every child can be educated to the limit of his/her capacity, and the brightest can soar. Their Rolls-Royce minds should be trained to cogitate on meaningful things that cannot be learnt from textbooks. Whilst knowledge may be limiting, imagination embraces the entire world and stimulates progress.

The most successful universities are intellectually diverse, gathering talented students and staff without regard to social background, ethnicity, or religious upbringing. Any obstacles to those candidates from disadvantaged backgrounds must be eliminated. Individuals should be the authors of their own destinies. The current higher education teaching model may have to change in the post-COVID world.

Silencing and censoring contentious views on university campuses will threaten free and open discussion of ideas and is unacceptable. One should not exclude people from the conversation simply because one disagree with them. Suppressing unconstrained expression, by shaming, ostracising or disinviting those deemed to have transgressed, ought to be forbidden. A "cancellation" attitude to public figures with a history of admirable work simply because of a clumsy comment or deviation from "approved" opinion is intolerable. Universities have long been a bastion of free speech and they have a duty to maintain their tradition of encouraging students and staff to express themselves candidly—challenging views and opinions, thereby cultivating untethered, enquiring minds. Universities are there to educate, not indoctrinate.

17.2 Will Society Be Supportive of Exciting Inventions, Current or Potential?

Our genius, as a nation, has been to ensure liberty under the law, to allow entrepreneurs to pilot and trial new ideas in a secure research environment, to allow risk to be properly rewarded (e.g., Watt, Brunel, Bessemer, W. Thomson and Bell). Achievement should be praised rather than belittled. We should never be squeamish about promoting our national interest.

Sadly, we have a long and inglorious history of rarely profiting greatly from our inventions. For instance, Sir Frank Whittle patented the jet engine in 1930, but it was Hitler's Germany that built the first jet plane. It was Berners-Lee who invented the World Wide Web and charged no royalties for its use, but it is the tech behemoths who 'share the spoils'.

Currently, fossil fuels are fundamental to our lives. However, looking to the future, industries connected with fossil fuels, the internal combustion engine and plastics are under threat and will decline. Britain is a nation of eco-conscious citizens, especially young people, and often localists who care deeply about their surrounding environs.

As well as 'vanity' projects, such as high-speed rail, HS2, other major engineering projects requiring changes to the surroundings that are visible to the local population are likely to be resisted by a combination of the Green movement and the not-in-my-back-yard (NIMBY) syndrome of the modern Briton. Any visual intrusion, and any effect on the countryside, plant, animal, insect and birdlife, no matter how small, is likely to be adamantly opposed.

There is overwhelming evidence that an economy based on the burning of fossil fuels for power generation will lead to the diminution of the world as we know it. This has resulted in some extreme forms of evangelical climate activism. Some radical climate protesters are obsessively anti-capitalist–against economic growth and the prosperity it generates—seeking to overthrow existing systems as penance for the "original sin" of the Industrial Revolution.

As a road to their 'climate salvation', there are calls to retreat to a utopian world, resembling our pre-industrial past. This is based on indigenous communities, rather than global markets, and a moneyless, subsistence economy where food, clothing and shelter are produced, only in sufficient quantities for survival, but not surplus for sale.

Undoubtedly, climate change is a serious threat to nature's wellbeing, but it will not be solved by abandoning our faculties and succumbing to the shrill catastrophism espoused by many of the climate warriors. The path to zero greenhouse gas emissions requires much innovation. Curtailing the free market and expunging multinational corporations will not halt global warming. On the contrary, capitalism, for all its flaws, will be an essential engine of innovation in the battle to achieve a carbon–neutral world, keeping the climate bearable and preserving the planet for future generations.

For instance, efforts are underway to capture greenhouse gases at the point of emission. Other research is seeking ways to remove, by sequestration, harmful gas molecules already in the atmosphere. Both approaches will help achieve net zero emissions by 2050 or earlier. Geo-engineering projects are being considered to spray calcium carbonate dust into the atmosphere to create a sun-reflecting aerosol. Alternative sources of power are being sought. Nuclear fusion is the holy grail of energy research.

Unsurprisingly, carbon-zero power sources, free at the point of delivery, are receiving particular attention. Visionary businesses have already devoted significant resources to solar, wind, wave and geothermal power generation. Since solar and wind power are inconstant, more effective ways to store the electricity generated from them are being developed. Power storage is being facilitated by advances in battery technology and hydrogen fuel cells. On those occasions when the national grid is overloaded with electricity from wind power, the excess could be used to produce 'green' hydrogen from the electrolysis of water. Given the opportunity, technical ingenuity will continue to help us live modern lives, whilst not despoiling the planet.

We are developing a habit of resisting any effort to improve infrastructure. Vociferous lobby groups opposing major civil engineering projects have the loudest voices and are given more attention by the media than do those muted citizens supportive of modernising projects. The latter have fewer activists, making their counter arguments less influential and effective. Noisy conviction almost always trumps apathy. Proposed developments are often stalled by disputes between builders, councils, and campaigners which reward lawyers and lobbyists. It appears that we are shifting from a nation with a 'can do' mentality to one of 'can't do'. We are becoming risk-averse, and expert in procrastination. Negativity is becoming endemic.

In 2001, Government ministers considered building a third runway at Heathrow. Whilst the debate about its suitability continues, particularly post-COVID, China continues at pace its airport construction programme. At the end of 2015, China had 207 certified general aviation airports which has grown to 299 at the closure of 2020. According to the Rhodium Group, in 2019, China's annual emissions of greenhouse gases (the six Kyoto gases) exceeded those of all OECD countries combined. The top-4 emissions were 27%, 11%, 6.6% and 6.4% for China, USA, India and EU-27 respectively.

Better means of transport invariably benefit society. In terms of transport infrastructure, the UK is slipping in the league of well-connected countries in the developed world. We invented the first successful railway engine and built, in four and a half years, the first intercity railway line which was thirty-five miles long. Whereas most major European countries have built high-speed railway lines, and are enjoying the benefits, twenty-five years after the Channel Tunnel opened, we have finally decided about the desirability of HS2. This will free up existing track for freight, helping to remove some of it from our roads.

In general, projects less visible to our citizens, such as subterranean Cross-rail and extra-terrestrial space travel, continue to be supported. Spin-offs from the space race include improved earth-observation imagery, helping with weather forecasting and assisting urban planners and farmers to optimise land use.

Powerful lobby groups have resisted the adoption of fracking and genetically modified foods in this country. Their protests have helped bring about the abandonment of the proposed transport corridor between the two centres of scientific and engineering centres of excellence, Oxford, and Cambridge, threatening the ambitious OXCAM arc.

These naysayers think it preferable to transport gas, liquefied at −160 °C, hundreds of miles in oceanic tankers than it is to extract gas from shale deposits beneath our feet, potentially giving us self-sufficiency for up to fifty years, and offering a pre-combustion carbon footprint about half that of imported gas. Shale gas, suitably extracted, may be the best option to plug the energy gap until eco-technologies catch up.

There is a growing interest in "nutraceuticals" and "functional foods"—nutrient-rich foods which are medicinally active. For example, rice, biofortified with beta-carotene, will combat vitamin-A deficiency in those countries where rice is a staple food. However, opposition to genetically modified organisms (GMOs) has blocked the roll-out of so-called 'Golden' rice in developing countries where a lack of vitamin-A causes blindness, even death to thousands of children each year.

Genetically modified foods have been marketed in America since 1994, apparently without ill-effects. However, it seems unacceptable to grow genetically modified foods in many other parts of the world to help feed the estimated 800 million, of the world's population of 7.3 billion, who are suffering from chronic under-nourishment.

This non-acceptance of GMOs remains, even though there are no reported major negative impacts on the environment, the animals which have been fed on GM crops and the humans who have consumed them. The global need to substantially increase food production can be achieved by a combination of sustainable farming, avoidance of monoculture crops, climate-resilient cultivars and judicious use of genetic engineering, controlled by robust regulation.

Industrial bioprocesses, especially fermentation, have been with us for centuries—extending the shelf-lives of some foodstuffs and creating new ones. For example, man has utilised micro-organisms (e.g., yeasts and bacteria), as well as extracted enzymes, to make beverages, cheese, yoghurt, bread and pickled products.

The next major leap in applied biology—biotechnology—will be the commercially viable production of meat analogues via *in vitro* cell culture of animal cells in bioreactors. Such 'cultivated' meat will mitigate the emissions of egregious greenhouse gases, methane and nitrous oxide—methane from enteric fermentation in belching, flatulent ruminants and their manure—nitrous oxide from synthetic and natural fertilisers. Cultivated meat will also eliminate the need for antibiotics and growth hormones in farm animals, as well as the risk of pathogenic bacterial cross-contamination, at their slaughter.

Whereas genetic modification involves the movement of whole functioning genes, from one organism to another, gene editing involves molecular "scissors" to make targeted changes to the resident DNA of an organism. Gene editing slightly modifies the DNA of organisms, thereby accelerating the selective processes that could happen naturally or take years for horticulturists or livestock breeders to achieve the desired outcome by traditional crossbreeding.

For instance, using this genetic "surgery", it might be possible to remove toxic ricin from the seeds of the castor oil plant; eliminate the protein in peanuts which cause food-induced anaphylaxis; reduce the amount of asparagine in wheat which combines with sugars, by the Maillard reaction to form carcinogenic acrylamide, when bread is toasted.

Rice crops feed about half the world's population. Historically speaking mutations occurring randomly, under natural conditions, have generated limited yield benefits. However, using genome editing technology, significant yield increases could be achieved, not only for rice, but also another staple food, the potato.

Gene editing could also be employed to create crops more resistant to pests and disease—lowering the need for artificial protective inputs. An example might be bananas resistant to the fungal Panama disease. Similar research may offer a pig breed that is resistant to PRRS. Porcine reproductive respiratory syndrome (PRRS) is a viral disease causing late-term abortions in sows, as well as respiratory problems in piglets. Any activity to reduce the use of antibiotics in farm animals, and the danger of transferring antimicrobial resistance from livestock to humans, is to be welcomed.

To mitigate the negative impact of chemical fertilizers on the natural environment, it might be possible to encourage cultivated crops to form associations with beneficial fungi that invade plant roots to deliver nitrogen and phosphorus. By genetic modification, research has begun to transfer the nitrogen-fixing capability of legumes to, say, cereal cultivars.

Dengue fever is spread by the Aedes aegypti mosquito. Luke Alphey, at Oxford University, has developed genetically modified male mosquitoes, the offspring of which do not mature into adults. This approach is a viable alternative to insecticides which are currently sprayed in the rainy season. Controlling the reproduction cycle of mosquitoes may also be used to stop the spread of Zika and malaria.

The industrial pendulum seems to be swinging away from factory mass production, back to cottage industries and artisanal, hand-crafted goods made in a modern way. More people work from home with a smaller work force but better internet access. This has been accelerated by the 2020 coronavirus pandemic. Working from home (WFH) was often a drudgery in Victorian times and the same pattern could re-emerge. Economic downturns have a history of spurring further automation. Technology is making jobs redundant at a faster rate than before, but inventiveness will never go out of style. The manufacturing sector must respond in a climate-friendly way.

In the past, business was generally seen as a powerful force for good, generating the growth which pays for a civilised society and powering the innovations that make our lives better. But capitalism may have to be modernised so that it behaves more humanely, ceasing the exploitation of the lower-skilled and vulnerable. Let us further harness our remarkable capacity for invention, adaptation, and cooperation. When considerately deployed, it will create practical remedies for the shortcomings in today's world.

We are gradually moving from an industrial age economy to a knowledge-based, digital, and virtual economy, less exploitative of the Earth's finite raw materials. In the future, challenging opportunities for scientists and engineers appear more likely in, for example, electronics and life sciences, particularly pharmaceutical research, genomics, and ecology.

The "Medicare" sector is aiming to give us a longer, healthier lifestyle. Body-wearing fitness trackers can continuously monitor our vital signs, including body temperature, heart rate and respiration rate. "Smart speakers" are being developed to detect the sound of a person's heartbeat, and to distinguish it from an irregular one.

Wearable technology may be developed to measure alcohol levels and track blood sugar levels for diabetics, as well as blood pressure. Sub-cutaneous sensors might be able to monitor oxygen saturation levels. These advances will allow medical professionals to diagnose both physical and mental conditions remotely, even allowing personalised, targeted therapeutic interventions with fewer negative side-effects. Other, implanted, microchip sensors could monitor nutritional and calorie intakes, again helping with health and wellness.

A greater array of biomarkers will form the basis of quick and simple diagnostic tests. Volatile organic compounds emitted by sufferers (from breast, lung and prostate cancers; even Covid) are being identified, catalysing the development of portable sensors/apps. "Liquid biopsies" are less invasive than tissue ones. The detection of DNA fragments—shed by tumours—in blood biopsies might help in the early detection of a variety of cancers. This is especially important for solid tumours for which no screening options currently exist (e.g., oesophageal, liver and pancreatic cancers). Prostate cancer accounts for a high percentage of cancers in men. A non-invasive test for gene expression is being trialled. Biomarkers in urine may provide a more informative preliminary diagnosis for prostate cancer than rectal examination.

Nanotechnology could be used to search for specific invasive bacteria and viruses, and to deliver drugs to precisely targeted tissues. It may be possible to use magnetic nanoparticles to bind to invasive bacteria and viruses in the blood and then remove them via a magnetic sieve. Nanopore sequencing and rapid identification of viral pathogens may, one day, alert us to the presence of pathogens, particularly zoonotic organisms, in our vicinity.

Currently, invasive, uncomfortable colonoscopy is used to detect bowel cancer and Crohn's disease. As an alternative procedure, pill-sized cameras small enough to ingest, are already being evaluated. Futuristic "four-legged" microscopic- and nano- robots may be able to roam the human body exploring, repairing tissue and detecting toxins.

Our microbiome will be monitored to control the balance between 'good' and 'bad' bacteria in our alimentary tract. New immuno-therapeutic drug will be used to control previously untreatable diseases. A broader understanding of virology will lead to more vaccines and anti-viral drugs to better protect us from pandemics such as COVID-19. In regenerative medicine, stem cell research has begun to generate healthy replacements for diseased or damaged tissues and organs.

In the UK, universal electronic health records (EHR's) could be used for assessing the effectiveness of specific treatments, monitoring the spread of infectious diseases, or discovering correlations in the development of new medical treatments. Naturally, there is concern about data breaches. Subject to citizens agreeing to EHRs, these records could hold precise information about our vaccination history; disease exposure; natural genetic resistance and known drug resistance. They could track our immune response, including unwelcome auto-immune and allergic reactions to a threat.

Harvesting personal information for the common good, whilst protecting privacy, could drive revelations and progress, not just in healthcare, including the COVID-19 challenge, but in every field of human endeavour. Citizens will need to partner voluntarily with government, so information flows freely, both bottom up, and top down. The Unique Identity Authority of India (UIDAI) uses biometrics (viz fingerprint and iris scans) to create individual IDs for healthcare benefits, financial services and agricultural subsidies.

Digital maternity records will bring maternity care closer to home. Apps will allow pregnant women with hypertension to monitor high blood pressure and proteinuria to avoid pre-eclampsia.

The UK is a world leader in genetics and genomics, making vast contributions to this rapidly evolving field—from the ground-breaking fundamental findings of Crick to its application in clinical practice—improving healthcare and patient outcomes.

The concept of genetic engineering has long been controversial. It is noteworthy, however, that when the lives of our global citizens were imperilled by SARS-CoV-2 virus, it appears that the vast majority are prepared to be vaccinated with material resulting from genetically modified technology. This is in the form of synthetic RNA, coupled with lipid nanotechnology.

The next generation of immunotherapies may find universal flu vaccines, as well as mRNA vaccines for malaria and allergies. It may be possible to make mRNAs for mutant proteins specific to a patient's tumour. There are visions of applying RNA therapeutics when key proteins are missing, or defective, and mRNA treatment could be used to express a functional copy of the said protein.

Our ability to understand and manipulate life at its most fundamental level has paid dividends so globally, millions of lives may be saved. Genetic medicine is advancing at pace because an individual's genetic make-up can now be analysed relatively easily and cheaply. This promises faster diagnosis and personalised medicine. Gene therapy is a major focus, to edit defective genes which cause specific types of cancer, and genetic defects such as cystic fibrosis and muscular dystrophy.

A better understanding of our epigenetic profiles will help us control health outcomes related to age—such as cognitive decline; disease; disability; frailty and mortality. Telomeres are protective caps on our chromosomes, and they are particularly important when cells replicate. They are like plastic caps on the ends of shoe lacers—disposable DNA that protects the helix from wear and tear. Damage to telomeres can signal to cells that it is time for apoptosis (controlled cell death) or senescence. Shortened telomeres are associated with some of the diseases and dysfunctions of ageing. If their rate of shortening can be reduced, resilience to disease and longevity may be enhanced.

The World's love affair with plastic may be over but research will continue into endlessly recyclable plastics that retain their original properties. Biodegradable film, made from re-organised plant protein (e.g., 'vegan spider silk') could replace some single-use plastics.

Other areas of interest include biomimicry for building construction; enhanced communication; further miniaturization; never ending computer upgrades; artificial intelligence (AI); automation; robotic carers for the aged and infirmed to facilitate more independent living; ever more sophisticated artificial body parts; humans living in floating space colonies; space travel to facilitate the colonisation of other planets.

The capacity of the human brain may be enlarged by physically linking it to computers. "Augmented intelligence" will develop whereby humans and machines will exploit each other's strengths. Neural interface technology is endeavouring to amplify signals from the brains of paralysed people, using micro-circuitry to digitise the signals so they can be transmitted wirelessly to receivers, capable of steering a wheelchair or directing a robotic arm.

Developments in artificial intelligence (AI) could free us from the drudgery of domestic tasks and conveyer belt working. It could augment medical diagnoses by human specialists. However, there are those who fear it will replace us in the workplace, ultimately enslave us, or even exterminate us.

The use of in-soil sensors and drones will help in the agriculture industry to better control soil health and plant disease. The use of robots as pickers of fruit and vegetables will raise productivity. There are opportunities for small-footprint fish farming, coupling aquaculture with hydroponics to give aquaponics. This may become a sustainable farming option in some parts of our world and beyond, offering fish fed on a vegetarian diet.

Agricultural research will continue into plant yields, nutritional value, and resilience to both pests and abiotic factors such as climate change and erratic weather patterns. A better understanding of plant ecology will result from improved knowledge of their molecular machinery. Soil erosion by heavy ploughing is to be avoided since it disrupts soil ecosystems, releasing carbon into the atmosphere, hindering the fight against climate change.

17.3   How to Overcome Resistance to Change, Drive Innovation, Accelerate Commercialisation and the Safe Adoption of New Technologies?

We have some world-beating universities, research institutes, teaching hospitals, academic and professional bodies and journals. Our rich cultural heritage, political and legal systems facilitate free thought and expressive individualism in a secure research and innovation environment. Our schools encourage children to grow with enquiring minds, providing the great British scientists and engineers of the future.

Our place in the world has been driven partly by inquisitive minds persevering with 'blue sky' (curiosity driven) scientific projects. Sadly, our dynamism as an innovative nation appears to be fading and we are submissively adopting the posture of a mature, politically correct, inflexible economy, over-burdened by big government and regulatory "nannyism". Performative virtue signalling has started to take pride of place.

Little creativity arises from a period of stasis. We need to rejuvenate a spirit of free enquiry and return to an enterprise-friendly economy. This nation has never suited modest ambitions, and there is no reason why we should not dream big again. By further unlocking a global UK innovation economy, advances in science and technology will take us to new levels of modernity.

We must guard against intolerant zealots strangling intellectual freedom and the will to progress. We can be better than our ancestors by ensuring everyone, regardless of their background and beliefs, has an equal chance to decide their own destiny; pursue their dream career; speak freely and contribute to intellectual endeavour. To achieve this, we must reinvigorate an environment where innovation thrives, and equip every citizen to succeed, thereby unleashing talent and enterprise.

We ask whether we are still a buccaneering nation of risk-takers? If our progress falters, it will not be because we lack for creative genius, technical prowess, vision, and motivation in these islands, it will be because we cannot set a clear direction and march collectively into the future. Patriotism, pride in our past, and ambition for that future should be the clarion call for us all. We should be neither reliant on past glories nor frightened of a bright, imaginative future.

We are reminded daily that we are inching towards a climate disaster. The crisis is not yet terminal, so we can rewind. We will be saved by a combination of collective action and technological breakthroughs. Many of the various obstacles facing us in the future will be solved, not by politicians or cynics, but by remarkable scientists and engineers with support from the public. We have the resources and the ingenuity to fix things. In general, science is the best warrior. We can work together to convert problems into opportunities and benefits.

Appendix A: Technical Comparison of Brunel's Great Oceanic Steamships

See Table A.1.

See Fig. A.1 and Table A.2.

© The Editor(s) (if applicable) and The Author(s), under exclusive
license to Springer Nature Switzerland AG 2022
Bailey, *Inventive Geniuses Who Changed the World*,
https://doi.org/10.1007/978-3-030-81381-9

Table A.1 Great Western and Great Britain

Ship	PS Great Western	SS Great Britain
Launched	1837	1843
Sponsor	Great Western Steamship Co	Great Western Steamship Co.
Builder	Patterson and Mercer Shipyard in Bristol	William Patterson Shipyard in Bristol
Overall length (m/ft) between perpendiculars	71.6 (235 ft)	98 (322 ft)
Beam – breadth (m/ft) at widest point	10.7 (35 ft)	15.4 (50.5 ft)
Beam + paddle wheels, m/ft	17.6 (57.7 ft)	N/A
Draught, m/ft	5.1 (16.7 ft)	4.9 (16 ft)
Bruto/gross registered tonnage BRT/GRT	1340	3443
Displacement at load draught (tons)	2300	3618
Steam engines	×2 Maudslay, each 375 hp	M.I. Brunel's V-engine, 5psi, 2 × 500 hp
Propulsion	Paddle steamship + 4 masts	Single screw steamship + 5 × schooner-rigged & 1 × sq.-rigged masts
Speed (knots)	8.5–9	10–11
Crew	57	130
Passengers	128 × 1st + 20 × 2nd	360 × 1st class
Noteworthy features	Iron-strapped, oak hull. Steam ship race with Sirius (British & American Steam Navigation Co.) to be the 1st to New York. Sirius won but took 3 days longer and had exhausted coal stock	1st passenger liner to combine wrought iron hull with screw propeller. Largest ship until 1854. 1st commercial use of 'silent' chain to minimise noise from engine room, using sprockets with wooden teeth (teak upper; lignum vitae lower)
Last act	Scrapped 1856 after service as a troop ship in the Crimean War	Dry dock, Bristol

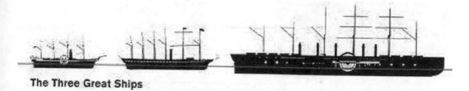

The Three Great Ships

Fig. A.1 Brunel's three Great Ships

Table A.2 Great Eastern

Ship	PSS Great Eastern
Launched	1858
Sponsor	Eastern Steam Navigation Co.
Builder	J. Scott Russel & Co. at Millwall, on R. Thames
Overall length (m/ft) between perpendiculars	211 (692 ft)
Beam − breadth (m/ft) at widest point	25 (82 ft)
Beam + paddle wheels, m/ft	36.6 (118 ft) over the paddle boxes
Draught, m/ft	7.6 (25 ft)
Bruto/gross registered tonnage BRT/GRT	18,915
Displacement at full load draught (tons)	32,160
Steam engines	In separate engine rooms, Boulton & Watt 6000 hp for single screw + Scott Russell 4000 hp 4-cylinder oscillating engine for paddle wheels
Propulsion	5 funnels + paddles + single screw steamship + 6 masts
Speed (knots)	14
Crew	418
Passengers	800 × 1st; 2000 × 2nd; 1200 × 3rd
Noteworthy features	Her gross tonnage was not surpassed until 1901. Coal storage sufficient for round trip to Australia. Safety features 50 years ahead of her time. Double, iron hull. Water-tight, longitudinal, bulkheads (×2) + crosswise bulkheads (×9), giving a total of 19 water-tight compartments
Last act	Scrapped 1889/90

Appendix B: Some Key Milestones in the History of Vulcanised and Synthetic Rubber Pneumatic Tyres

See Tables B.1, B.2, B.3 and Fig. B.1.

Table B.1 Period 1820–1895

Date	Event
1820	Englishman, Thomas Hancock, invented a mastication machine to mix and soften natural rubber, enabling it to be shaped. He built a factory in London. His machine and spreader were not patented until 1837 (UK patent 7344)
1823	Glaswegian, Charles Macintosh, developed a process of coating cloth with a solution of rubber in benzene and started manufacturing water-proof mackintoshes. He was granted UK patent 4804
1839	American, Charles Goodyear, started experimenting with the vulcanisation of rubber using sulphur and heat but it took 5 years to perfect the process
1843	Englishman, Thomas Hancock, now an associate of the raincoat firm, Charles Macintosh, filed for a UK patent, in November 1843, for the vulcanisation of rubber using sulphur
1844	Goodyear filed his US patent in January 1844 and was granted US patent 3633 in June 1844. Goodyear failed in British courts to get Hancock's patent revoked
1845	Robert William Thomson was granted British patent 10,990 for a rubberized fabric tube (bladder) inflated with air and encased in a thick leather outer casing which was bolted to the wheel hub of a horse-drawn carriage. Such pneumatic tyres were meant for carriages, rendering their motion easier to lessen the power required to pull them, and diminishing the noise they made in motion. Successful tests were performed in Hyde Park, London, in 1847
1846–7	Patents for Thomson's 'pneumatic tyre'/ 'aerial wheel' followed in France, in 1846, and in USA, in 1847. The aerial wheel was never commercially successful because of the poor state of roads and the high cost of the rubber for the bladder

(continued)

© The Editor(s) (if applicable) and The Author(s), under exclusive
license to Springer Nature Switzerland AG 2022
Bailey, *Inventive Geniuses Who Changed the World*,
https://doi.org/10.1007/978-3-030-81381-9

Table B.1 (continued)

Date	Event
1876	Seeds of Hevea brasiliensis rubber tree were smuggled to Kew Botanic Garden in London and used to develop more resistant varieties. These were sent to Ceylon, Singapore, and Malaysia where massive, well-cultivated and well- managed rubber plantations were established, bringing down the price of natural rubber
1878	Nikolaus Otto constructed a 4-stroke engine with a cycle: induction → compression → firing → exhaust which led to the modern internal combustion engine. He collaborated with Gottlieb Daimler
1886	Carl Benz's Patent-Motorwagen is regarded as the 1st production automobile propelled by an internal combustion engine. Benz was preoccupied with a 2-stroke engine. His automobile patent application (37,435) was filed in January 1886. His 3-wheeled vehicle had steel-spoked wheels and solid rubber tyres
1888	In July 1888, Dunlop was granted patent 10.607 for an improvement in tyres of wheels for bicycles, tricycles, or other road cars
1890	Dunlop's patent was officially declared invalid because of the belated discovery of prior art by fellow Scot, Robert William Thomson, in 1845
1890	The Welsh and Bartlett patents for demountable pneumatic tyres were granted
1891	Édouard Michelin filed a patent (France, 216,052) for his version of a detachable pneumatic bicycle tyre which allowed the rider to fix a puncture quickly, rather than having to rely on a mechanic. It consisted of an inner tube and outer tyre, pierced with bolt holes so that a circular metal strip could secure it to the wheel rim
1895	The 1st pneumatic tyres for an automobile were made by the brothers, Édouard and André Michelin, for their L'Éclair which was entered for the Paris to Bordeaux road race. After early reservations from the French public, the Michelin Company started to produce tyres for the expanding car industry in France

Table B.2   Period 1898–1942

1898	Frank A. Seiberling and his brother founded the Goodyear Tyre Co., named after Charles Goodyear
1900	The Dunlop Tyre Co. produced its 1st automobile tyre
1904	Demountable rims were introduced that allowed car drivers to fix their own flat tyres. After removing several nuts and lugs, a flange ring could be removed from a wooden wheel, followed by the tyre and inner tube
1908	Mechanic, W. C. State together with Seiberling of Goodyear invented the Seiberling-State tyre-building machine. Once the tyre has been assembled, it is subject to a curing process in a metal mould. As it expands, in the mould, the tread pattern is formed. A machine cut groves into the surface. Overall, the tyre offered better road traction, anti-skid in wet conditions, and improved driver safety
1910	B.F. Goodrich added reinforcing agent, carbon black, to tyre rubber to increase stiffness and tensile strength and to colour the tyres, black
Pre-1911 and beyond	In the early pneumatic tyres, chafing of the inner tube occurred due to the scissor-like action of the warp and weft threads of the reinforcing woven canvas. To minimise this, high tyre pressures were adopted
	Tyre engineering and manufacture evolved (see developments from Hardman, Dunlop, Palmer and Michelin below) to eliminate the wefts, so lower pressures could be applied. Sheets of fabric cord canvas in so-called 'cross-ply' or 'bias-tyres' were developed. Layers of "plies" of fabric cord embedded in vulcanised or, later, synthetic rubber encased the inner tube. The plies are laid across each other, running from edge to edge, in alternating diagonal layers, giving a criss-cross pattern
	Cotton fibres were eventually superseded by man-made fibres such as polyester, and Nylon. The finer the fabric (measured in threads per inch, tpi) the better the quality
1911	Philip Straus worked for the Hardman Tyre and Rubber Co. In 1894, his father, Alexander, had invented a process which allowed fabrics to stretch in one direction while not yielding in another (US patent 526546A). Philip recognised their usefulness to tyre strength. The Company produced the 1st successful motor car tyre which was a combination of fabric-reinforced, hardened rubber tyre and an air-filled inner tube
1914	The Dunlop Tyre Co. developed a process of spinning and doubling cotton for a new fabric for its pneumatic tyres
1915	Detroit's Palmer Tyre Co. developed the cord tyre. The fabric was not woven. Cords were kept parallel to one another and then pressed into the rubber sheet. The tyres were cast using these sheets. Each ply was separated by its own coating of rubber
1922	The 1st Dunlop tyre using steel rods together with a canvas casing was developed. It provided thrice the service-life of other tyres
1937	B. F. Goodrich manufactured the 1st synthetic rubber tyre, made from a synthetic rubber, Chemigum, patented by Ray P. Dinsmore of Goodyear, in 1927, and mass-produced in 1935
1942	At the onset of America's involvement with WW2, the USA consumed 600,000 tons of natural rubber. With the Japanese invasion of Southeast Asia, its supply was extinguished. The US Government, together with American rubber and oil companies agreed to produce Government Rubber-Styrene (GR-S). This was derived from monomers butadiene (75%) and styrene (25%). By 1945, America was producing 920,000 tons annually of synthetic rubber, 85% of which was GR-S

Table B.3   Period 1948–1973

1948	Michelin's creation of the steel-belted, radial tyre in 1948, offered superior road grip, durability, and increased fuel efficiency. It was so named because the carcass ply forms arcs, from one bead to the other, running radially as viewed from the centre of the tyre. The side wall is soft, and the crown is strengthened. Under the thread, sitting on the carcass ply, are belts composed of several cord plies, reinforced with metal ply. Radials were quickly adopted in Europe but not America until after the gasoline shortage of 1973
1952	From 1903 onwards, there were many patents filed for tubeless tyres, but technical problems meant that production of such designs was limited
	In 1947, B.F. Goodrich announced the development of a tubeless tyre, eliminating the need for an inner tube. Pressurized air was trapped within the wall itself. This was achieved by reinforcing the tyre wall. They have continuous ribs moulded integrally into the beads, so that they are forced, by the pressure of air inside the tyre, to seal with the flanges of the metal rim of the wheel
	Synthetic butyl rubber was used in the inner liner of tubeless tyres because of its low permeability to gases, offering better leak resistance. There may also be incorporated, a soft inner layer of highly viscous material, sufficiently flowable to seal minor punctures
	US patent 2,587,470 was awarded in 1952 to Goodrich engineer, Frank Herzegh. The 1st Goodrich tubeless was used on the Packard Clipper in 1954
	The credibility of tubeless tyres progressively advanced in the 1950s
1973	Dunlop produced a self-supporting, run-flat, Total Mobility Tyre, called the Denovo

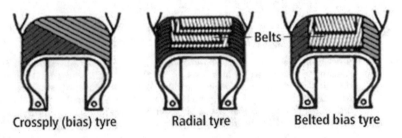

Crossply (bias) tyre   Radial tyre   Belted bias tyre

Fig. B.1.   Cross ply versus radial versus belted bias tyres

Appendix C: Chronicle of Early British (German and American) Jet Engines. Frank Whittle's Engine Lineage

See Table C.1.

© The Editor(s) (if applicable) and The Author(s), under exclusive
license to Springer Nature Switzerland AG 2022
Bailey, *Inventive Geniuses Who Changed the World*,
https://doi.org/10.1007/978-3-030-81381-9

Table C.1 Chronicle

Event/engine/model	Key features	Significance
RAF flying instructor, Frank Whittle, aged 22, applied for patent on 16/01/1930; numbered 347,206 for improvements to the propulsion of aircraft & other vehicles. Granted Oct 1932	Air passes through' a 1-stage axial-flow turbine, followed by a 1-stage centrifugal compressor. After passage through a diffuser, the compressed air enters a series of radially arranged combustion chambers. The hot gas is guided through a 2-stage axial-flow turbine on the same shaft as the compressor. Gas expansion/pressure drop occurs through turbine and propelling nozzles creating thrust	1st patent in the world to outline the fundamental design of a turbojet with sufficient thrust to propel an aircraft by way of its exhaust gases. Covered both axial and centrifugal options. Not having the means to pay the £5 renewal fee, he allowed the patent to lapse in 1935
von Ohain deposited patent application in 1935	Like Whittle's patent, the design featured both axial and centrifugal compressors	Application initially refused due to Whittle's prior art. May have been granted later, in secret
Whittle unit WU, models 1, 2 and 3. Experimental bench engines. Patents for double-sided compressor: 456,976 and 456,980 on priority dates 16/5/35 & 18/5/35	Patent 456,980 filed for a single-stage, double entry 19" centrifugal compressor driven by a 16.4" turbine, directly coupled to it. The compressor had 30 blades to minimise blade loading and to avoid stalling. WU models 1 & 2 had single, large combustion chamber ex Laidlaw Drew & Co. WU model 3 had 10 combustion chambers with a reverse-flow layout. The 16.4" turbine was a single axial rotor with 66 × Firth-Vickers blades made from austenitic steel, Stayblade. A series of modifications were made to the diffuser system to raise pressure from the compressor. In May '39, a new compressor with 29 (not 30) blades (Hiduminium RR59) was installed to avoid resonant coupling with 10 blade diffuser system. On 9/10/40, Model 3 was run with Lubbock's controllable atomizing burners + Shell's combustion chamber 75	Models 1–3 built by British Thomson-Houston. World's 1st successful ground test of liquid-fuelled, turbojet, bench engine on 12/04/1937. Models 2 and 3 tested on 16/4/38 and 26/10/38, respectively

Event/engine/model	Key features	Significance
von Ohain's jet engines, He-S series incl. He-S.1 & 2 and He-S.3B. Heinkel He-178 jet aircraft	16-blade centrifugal compressor. 1st experimental device was not self-contained, requiring a separate electric motor to power the compressor. Performance figures for the 3 models are unconfirmed	According to E. Heinkel, He-S.1 bench tested in Sept 1937 fuelled by hydrogen, served to assess flow characteristics. Six months later, He-S.2 ran under its own power, fuelled by gasoline. 1st jet aircraft, He-178, piloted by E. Warsitz, was flight-worthy, in Germany, on 27/08/1939, with the engine permitted to run at flight-thrust for only a few minutes. Development of von Ohain's jet engine design was finally abandoned in 1941
Power Jets W.1X/BT-H. Experimental version of flight engine W.1. Became demo model for GE in USA	Early edition of W.1. Cannibalized spare parts to create non-flying, experimental engine. 1st run in Nov 1940 @16,500 rpm at the Ladywood Works test facility	1st used as a mock-up for the installation of the W.1 in the E.28; then used for taxiing trials on 7/4/41. In Oct '41, after 132 h running, over 9 months, W.1X shipped to USA to assist GE produce jet engines, leading to GE 1 & 1-A. American version 1st run on 18/03/1942. 1,250 lbf thrust. W.1X is displayed at the Smithsonian National Air & Space Museum in Washington. Knowing its raison d'être, it is understandably dirty and dented

(continued)

Table C.1 (continued)

Event/engine/model	Key features	Significance
Power Jets/BT-H W.1 (differed in minor ways from WU model 3) & W.1A Both W.1 & W.1A were flight engines	Double sided centrifugal compressor. Reverse flow 10-can combustor positioned around the turbine area, making for shorter engine. W.1 first run on 12/4/41. Rated at 1240 lb thrust @17,750 rpm	W.1 turbojet propelled single engine Gloster E.28/39 on 1st demonstration flight on 15/5/1941, piloted by G Sayer. When rpm-restrictions were lifted, the thrust increased dramatically, as did the speed of the E.28. In Feb 1942, flight trials of the E28 powered by twin W.1A engines began at Edgehill in Warwickshire. The extra power of the W.1A facilitated a more rapid climb so the fuel available for cruise-at-height was extended, giving prolonged endurance. Speeds of 430 mph @ 15,000 ft were achieved by March 1942. In the autumn, the twin W.1A engines completed a 100-h test—the 1st turbo engines to do so. It is justifiable to say that the E28/39 was the 1st aeroplane to fly, remarkably trouble-free, using a gas turbine jet propulsion engine. The W.1 became the parent of a series of successful British & American jet engines
In 1939, patent 577,971 for general arrangement of W1. Patent 577,972 for combustion vaporizers	Incorporated I. Lubbock's atomiser burners + Shell Petroleum Co. combustion chamber 75	
	1-stage axial turbine with 72 (not 66) Firth-Vickers Rex 78 turbine blades, secured by fir-tree root fixing	

Event/engine/model	Key features	Significance
	Convergent nozzle	
	Cleared for flight after a 25-h special category test subject to de-rating giving 850 lbf @ 16,500 rpm	
	W.1A incorporated several features of the planned W.2 incl. a turbine cooled only by air from fins of a cooling fan	
	When W.1A blade profiles corresponded with the inter-blade channels, the actual thrust increased to 1,340 lbf	
American GE 1-series → GE J31	Based on W.1X + Rover's W.2B/23 drawings	Mass produced model General Electric J31 (1650 lbf thrust) powered Bell XP-59A Airacomet which taxied on 1/10/1942; maiden flight 2/10/42. It achieved a top speed of 380 mph
Junkers Motoren (Jumo) 004 A & B. Messerschmitt Me-262 jet fighter	The 004 had a single-spool, 8-stage axial compressor. Model 004B-1 had a thrust of 1980 lbf (8.8 kN) @ 8700 rpm	Slung from swept-wings via under-wing pods of Messerschmitt Me-262. 1st flown on 18/07/1942. Entered squadron service on 30/6/1944. Jumo 004.B engine came off supply chain in Sept, ready for active operational service with Luftwaffe in Oct '44. With advanced aerodynamic characteristics, it was the fastest operational jet in the world. Top speed 530 mph (850 km/h)

(continued)

Table C.1 (continued)

Event/engine/model	Key features	Significance
Power Jets W.2	Same configuration as W.1 with lower frontal area	Ordered in 1940 and built mainly by Rover. More powerful than W.1 but exhaust velocity reached a critical value. Surging of the blower and high jet pipe temperatures resulted. Whittle accepted W.2 was a failure and devoted his attention to W.2B
Planned as 2nd generation turbojet	Novel compressor inlet, having stationary guide vanes giving the air a pre-whirl	
	Air cooling of the turbine disc	
Power Jets W.2B	Calculations by Whittle made him feel uneasy about aspects of W.2	Rover's starting point for its W.2B/23. Parent design for Rolls-Royce Welland and GE Type I
Rover's W.2B Mk II/23	Centrifugal flow design retained	In Jan '43 a W.2B was run for 400 h. On 7/5/43, a 100-h development test delivered 16,000 lbf thrust at full power. In late June 1943, 50 h of flying in a Gloster/Whittle E.28
	10-vane diffuser with fewer but stiffer blades	
	In 1941, to overcome the frequency of turbine failure, Nimonic 80 turbine alloy was trialled. The turbine blades were twisted through 5 degrees to reduce the pressure differential between the blade-root and blade-tip	

Event/engine/model	Key features	Significance
Rover W.2B/23 → R-R RB.23 → R-R Welland 1	Single-stage, double-sided centrifugal compressor. 10-can reverse-flow combustion chambers	Adopted by Rolls-Royce becoming Welland 1. 1600 lbf (7.1kN) thrust @ 16,000 rpm on 7/05/1943. Rolls-Royce built 100 × Wellands (production W.2B's)
Rolls-Royce 1st production of a jet engine. Military turbojet based on Whittle's W.2B		Early versions propelled Gloster F.9/40 (prototype to the Meteor) first flown on 5/3/43, from Cranwell. Delivered to RAF as twin-engine interceptor jet fighter, Meteor 1, in May 1944. Had mid-wing nacelles for the engines. At 10,000 ft had max speed of 480 mph. 1st allied jet aircraft to fly on combat mission on 27/07/44, to intercept V-1's. On 4 Aug, Meteors downed 2 × pulsejet-powered V-1 flying-bombs
Power Jets W.2/500	On 13/3/42, started complete redesign of W.2B. Increased size of turbine blades. New blower case	On its 1st run in Sept 1942, attained thrust of 1755 lbf. Flight-tested in prototype Meteor in Nov 1943
Rolls-Royce RB.37 → Derwent 1. R-R 2nd production jet engine	The RB.37 (Derwent I) displayed some features of Whittle's W.2B & W.2/500 but was not reverse flow. Straight-through design based on Adrian Lombard's (Rover) B.26. It had a 1-stage, dual entry compressor with a 2-sided 20.68" impeller. 10-can combustion chambers with ignition plugs in chambers 3 & 10. Single-stage axial flow turbine with 54 blades	Compressor + turbine air & gas flow increased by 25% giving 2000 lbf static thrust. Passed a 100-h test in Nov 1943. Power plant for Meteor III which entered service in 1945 and attained a speed of 460 mph
Power Jets 'by-pass jet'. Thrust augmenter could be tacked onto jet engine to increase mass flow of gases via aft-fan turbine (patent 471,368; priority date 4/3/36). High Court gave a patent extension, from 1952 to 1962, on the grounds of 'exceptional merit'	It established the thermodynamic principles for all subsequent 'ducted fan', 'by-pass fans' and 'turbofan' aeroengines. It traded exhaust velocity for extra mass-flow, giving the required thrust with lower fuel consumption. In the case of aft fans, some air by-passes the combustion chamber & main turbine to drive an additional turbine positioned in the exhaust of the main engine, thereby augmenting the jet thrust from the main turbine	Distinct engine performance and economic advantages of by-pass technology not fully exploited until the early 1960s. No. 2 thrust augmenter formed the basis of GE aft-fan design which became part of the GE power plant for the Conair 990A—the fastest subsonic commercial airliner in 1961

(continued)

Table C.1 (continued)

Event/engine/model	Key features	Significance
Power Jets W.2/700. On 2/3/40 & 17/11/42 aft-fan (augmenter) patents 583,111/2; 588,918 and 588,085. After-burner/reheat jet-pipe developed 1944–5	Improved version of W.2/500 via compressor impeller, completely new (Type 16) diffuser and blower casing. Aiming for supersonic speeds, a version of W.2/500 was trialled with a No.4 aft-thrust augmenter coupled with an afterburner. Fuel injected into the exhaust gases was burnt with residual oxygen, after which the reheated gases exited the tail pipe. Offers transitory thrust boost	Attained Whittle's aim of 80% compressor efficiency at pressure ratio of 4:1. Rated at 2,500 lbf thrust. The modified version of the W.2/700, in combination with the No. 4 augmenter + reheat, was intended for the M.52 supersonic aircraft. In 1943, the Air Ministry contracted the Miles Aircraft Co., in Reading, to build a plane able to fly @ 1,000 mph @ 36,000 feet. With the prototype almost ready, the contract was cancelled in Feb 1946. On 14/10/47, the Bell X-1, powered by a rocket engine, made the 1st manned, level-flight at a speed exceeding the speed of sound
Power Jets W.2/850	By 1944, produced 2,485 lbf thrust @ 16,500 rpm	
Rolls-Royce RB.41, Nene. 3rd R-R jet engine to enter production. Engine for both military & civilian use	Enlarged Derwent to achieve 5000 lbf @ 12,400 rpm. Double sided centrifugal compressor with 28.8″ impeller; 9 × new (Lucas) combustion chambers + single-stage turbine. Manufactured in 1944	Most powerful engine of its era. Atlee's post-war Labour government sold 35 Derwent's + 25 Nenes to the Soviets for 'non-military purposes'. The Soviets reverse engineered the Nene, building 39,000 without licence. The technology migrated through Eastern Europe to China. The Soviet derivatives, Klimov RD-45 & VK-1, powered MiG-15's used against USA, in the Korean War, in 1950. Bizarrely, jet aircraft for all combatants were powered by developments of Whittle's designs

Event/engine/model	Key features	Significance
Nene → Rolls-Royce RB.55 (Derwent V). Scaled-down Nene	The Nene engine had an overall diameter of 49.5″. Being larger than a Derwent, it could not fit in the nacelles of a Meteor. Therefore, a Nene was scaled down (by a factor, 0.855) to the 43″ diameter of a Derwent, creating the RB.55, confusingly called the Derwent V, but having no relationship with Derwents I to IV	It had a service rating of 3500 lbf (15.6 kN) thrust. In a twin-engine Meteor F4, it flew over 600 mph (976 km/h; 0.8 Mach)—obtaining an official world air speed record of 606 mph, in 1945
Power Jets W.3X	Straight through design concept	Power Jets had been contracted to build an experimental engine like Rover's B.26, comprising a long shaft with extra bearings and couplings to allow for thermal expansion
Power Jets Long Range 1 (LR1). Front fan turbofan	3-stage turbofan ahead of, and supercharging, the core engine which comprised an 8-stage axial compressor, followed by a centrifugal compressor. Would have been the world's 1st turbofan	Design prompted by Whittle's appreciation of the need for a long-range (4,000 miles), 4-engined bomber (10,000 lb bomb load) for the Far East/Pacific. LR1 engine was half built when the project was cancelled because Japan had been defeated (1945) and the National Gas Turbine Establishment was formed (1946) precluding further engine design & development by Power Jets

Appendix D: Key Properties of, and Conditions for, the Formation of Electromagnetic Radiation (EMR)

See Table D.1.

) The Editor(s) (if applicable) and The Author(s), under exclusive
:ense to Springer Nature Switzerland AG 2022
 Bailey, *Inventive Geniuses Who Changed the World*,
:tps://doi.org/10.1007/978-3-030-81381-9

Table D.1 Properties and generation of EMR

Type of EMR	Wavelength range Distance (m) between adjacent wave crests λ	Frequency range (Hz) No. of crests passing a given point per second Speed of wave, $\nu = f\lambda$	Quantum energy range (eV) Amount of work done in moving an electron through 1-V potential $E = hf = hc/\lambda$	Examples of EMR generation caused by an accelerating charge outside the nucleus or inside (γ-rays)
	Longer waves to shorter	Lower to higher frequency	Lower to higher energy quanta	
Radio waves/TV/Radio Detection & Ranging (=RADAR)	1×10^5 down to 1×10^{-3} Median long wave = 1000 m Short wave = 85–10 m RADAR = 1 m to 0.02 m	3×10^3 to 3×10^{11} Shortwave freq. = 3.5–30 MHz Longwave freq. = 30–279 kHz VHF TV = 54–88 and 174-222 MHz UHT TV = 470–1000 MHz RADAR = 3×10^8–1×10^9 Hz	1.2×10^{-11} to 1.2×10^{-5}	The oscillation of electrons in a conductor (a radio antenna), driven by a time-varying voltage, generates both electric and magnetic fields and the combined field is radiated as a radio wave
Microwaves	1×10^{-1} to 1×10^{-3}	3×10^9 to 3×10^{11}	1×10^{-5} to 1.2×10^{-3}	Emission results from rotational transitions and torsion. Caused by (i) thermal motion of atoms and molecules in any non-metallic substance, at a temperature above absolute zero (ii) currents in metallic conductors/macroscopic circuits and devices (e.g., a microwave oven has a concealed antenna that has an oscillating current imposed on it)

Type of EMR	Wavelength range	Frequency range (Hz)	Quantum energy range (eV)	Examples of EMR generation caused by an accelerating charge outside the nucleus or inside
	Distance (m) between adjacent wave crests	No. of crests passing a given point per second	Amount of work done in moving an electron through 1-V potential	accelerating charge outside the nucleus or inside (γ-rays)
Infrared (thermal radiation)	1×10^{-3} to 7×10^{-7}	3×10^{11} to 4×10^{14}	1×10^{-3} to 1.7	Emission results mainly from vibrational transitions. When an object is not quite hot enough to radiate visible light, it will emit in the infrared region. Common natural sources are the Sun and fire. Common artificial sources are domestic heating appliances and infrared lamps
				Caused by rotational movement (=gas phase molecule) or vibrational movement (=gas, liquid, or solid phase molecule). The energy released when excited vibrational states relax depends on the chemical structure and type of chemical bond

(continued)

Table D.1 (continued)

Type of EMR	Wavelength range Distance (m) between adjacent wave crests	Frequency range (Hz) No. of crests passing a given point per second	Quantum energy range (eV) Amount of work done in moving an electron through 1-V potential	Examples of EMR generation caused by an accelerating charge outside the nucleus or inside (γ-rays)
Visible light (optical waves) See Newton Section 6.3.2 and Maxwell 10.4.8	7.5×10^{-7} to 3.9×10^{-7} Red = 800 nm Violet = 400 nm	4×10^{14} to 8×10^{14} Red = 3.7×10^{14} Violet = 7×10^{14}	1.65–3.17	Emissions in the visible and UV regions result from electronic transitions. Electrons can be excited to higher energy orbitals lying further from the nucleus. The bigger the transition, the more energy is emitted (higher frequency) when the electron falls back to a lower level. Excitation is caused by (i) subjecting a substance to heat stimulation or (ii) extreme excitation using an electric arc, spark, or discharge. In the visible region, typical quantum jumps involve photon energies of 2–3 eV. UV emissions involve more energetic quanta In the case of atoms, only excited electron states exist, so only electronic emission can occur. In diatomic or polyatomic molecules, one or more series of quantized vibrational and rotational states are superimposed on each electronic state. Consequently, when molecules are in excited states, EMR emission can result from relaxation to electronic, rotational, or vibrational ground states A variety of emission spectra are obtained. Line spectra are characteristic of atoms, band spectra of molecules

Type of EMR	Wavelength range	Frequency range (Hz)	Quantum energy range (eV)	Examples of EMR generation caused by an accelerating charge outside the nucleus or inside (γ-rays)
	Distance (m) between adjacent wave crests	No. of crests passing a given point per second	Amount of work done in moving an electron through 1-V potential	Warm objects emit invisible infrared light. Incandescence is the emission of light by a solid heated until it radiates light. A red-hot object emits the lowest energy light—red. As the object gets hotter it emits red and yellow light and looks orange. A white-hot object emits red, yellow and blue light
				In an incandescent light bulb, electrons travelling along a tungsten filament, collide with the metal's atoms, boosting their electrons to higher energy levels. When they fall back to lower levels, radiation occurs. If the filament is at ca. 2200 degC, a mixture of IR (90%) and visible light (10%) is emitted. A continuous spectrum of all visible wavelengths is achieved with tungsten. Argon is used to replace air in the light bulb to prevent combustion/evaporation of the abnormally high melting tungsten

(continued)

Table D.1 (continued)

Type of EMR	Wavelength range	Frequency range (Hz)	Quantum energy range (eV)	Examples of EMR generation caused by an accelerating charge outside the nucleus or inside (γ-rays)
	Distance (m) between adjacent wave crests	No. of crests passing a given point per second	Amount of work done in moving an electron through 1-V potential	
Ultraviolet light	4×10^{-7} to 1×10^{-8}	7.5×10^{14} to 3×10^{16}	3.1 to 1.2×10^2	On the Sun, the energy released by nuclear fusion of hydrogen, to helium, is radiated away by EMR. In the Sun's photosphere (4200–6000 °C) the radiation is mainly radio wave and IR. In the chromosphere (ca. 10,000 °C) radiation is UV. The corona (1 m °C) emits X-rays. Solar flare eruptions cause very highspeed electrons and protons to strike ordinary atoms with such force to generate both X-rays and γ-rays. The ratio of IR to visible and UV radiation is roughly 50%:40%:10%. The Sun is a source of UV-A, UV-B and UV-C. The latter is the most energetic and harmful. Fortunately, UV-C is almost completely absorbed by the Earth's atmosphere. On Earth, UV can be produced by de-excitation of atoms that may be part of a hot solid or gas (e.g., carbon arc; mercury vapour lamp)

Type of EMR	Wavelength range Distance (m) between adjacent wave crests	Frequency range (Hz) No. of crests passing a given point per second	Quantum energy range (eV) Amount of work done in moving an electron through 1-V potential	Examples of EMR generation caused by an accelerating charge outside the nucleus or inside (γ-rays)
X-ray	1×10^{-8} to 1×10^{-11}	3×10^{16} to 3×10^{19}	1.2×10^2 to 1.2×10^5	Emitted when (i) high energy electrons (50 keV) are instantly decelerated by smashing into a heavy metal target whereupon their kinetic energy is transferred to X-ray photons, and (ii) when electrons from an outermost orbit of an atom drop to an inner shell. Our Sun's radiation is mainly in the IR and visible ranges, but the Sun's corona is much hotter and radiates mainly X-rays
γ-rays	Smaller than 1×10^{-11}	3×10^{19} to 3×10^{26}	1.2×10^5 to 1.2×10^{12}	They are produced naturally by the hottest and most energetic objects in the universe (e.g., solar flares, neutron stars; pulsars). Unlike X-rays, γ-rays are produced by nuclear processes and their wavelengths are characteristic of the nuclide producing them. On Earth, they are emitted by an excited daughter nucleus undergoing a 'settling process' after radioactive decay, by making a quantum jump to lower energy nuclear states. For instance, in the case of radioisotope, potassium-40, about 11% of it decays by electron capture to argon-40, accompanied by γ-ray emission with energy 1461 keV. Some nuclides produce γ-rays of more than one wavelength
Ultra-high energy γ-rays	1.2×10^{-25}	2.4×10^{33}	1×10^{19}	

Appendix E: Biological and Global Consequences of Exposure to Photons

See Table E.1.

Bailey, *Inventive Geniuses Who Changed the World*,
https://doi.org/10.1007/978-3-030-81381-9

Table E.1 Impact of exposure to photons

Impact of EMR on atoms and molecules in living tissues. Representative energies associated with atomic and molecular rotation, vibration, electron transition, chemical bond breakage, and ionisation

Energy associated with	Representative energy (eV)	Biological and global consequence of photon exposure
Rotation of molecules	1×10^{-4}	Photon energies range from a harmless billionth of an eV to harmful trillions of eVs
Vibration of molecules	0.5	If there are no available quantized energy levels, with spacings that match the quantum energy of the incident radiation, then the material will be transparent to that radiation, and it will pass through
Descent from outer electron shells in atoms	1	
Binding of weakly bound molecules	1	If the EM energy is absorbed but cannot eject electrons from the atoms of the irradiated material, then it is classified as non-ionizing, and will normally just heat the material
Binding of tightly bound molecules	10	
Ionisation of an atom or molecule	10 to 1000	As a generality, IR and visible light influence biologically significant habitat, and nutrition. Ionising radiation, on the other hand, contribute to morbidity and mortality, and influence the course of evolution

Impact of EMR on atoms and molecules in living tissues. Representative energies associated with atomic and molecular rotation, vibration, electron transition, chemical bond breakage, and ionisation

Extremely low frequency (ELF) radiation Electric powerline (50 Hz; λ = 6000 km) Mobile phone (800 Hz; 4G; λ = 374 km)	Powerline = 2×10^{-13} Mobile = 3.3×10^{-12}	In our living environment, electric powerlines and appliances are the most common sources of extremely low frequency (ELF) radiation. Most electric power operates at 50 Hz (UK) or 60 Hz (USA). The strength of an electric field is measured in volts per meter (V/m). For a computer screen and powerline, the electric field strengths are 10 and 10,000 Vm^{-1}, respectively Unlike ionizing radiation, non-ionizing EMR, including ELF radiation, is too weak to damage DNA or cells directly. Research continues to better inform us about the possibility of any other effects of ELR radiation on the human body
Radio waves AM (medium wave) = 530–1600 Hz; λ = 565 to 187 km RADAR = 300MHz–15GHz	2.1–6.6×10^{-12}	Everyday sources of radio frequency EMR are telecommunications, broadcasting antennae and microwave ovens Solid objects are transparent to radio waves
	RADAR = 1.2×10^{-6} to 6.2×10^{-5}	Close exposure to radar sources can cause heat injury. For RADAR at 10GHz, penetration to inner tissues is low, but the high-power density can cause skin burns and cataracts. Protective measures are enforced at RADAR installations so human exposure, immediately in front of a RADAR antenna, s avoided

(continued)

Table E.1 (continued)

Impact of EMR on atoms and molecules in living tissues. Representative energies associated with atomic and molecular rotation, vibration, electron transition, chemical bond breakage, and ionisation

MW → IR → visible Microwave oven (2.45 GHz; λ = 12 cm)	MW oven = 1×10^{-5}	EMR is a carrier of energy, so as the frequency of the radiation increases, the amount of energy absorbed rises, and the irradiated object becomes warmer Penetration depth increases with increasing wavelength. Thus, IR penetrates the skin further than visible light The Earth's climate results from the relationship between the incoming energy from the Sun, and the energy radiated from the Earth, back into space. Anthropogenic emissions of greenhouse gases are increasing. They absorb some of the IR radiated back to space, and then radiate some of it back to Earth, causing global warming
Red light	2	Red light can be transmitted through a fold of the skin, showing that the red end of the visible spectrum is not absorbed as strongly as the violet end The epidermis acts, not as a sun block, but as a sun filter, facilitating the passage of sufficient UV(B) for vitamin D manufacture, but limiting, to a degree, cellular damage by UV light which could lead to skin cancer Red light gently stimulates the red colour receptors in the retina

Impact of EMR on atoms and molecules in living tissues. Representative energies associated with atomic and molecular rotation, vibration, electron transition, chemical bond breakage, and ionisation

Red and blue visible light	1.6–3.3	Plant leaves are natural solar panels, harvesting light energy, using it to make sugars by the bioprocess of photosynthesis, becoming a primary source of food for many animals, and the origin of the food chain. The procedure is achieved without the need for quanta being large enough to break chemical bonds, whereupon molecules would decompose into free atoms, or ionised radicals
		Green is the colour that reigns over much of the plant kingdom. This is because a variety of pigments effectively absorb visible light in the violet/blue and red/orange regions. They transmit light in the middle range of the visible spectrum, giving leaves their green visual colour. Chlorophyll "a" is the primary light-capturing pigment. In methanol, it has absorption peaks at ca. 422 nm (blue/violet) and 670 nm (red)
		Light absorption is extremely fast, occurring within femtoseconds (10^{-15} s) causing transition from the electronic ground state to an excited one. Within 10^{-13} s, decay by vibrational relaxation occurs to the 1st excited singlet state. The pigments optimise solar power conversion, passing excited electrons to an electron transport chain, leading to the synthesis of NADPH and ATP which serve as energy stores for plant metabolism

(continued)

Table E.1 (continued)

Impact of EMR on atoms and molecules in living tissues. Representative energies associated with atomic and molecular rotation, vibration, electron transition, chemical bond breakage, and ionisation

UV with wavelength 1×10^{-7} m	12.4	UV photons below the ionizing energy boundary are absorbed in producing electronic transitions. UV disrupts atoms and molecules with which it interacts. Could break up a single, tightly bound molecule and destroy about a dozen, weakly bound molecules
		Lower UV absorbed by outer layer of skin. The most harmful wavelength for producing skin burns, in the UV-B range, is said to be about 300 nm
X-rays with wavelength of 3×10^{-10} m	4.14×10^3	Pass readily through skin, fat, muscle, and air-filled lungs. Absorbed by calcium-rich bones. X-rays are capable of completely knocking out an electron from an atom, so are classified as ionising radiation. This means that repeated exposure can cause harm to living tissues
Gamma rays with frequency 1×10^{21} Hz	4.14×10^6	Each photon causes violent ionisation events. Can ionise thousands of atoms and molecules, resulting in irreparable damage to biological tissues
		Damage occurs to DNA inside cells, leading to mutations and uncontrolled cell growth (cancers)

Index

Printed in the United States
by Baker & Taylor Publisher Services